Pennsylvania
Landscapes and People

Charles Geiger
Millersville University

KENDALL/HUNT PUBLISHING COMPANY
4050 Westmark Drive   Dubuque, Iowa 52002

The following photos were taken by Lydia Geiger: Figures 17.2; pg. 103, 17.3; pg.104; 20.1 pg. 122; 20.2 pg. 123; 20.3 pg. 124; 21.1 pg. 127; 21.3 pg. 127; 24.1 pg. 143; 24.2 pg. 144; 24.5 pg. 147; 29.1 pg. 174; 36.1 pg. 214.

The following photos were taken by Charles Geiger: Figures 26.3 pg. 159 and 37.1 pg. 219.

Map credits can be found on page 237.

Cover image provided by the author.

To Dad,
who traveled these roads
a lot more than I ever will.

Contents

Acknowledgments vii

Chapter 1:   Pennsylvania: A Crossroads 1

Section A: Chapters 2 to 12:   Pennsylvania's Natural Landscapes and Regions 5

Chapter 2:   The Earth Beneath Your Feet 7
Chapter 3:   The Geological Story 17
Chapter 4:   Mineral Resources and Hazards 23
Chapter 5:   Winter Weather 33
Chapter 6:   Summer Weather 39
Chapter 7:   Surface Water Resources 45
Chapter 8:   Groundwater Resources 51
Chapter 9:   Soils 55
Chapter 10: Forests 61
Chapter 11: Vegetation and Soil Erosion 67
Chapter 12: Wildlife 71

Section B: Chapters 13 to 19:   Pennsylvania's Human Landscapes and Regions 75

Chapter 13: People, By the Numbers 77
Chapter 14: Ethnicity and Immigration 83
Chapter 15: Urbanization 91
Chapter 16: Ethnicity and Cultures 97
Chapter 17: Pennsylvania Cultures 101
Chapter 18: Government 107
Chapter 19: Politics 113

Section C: Chapters 20 to 31:   The Structure of Economic Activity 119

Chapter 20: Traditional Agriculture 121
Chapter 21: Modern Agriculture 125
Chapter 22: Forestry 133
Chapter 23: Mining 137
Chapter 24: Early Transportation 143
Chapter 25: Modern Transportation 149
Chapter 26: Energy in Pennsylvania 155
Chapter 27: Manufacturing 161
Chapter 28: Iron and Steel 167
Chapter 29: Retail Economy 173
Chapter 30: Services 177
Chapter 31: Recreation and Tourism 183

Section D: Chapters 32 to 37:   Urban and Environmental Issues 189

Chapter 32: Uniqueness of Urban Places 191
Chapter 33: Pittsburgh 195
Chapter 34: Philadelphia 201
Chapter 35: Air Pollution Issues 207
Chapter 36: Water Pollution Issues 213
Chapter 37: Waste Management 217

Bibliography 223
Index to Places in Pennsylvania 233

Acknowledgments

When I began this project, I thought I had everything I needed. I had lived (or spent many months over the years) in West Chester, Montrose, Edinboro, Franklin, Slippery Rock, The Main Line, Clarion and Lancaster. I've lived in Pennsylvania for all but about seven out of my fifty-some years. However, research teaches you that you never know enough. The one thing I have enjoyed the most about this project is that I have learned to listen differently to people.

Several students, whom I have had the pleasure of teaching in my Geography of Pennsylvania and other courses, stand out in this regard. Stephen Scanlon, Jon Egger, Mike Sowizral and Tom Hetter all have made astute observations or led me to excellent sources that have inspired some of the contents of this book. I especially thank two students, Jon Egger and Rachel Ralls, as well as my sister Joanne Mark, who provided excellent photographs. Of the rest of the photos, I took a couple, but most were taken by my wife, Lydia, a true artist.

I was never trained as a regional geographer, but conversations with some who have been, such as Joe Glass, Mario Hiraoka, Frank Pucci and the late Art Lord from my own department, have helped me to see what it takes. Many authors' written works have also been influential, of course, especially the Penn State group including E. Willard Miller, Wilbur Zelinsky and Peirce Lewis. On specific topics I have also drawn on past work with Kent Barnes, Jay Parrish, Charles Scharnberger, Steve Thompson, and countless others.

I created all of the maps, graphs and tables in Pennsylvania Landscapes and People, and all but one of the diagrams, using the wonders of computer technology. Data sources are noted where appropriate, but I excluded from that process the different sources of digital files containing elements that became parts of the maps. The software I used to create the maps is ESRI's ArcGIS. One Internet site which is a rich source of digital map files is titled "Pennsylvania Spatial Data Access" (http://www.pasda.psu.edu), hosted by the Bureau of Geospatial Technologies of the Office of Information Technology of the Pennsylvania Governor's Office of Administration, and the Penn State Institutes of the Environment of the Pennsylvania State University. Other map files, or the data to create them, came from mostly federal government Internet sites.

There is one more form of assistance that I most happily acknowledge, and that is the support my family has provided. My wife, Lydia, and daughters, Sarah and Emma, have tolerated my spending a lot of time focused on maps and paragraphs. As we all know, love sustains all.

Chapter 1

Pennsylvania: A Crossroads

Pennsylvania needs well-informed citizens. The purpose of this book is to enhance Pennsylvanians' understanding of Pennsylvania in order to help them become better informed. I assume that, whether you were born here in Pennsylvania or not, you have lived here long enough to be interested in learning more about the state. Every one of us is part of a much larger whole. Your neighborhood, your home town, your county, Pennsylvania and the United States are successively larger and more complex parts of that whole. You are already here, influencing and being influenced by this whole and these parts.

Three principles have guided the selection and presentation of content. The first principle is to consider issues and information in the context of the challenging question, "What does an informed citizen of Pennsylvania need to know in order to be a valuable contributor to the affairs of the Commonwealth?" This may sound like a political, or at least governmental, question but it is not.

Being a valuable contributor requires that we examine a wide range of issues and opportunities occurring in Pennsylvania. We will not be *answering* the question so much as restating the question in a variety of contexts. The emphasis here will be on those questions that are appropriately considered at the statewide level or at the level of one or several major areas of the state. This includes questions that may appear to be about one place, but end up applying to many comparable places. Questions of national or federal importance will be avoided unless Pennsylvania's position is somehow unique. At the other end of the scale spectrum, questions of relevance to one community only (there are relatively few such questions) will not be considered unless it is an illustration of Pennsylvania's great variety.

The questions that need to be resolved at the state level ask what our natural and human resources are, how we are using them, and what problems or issues are known about them. They ask what various natural and human-built landscapes look like, how they are changing and, again, whether that has become problematic or controversial. The questions also ask how best to organize discussions about these landscapes and resources; in this book we will especially emphasize the value of a geographical perspective.

The second guiding principle is that Pennsylvania today is the product of a long and colorful history, and yet that history does not control or predict our future. This principle leads to questions that ask how our landscapes, resources, population and economy arrived at their present condition, emphasizing past changes and movement. Many "firsts," "mosts," and other "great things" have happened in Pennsylvania, while many others have never or hardly ever happened here.

Both of these first two principles, the desire to contribute to a more knowledgeable citizenry and the attempt to explain how things got to be the way they are, are best explained from a geographer's perspective, which will serve as our third guiding principle. Geography is the discipline which, above all others, understands the relevance of considering nature and humans together.

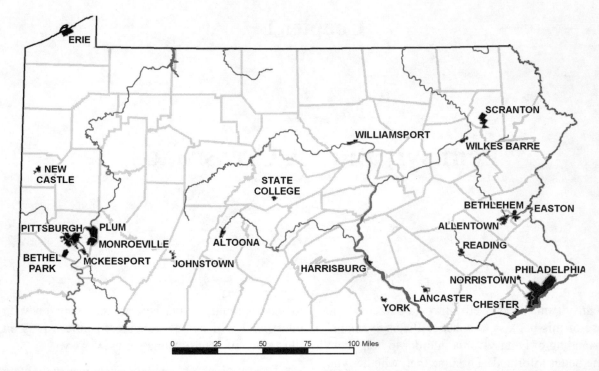

Figure 1.1 **Pennsylvania's 67 Counties:** County boundaries will be repeated on most of the maps in this book. In order to relate the many topics we will be examining to your own experience or research, make sure you can locate particular places within their counties.

This third principle guiding this book, the need to convey a sense of the perspective and value of Geography as a discipline, is behind the "level" of writing used. Unfortunately, it is still the case that most readers will never have had a class whose name included the word "geography" and whose teacher had a strong background in geography. For those who have, as often as not, the curriculum guiding your teachers was built around the conception that geography is map work only, or that geography is trivia (capitals, largest rivers, etc.). Geography, especially at the university level, is often nothing like that, though such background can help.

Why Pennsylvania?

Given the range of contexts that we are part of, from the local to the global, how important is it to study your state? To put it geographically, what is the significance of this *scale* of discussion? Early in the existence of our country, and even more so when it was one of Britain's North American colonies, it made a lot of difference. Each colony or state was significantly different culturally, economically and politically from its neighbors. Today these differences are reduced to more subtle differences such as tax rates and road conditions, as anyone who lives near one of the state's borders is well aware of.

Indeed, all of the content areas to be included in this text would be included in a book about any state, or even about the whole country. However, at the scale of Pennsylvania as a state we can simplify many of these discussions somewhat and bring them closer to our actual experience. This is very much a case of learning based on experience, except that we will be working at organizing those experiences.

Two personal observations also have convinced me that this state scale is highly appropriate. One is that my students have spent about as much time traveling around Pennsylvania as they have around the rest of the country. Their identification with their state in making travel decisions does matter to them. My second observation is that more than half of those students expect to live in Pennsylvania after completing their education. They say that it is because both their real and virtual travels (via television, movies, etc.) have convinced them that Pennsylvania has the "right" variety, whether of weather conditions or cultural diversity.

The question of "Why Pennsylvania?" can be addressed from a historical perspective also. How significant has Pennsylvania been in US and even world history? In our early history, until the late 1800s, we were a leader in the US economy and in North

2

American culture. When the US emerged onto the world scene starting in the late 1800s but especially after World War I, Pennsylvania was still its industrial and economic heart. Even if that is now history, and we have gone through a lot of "down-sizing" since then, it still helps to explain a lot about the population, landscape and issues that surround us.

Peirce Lewis, a well known geographer who has taught geography at the Pennsylvania State University for many years, and several others, have gone so far as to assert that Pennsylvania's culture is the largest part of US culture (Lewis 1995, 1-4). Once again this idea links past with present: the most common pronunciations of American English, the food preferences and agricultural methods, and values such as tolerance, hard work and adaptability are all characteristics of Americans that first became dominant in southeastern Pennsylvania. Because of that location and early settlement routes, it was also usually the first set of cultural traits and values in the American heartland and the far west as well.

Getting Started

Even though most of us have traveled elsewhere within Pennsylvania, few of us have traveled in all parts of the state. Thus, we will inevitably be talking about places you are unfamiliar with. The maps in this chapter represent a geographical context for locating unfamiliar places.

The first map (Figure 1.1) shows Pennsylvania's counties. Perhaps related to the extent of our intrastate traveling, a third observation I can make is that Pennsylvanians seem to identify themselves as much by their home county as by their home town. This is remarkable indeed given that county government has probably the least influence of any level of government in our day-to-day lives. In keeping with my observation, though, I will usually refer to cities, towns and even smaller natural features with their county as well as their place name. In addition, county boundaries will appear on most maps of other phenomena, in order to provide a geographical frame of reference.

The map in Figure 1.2 shows Pennsylvania's major cities and rivers. Early in history the two were strongly linked: notice that only a few of the cities on that map are not on major rivers. Even those few, though, are located on or very near a more locally significant river. The rivers provided water for domestic uses and for transportation. Some are still important in the same ways.

Figure 1.2 **Pennsylvania's Major Cities and Rivers:** Each city was located, and grew best, where there was some advantage to its site or situation. Often, in early times, that advantage grew from its access to water.

3

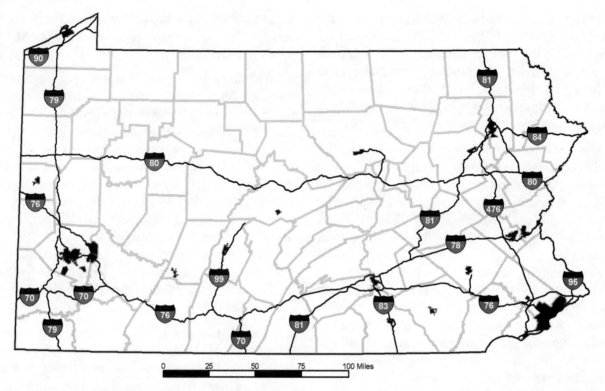

Figure 1.3 **Pennsylvnia's Major Highways:** The original Interstate Highway System was built with federal funding from the late 1950s through the 1980s. It carries much more traffic now than original designers envisioned, and has been expanded several times. The building of the Pennsylvania Turnpike (now Interstates 76 and 476) was an important development leading to the Interstate system.

The final map (Figure 1.3) shows the major part of our modern transportation preference: driving an automobile or truck on the interstate highways. Because you are likely to have driven on at least some of the roads, they again give us some spatial reference. Hopefully they also give you some ideas for routes to take to an as yet unvisited part of Pennsylvania on your next trip.

The Organization of this Book

Regional geography textbooks such as this are generally organized either spatially, covering a new region in each chapter, or topically, as this book is. This topical organization reflects the variety of ideas commonly of interest to geographers, and the structure usually used to organize those ideas.

First, the discipline of geography, like academia in general, recognizes a distinction between questions about nature, covered in this textbook in Part A (Chapters 2-12), and questions about people, covered here in Parts B and C (Chapters 13-31), and also the need to bring those questions to bear on each other, covered here in Part D (Chapters 32-37). Each of the Parts, A-D, is preceded by a short introduction that lays out the context and organization of the ensuing chapters.

Secondly, the perspective of the geographer is best represented by his or her most common tool: the map. That is to say that all of those questions include wondering about the significance of <u>where</u> an item or event is located or distributed or moving. Maps are abundant in this textbook.

The treatment given to each topic is relatively brief, compared to the amount of questioning and investigating that geographers have engaged in. Any topic that piques your interest can be delved into much more deeply. The bibliography at the end of each chapter will give you some tools to start such a search, including many windows on the Internet, but you should also include the living, breathing geographers in your university and community. We tend to be very passionate about our discipline!

Bibliography

Lewis, Peirce 1995. "American Roots in Pennsylvania Soil." Chapter 1 in Miller, E. Willard (ed.) A Geography of Pennsylvania. pp. 1-13.

Section A: Chapters 2 to 12

Pennsylvania's Natural Landscapes and Regions

It has been said that change is the only constant. It is helpful, though, to appreciate the types of changes that are happening all around us and the rates at which those changes occur. Why? It helps us to appreciate how things got to be as they are, and that they will continue to change without, or even despite, any effort on our part. In many cases, it also helps us to see how much of an influence on our surroundings we humans can have.

This first set of chapters describes the physical landscapes of Pennsylvania and the forces and processes that created them. The forces range from the tremendously slow but powerful ones that move continents and create mountain ranges, to the quick and subtle ones that remove small amounts of topsoil over a year. Some are lasting (at least compared to our life spans), and others are temporary like those that flood stream valleys after a heavy rain.

The processes that built up our landforms are the result of <u>tectonic</u> forces that primarily work from within the Earth. These forces, just like our planet's spinning on its axis and rotating around our sun, are the after-effect of the swirling forces that formed the Earth and our solar system from clouds of interstellar debris. The tectonic forces that work to shift the North American continent relative to its neighbors are measured in inches (or centimeters) per year, and persist for millions, even hundreds of millions, of years before altering direction. They have created huge mountain ranges in the vicinity of Pennsylvania that would rival the Himalayas of today. They have compressed, torn and contorted solid rock.

Equally powerful, at least in its effect, has been the process of <u>erosion</u> due to rainfall, temperature variations, chemical reactions, and gravity. By the very slow process of separating particles of mineral, one at a time, from their parent rocks, and the relatively quicker process of carrying it downstream to the ocean or other inland water basins, this force has worn those same mountain ranges down to sea level again. Such is the power of water, including the often tranquil waters of Pennsylvania's abundant lakes and streams.

Those particles are a principal component of our soils, which again illustrate the idea of the permanence of change. On their way downstream, the particles contribute to the growth of vegetation; once they reach their destination, they become part of future sedimentary rocks. The multiple layers of soil and rocks beneath the surface create a complex three-dimensional area that is a challenge to show in map form.

Another natural feature of our landscape, which plays competing roles in the erosion process, is vegetation. In Pennsylvania that means forests. If not for human interference, Pennsylvania would indeed be Penn's Woods, nothing but forest. Again working at rates that make our lives brief by comparison, trees and other vegetation add the necessary organic component to our otherwise mineral soils. Individual trees may live to be hundreds of years old, though few trees in Pennsylvania today are that age. More importantly, forest areas consisting of large numbers of trees within a limited range of species can persist for tens of thousands of years.

Wildlife represents an even shorter time horizon, considering the life expectancy of most creatures. Once again, humans have had a major impact on the assortment of species and their range and sustainability. Some animal populations have increased due to our presence, but many more have seen their habitat reduced and food supplies altered.

Key periods in Pennsylvania natural history include:

Table A.1 Geologic Time Periods (Source: Barnes and Sevon 2002, 32 and back cover)

Time Period	Geological Name	Activity
4.5 to 3.8 billion years ago	Pre-Archean Eon	Earth and solar system formed. No life.
3.8 to 2.5 billion years ago	Archean Eon	Meteors created the continental crust. Bacteria is the only form of life.
2.5 billion to 570 million years ago	Proterozoic Eon	Laurentia, the base of North America, formed. Life forms consist of algae, jellyfish and worms.
248 to 570 million years ago	Paleozoic Era	Three "orogenies" were each followed by the erosion of mountains. Deposition of biomass led to the eventual development of coal and oil. Life progressed from trilobites to insects, amphibians, reptiles, and land plants.
65 to 248 million years ago	Mesozoic Era	North America separated from Africa. Igneous rocks intruded into the northwest Piedmont area. Dinosaurs, birds, and early mammals were alive.
0 to 65 million years ago	Cenozoic Era	The present landscape was created by weathering, erosion and glaciers. Three different glaciers entered Pennsylvania within the last 800,000 years. Life includes mammals, grasses, and humans.

Notice the time scales we have considered: hundreds of millions of years to create and remove mountain ranges; thousands to millions of years to see species evolution; tens to hundreds of years in the lifespan of individual animals and plants or trees.

Understanding the context of distant past events is important in both human and natural history. Even though such natural events as earthquakes and tornadoes happen quickly and can be violently forceful, the vast majority of substantial change in our Earth's history has been gradual. The emphasis in this section will be on identifying those on-going processes and their visual impacts on the landscape.

Bibliography

Barnes, John H. and William D. Sevon 2002. The Geological Story of Pennsylvania (3rd edition). Harrisburg, PA, Pennsylvania Geological Survey. 4th series, Educational Series 4.

Chapter 2

The Earth Beneath Your Feet

The earth's surface is the point at which soil, vegetation, minerals, atmosphere and flowing water all merge. That means that most of the essentials upon which we depend come together on and in the ground beneath our feet. All of these natural phenomena are studied extensively by a variety of scientists. More importantly, for most of us who are not scientists, this common ground is also our primary landscape, the vista upon which we look every day and which gives us a feeling of familiarity. We alter it by constructing artificial landscapes, like cities and huge reservoirs.

When the natural landscape is studied by earth scientists (geographers, especially physical geographers, and geologists), one common approach is to characterize the un-evenness of the terrain, while another is to identify regions of similarity. The earth scientists will refer to the landscape's relief, topography, terrain, landform or physiography. The regions of similar landscapes are known as physiographic provinces or landform regions; the latter term will be used in this text. As with many topics of study in geography, the amount of detail and variety depends on the scale at which you are studying areas.

Pennsylvania's landscape, from the perspective of the world or of the US as a whole, is consistent and continuous: it is characterized by its relatively low relief, with elevations ranging only from about sea level on the Delaware River to 3213 feet above sea level at Mount Davis in Somerset County, and what could be a virtually complete forest cover (if not for humans). However, from our perspective within Pennsylvania, there are significant differences, for example between the southeastern lowlands and the

north central plateau. Earth scientists, both geologists and geographers, consider a variety of factors in order to identify and characterize landscape regions.

The Importance of Landforms

Why is it important to understand landforms and their regions? In this book, they will serve as the reference point from which all of the natural phenomena of the state will be studied. The landforms give clues to what bedrock lies beneath the surface (and to what mineral resources the various layers of bedrock might contain). Landforms also influence water flow and atmospheric conditions. Vegetation types and soil characteristics correlate very strongly with landforms.

The character of the surface of the Earth, its elevation and slope, for example, is most strongly influenced by the top layer of bedrock, which is the first layer of solid rock encountered in a deep excavation beneath the soil. Beneath this surficial layer, though, are many other layers. Sometimes these lower layers are exposed at the surface, in an open pit mine or along a deep valley eroded by a river. Once geologists know the order and other characteristics of these layers of bedrock in one place, they can often use that information to find particular minerals in any of the deeper layers at other locations within a range of hundreds of miles.

The surficial layer of bedrock usually contributes the largest proportion of the mineral content of any soil. As we shall see in Chapter 9, soil quality and therefore agricultural success are very dependent on

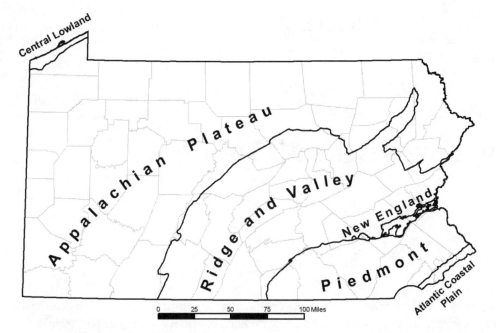

Figure 2.1 **Pennsylvania's Landform Regions**: The black boundaries (against the counties in gray) represent both structural (sub-surface) and visible patterns and characteristics of the Earth. Their diagonal orientation across the state is actually parallel to the Atlantic coast. Notice that Pennsylvania's "mountains" are actually ridges and a plateau.

that mineral content. In non-agricultural areas, success in forestry or road-building can also be affected.

The elevation of a particular place can affect its temperature, and differences in elevation between that places and others surrounding it can affect wind conditions and precipitation totals, not to mention the scenic view. Higher elevations relatively close to (such as within several miles of) lower elevations will probably show corresponding differences in soil, vegetation, water availability and, therefore, in economic potential.

Elevation differences also translate into the slope of a particular area's surface. Steeper slopes speed up overland water flow, and also make transportation difficult. The most challenging slope in Pennsylvania, especially historically speaking, was the Allegheny Front, a line along which Pennsylvania's two largest

landform regions meet. The challenges faced in surmounting this obstacle will also be discussed later in this chapter.

The Landform Regions

The **Atlantic Coastal Plain** in Pennsylvania is a small portion of a much larger region that includes most of New Jersey and Delaware and beyond. Pennsylvania's portion is the result of deposition of sediment both washed in by the Delaware River and its tributaries, and eroded from Piedmont areas to the immediate northwest (see Figure 2.2). The shores of the Delaware River, still part of the tidal Delaware Bay, are at sea level, and the highest elevation is approximately 180 feet above sea level (Potter 1999, 348). At 180 feet of relief, steep slopes are rare, most occurring at several higher points along the Delaware and Schuylkill Rivers.

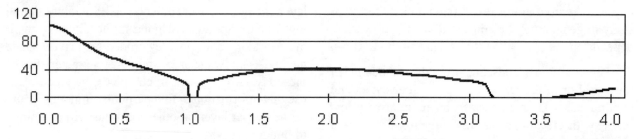

Figure 2.2 **Atlantic Coastal Plain Landform Region**: This is a profile along the line of Market Street in Philadelphia, starting west of the Schuylkill River, and finishing just east of the Delaware River in New Jersey. The markings across the bottom show ground distance in miles; those on the vertical scale show elevations in feet. The vertical scale is exaggerated about 28 times the distance scale, three to four times the exaggeration of the other profiles in this chapter, in order to more clearly show the small elevation differences of the Atlantic Coastal Plain.

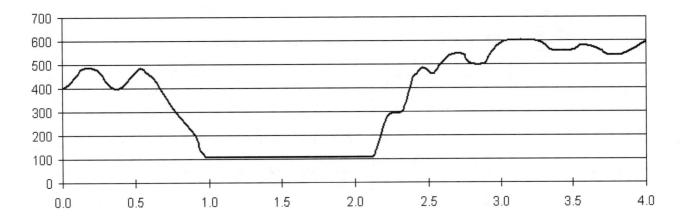

Figure 2.3 **The Piedmont Landform Region:** The profile shows elevations (in feet) on both sides of the Susquehanna River, exaggerated 10 times (10x) the length scale (in miles). Note the low river elevation (109 feet) and hilly terrain of this Piedmont Upland area. The map shows regions described in the text. The aerial photograph (USGS 1999a) shows an area in the Piedmont Lowland southeast of Lititz in Lancaster County. The highest and most steeply sloped area is still forested, and the hillier farmland has the more complex plow patterns.

Because it is primarily a depositional landscape, the parent materials for its soils and the upper layers of bedrock contain assorted minerals, of greatest value for the fact that they are easily returned to their original sandy and gravely states. Below these depositional layers are some of the oldest rocks in North America, which have been greatly deformed, as we shall see in the next chapter. Most of the area has no more than scattered trees and smaller wood patches. It has been settled the longest, and provided an ideal setting for Pennsylvania's largest city.

The **Piedmont** landform region is characterized by its rolling landscape (see Figure 2.3). It is divided into smaller sections, generally characterized as lowlands, ranging in elevation from about 300 feet up to 550 feet, or uplands, some of which are greater than 1000 feet in elevation (Potter 1999, 347-8). The transition from the coastal plain to the Piedmont is called the Fall Line. It is a relatively short climb that created non-navigable rapids and falls on all streams and rivers that cross it. Those falls helped establish Philadelphia as an industrial city in the early 1800s by providing an ideal source of mechanical power for the first factories there. The Piedmont remains a

9

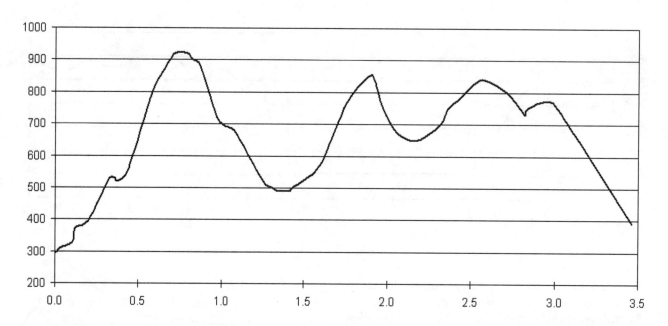

Figure 2.4 **New England Landform Region:** This profile runs across the New England landform region with a vertical exaggeration of 10x. The terrain is again hilly, but reaches higher elevations than the Piedmont and is based on an older bedrock than adjacent lowlands.

strongly agricultural region, and has been culturally the most influential area of Pennsylvania.

The lower elevations are the result of erosion that is exposing lower layers of bedrock, some of the oldest visible at the surface in Pennsylvania. Lowland areas include the Piedmont Lowland and the Gettysburg-Newark Lowland. The Piedmont Lowland is underlain by bedrock rich in carbonate rocks that weather into very rich soils, which support very intensive agriculture. As desirable as the carbonates' limestone content is, it does produce one hazard. It is prone to being dissolved by water that seeps into and through the ground, creating underground caverns that occasionally collapse into sinkholes. The Gettysburg-Newark Lowland is underlain by younger sandstone (readily eroded) and shale (more resistant) (Potter 1999, 347-8). The soils are not as rich as those in the Piedmont Lowland, but still support very productive agriculture. The Gettysburg-Newark Lowland was at one time a weak area of the earth's crust. Into some crack-like weaknesses volcanic material forced its way, and today appears as locally significant forested ridges a few dozen feet higher than the surrounding lowland. A few, such as those around Gettysburg, reach over 1000 feet in elevation, hundreds of feet above the lowland (Potter 1999, 347).

The higher elevations are the result of their upper bedrock layers being more resistant to erosion, with their shapes often reflecting geological forces which deformed the landscape hundreds of millions of years ago. Steep slopes occasionally occur in the region, and those slopes and the higher-elevation hilltops are more likely to be wooded.

North of the Piedmont is a smaller landform region called the **New England** region, better known locally as the Reading Prong. The New England region has some characteristics similar to the upland areas of the Piedmont, topping out at around 1300 feet (see Figure 2.4). It is an extension of a highly deformed landform of resistant rock older than the Piedmont bedrock. The New England landform region appears as several smaller areas within Pennsylvania, which is its southernmost end disappearing beneath the Piedmont. The eastern end of this landform in Pennsylvania is several smaller sections near the Delaware River in Northampton and Bucks Counties. The largest section covers a substantial part of Berks and Lehigh Counties. The westernmost section is a hill known locally as "Little South Mountain," which rises where Berks, Lancaster and Lebanon Counties meet. Many of these areas support locally important mining operations (Potter 1999, 346-7).

A Pennsylvania landform region which has been studied extensively is the **Ridge and Valley** region (see Figure 2.5). On relief maps its dramatic parallel ridges and valleys are very obvious. Even in aerial

10

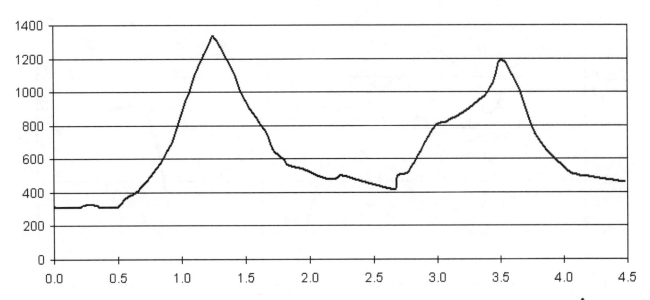

Figure 2.5 **The Ridge and Valley Landform Region:** The profile shows elevation changes perpendicular to the orientation of the ridges. The vertical exaggeration of 7x emphasizes the 1000+ foot elevation differences between ridge tops and valley bottoms. The map shows that the majority of the ridges and valleys are considered part of the Appalachian Mountains. The air photo (USGS 1999b), of an area northeast of Harrisburg, shows the urban and agricultural developmment of the valleys, in contrast to three forest-covered ridges. The paler lines crossing the ridges are clear-cut paths through the forest for electric power lines. Note also the stream winding along the more northerly valley.

photographs the topography is apparent because land clearing, rivers and streams, and settlement have occurred in the valleys, while the ridges are almost entirely forested.

It is easiest to describe the appearance of the region as the result of forces that horizontally compressed the land toward the northwest, which raised the ridges as if someone was pushing furniture across a loose rug. The pattern of ridges and valleys is regular in places, but most of the ridges are of limited length. The actual process was much more complex. The ridges are either the broad arched tops of resistant rock layers or narrow edges where those resistant arches have been split open lengthwise. They generally rise about 1000 feet above the neighboring valley floors, and reach over 2000 feet in elevation (Marsh and Lewis 1995, 27).

Like the Piedmont, the Ridge and Valley contains several noteworthy subsections. Its easternmost ridge along the Maryland border is the Blue Ridge, known

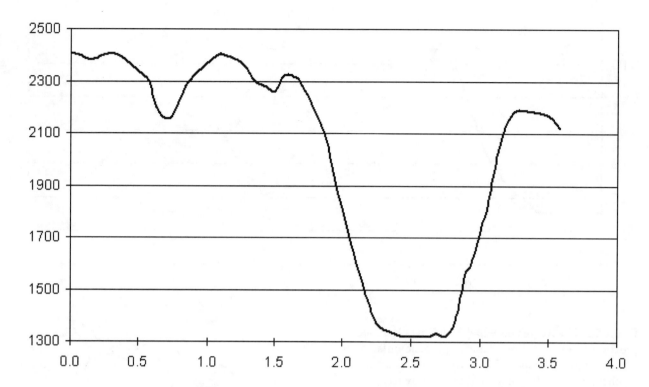

Figure 2.6 **Appalachian Plateau Landform Region:** The profile, again at a 10x vertical exaggeration, shows the more consistently higher-elevation plateau and the 1000-foot deep valleys cut by rivers. The map shows zones of the plateau that have different palteau elevations, divverent valley depths, and different forces that have worked to create the landscape. The aerial photograph shows the town of Honesdale in Wayne County on the Lackawaxen River (USGS 1999c). The area is in the Glaciated Low Plateau, and despite the lower elevations the tendency to confine development to the river valleys is still apparent. In this area, greater development pressure from Scranton and New York is encouraging farming on the plateau despite relatively weak soils.

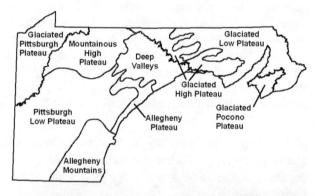

locally as South Mountain, an extension of the Blue Ridge Mountains of Virginia and North Carolina. Until recently, this feature was given the status of a separate landform region by Pennsylvania geologists, and not much longer back it was considered to be closely related to the New England landform region. The South Mountain sub-region is much more like the other ridges of the Ridge and Valley. It is the result of an upward fold in the crust that reaches up to 2000 feet in elevation, again mostly forested (Potter 1999, 346).

The first valley of the Ridge and Valley region is called the Great Valley, but is known locally in its northeastern end as the Lehigh Valley and in its

12

southwestern end as the Cumberland Valley. This valley is up to twenty miles wide and has generally fertile soils where it is underlain by carbonate rocks. Like the Piedmont Lowland, these have created a significant sinkhole problem. Within this valley are some significant cities and agricultural areas, such as Allentown, Bethlehem and Easton to the east, Harrisburg along the Susquehanna, and Chambersburg and Shippensburg to the southwest.

The first ridge of the Appalachian Mountains section is known as Blue Mountain. In the early years of Pennsylvania's settlement, this ridge turned out to be a significant barrier to westward travel, and guided many settlers south through the Great Valley into the Cumberland Valley and further into Virginia's Shenandoah Valley. Only by following the Susquehanna River north of modern Harrisburg did those settlers find access to Pennsylvania's interior.

Most of the other valleys of the Ridge and Valley region are narrower, more varying in their fertility, and much less populated. A few are well known, however, including Nittany Valley, home to State College and Pennsylvania State University, and the Wyoming Valley, which contains Scranton and Wilkes-Barre and the North Branch of the Susquehanna River. Many of the northeastern valleys, including the Wyoming Valley, are North America's most significant source of anthracite coal. Similarly, only a few of the other ridges are as well known as Blue Mountain;

possibly the best known of these in the southwestern part of this region is Broad Top, also known for its coal production.

The largest of the landform regions is the **Appalachian Plateau**, occupying well over half of Pennsylvania (see Figure 2.6). The plateau surface ranges in elevation from a little less than 1000 feet where it meets the Central Lowland near Lake Erie, to Mount Davis's 3213 feet, but is frequently deeply cut by rivers (Briggs 1999, 363-5). Again, though, it is much more meaningful to describe sub-sections of the plateau since different forces have been at work in different areas, and have provided for different resources.

The transition to the Appalachian Plateau from the Ridge and Valley region is a steep escarpment known as the Allegheny Front. If Blue Mountain was a significant first barrier encountered by potential settlers, then this obstacle must have seemed nearly insurmountable. It rises about 1000 feet above the final valley of the Ridge and Valley region, to elevations ranging from 3000 feet in the southwest to a few hundred feet in the northeast.

The areas of highest elevation are the Allegheny Mountain Section, adjacent to the southwestern end of the Ridge and Valley region, and the High Plateau and Mountainous High Plateau Sections of north central Pennsylvania. Both include large areas that are over 2000 feet above sea level. The larger rivers,

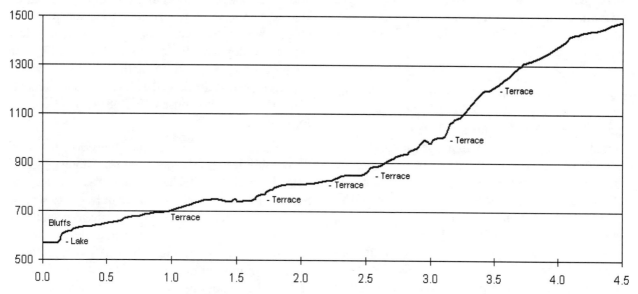

Figure 2.7 **Central Lowland Landform Region:** Notice the significant slopes of the region as it reaches from Lake Erie at just over 570 feet above sea level toward the Appalachian Plateau. The profiled site, again shown at a 10x vertical exaggeration, is perpendicular to the lakeshore, and located just northeast of the town of North East, about twelve miles northeast of Erie.

such as the Allegheny in the north and the Youghiogheny in the south, have cut valleys 500 to over 1000 feet deep. The western Pittsburgh Low Plateau Section is mostly less than 1600 feet in elevation with valleys mostly less than 1600 feet deep (Briggs 1999, 366-8).

Glaciers invaded the northeastern and northwestern corners of Pennsylvania during several different ice ages. The glaciers reshaped some plateau areas while leaving glacial deposits of mixed rock debris in others. These actions created some locally dramatic landscapes and also some poorly drained swampy areas, especially in the eastern Endless and Pocono Mountains areas.

Most of the developed land is located in the steep-sided river valleys. The stream-side flood plains are attractive sites because they are flat, easy to reach (most transportation routes follow the rivers also), and have easy access to water, but they are also at risk from flooding. Pittsburgh is the largest city in the region, occupying an ideal site for water-born transportation. Pittsburgh's other advantage was that much of the western areas of the plateau are underlain by rich coal beds (see Chapter 4).

Finally, as if to create a sense of landscape symmetry, the northwest corner of the state contains a narrow coastal plain adjacent to Lake Erie and extending into neighboring states just like the Atlantic Coastal Plain does in the southeast. This Lake Erie coastal plain is known by geologists as the **Central Lowland** landform region, because it is an extension of the entire Mississippi River valley and Great Lakes area (see Figure 2.7).

The inland extent of the Central Lowland region reflects an ancient shore of Lake Erie following the ice ages. Its terraced appearance is the result of extended periods of higher lake levels which have lowered to today's 571 feet above sea level (Briggs 1999, 376). The dominant features of the shoreline today are the Presque Isle "sand spit" and the shoreline bluffs. The sand spit is constantly being re-formed by erosion and deposition due to lake waters carrying sediment toward Niagara Falls. The bluffs, 15 to 170 feet high, are caused by erosion undercutting the old lakebed sediment and glacial deposits (Delano 1999, 781-2). Much of the Central Lowland's land is cleared to create locally significant farmland, and to make space for our Lake Erie port city of Erie.

Case Study: Altoona and the Horseshoe Curve

In the 1820s the US was still a young nation and our Midwest states, Kentucky, Tennessee and Ohio, were being joined by Indiana and Illinois in the Great Lakes area and by Missouri, Louisiana, Mississippi and Alabama in the lower Mississippi River and Gulf of Mexico area. These areas needed many goods not yet locally available. When New York and Maryland opened canal systems in the middle 1820s, a tremendous amount of trade destined for those areas was taken away from Philadelphia. The Pennsylvania legislature supported two efforts to gain back Philadelphia's former status as the nation's leading city: the Main Line of Public Works canal and railroad system, and the Pennsylvania Railroad.

The first project, built between 1826 and 1834 used an inclined plane rail system to haul canal boats from the east 1400 feet up the Allegheny Front and then 1170 feet down the sloping plateau to Johnstown, where the canal continued (see Chapter 24). The Pennsylvania Railroad chose a similar route through Johnstown. After rejecting (as beyond the engineering capabilities of the day) the idea of a four-mile-long tunnel through the Front, they arrived at another way to make the grade up the slope manageable.

The Allegheny Front is cut into by valleys eroded by relatively small streams. The engineers decided to follow one side of such a valley upstream into the face of the Front, cross that stream and reverse direction downstream while continuing up hill (this is the U-shaped "horseshoe"), turn and continue to climb upward across the face of the front, and then follow the next valley over the crest (see Figure 2.8). The project still required earth moving on an unprecedented scale, as well as several smaller tunnels (Thompson and Wilshusen 1999, 814-5). It opened for business in 1854.

Despite the accomplishment, the slope still taxed the capabilities of early locomotives. Altoona, located at the base of the Allegheny Front, was founded and built by the Pennsylvania Railroad in 1849 as a service center. The location proved strategic, especially as the western Pennsylvania iron and steel industry developed. By the middle 1850s Altoona factories were building locomotives, freight cars and cabooses, and by the 1880s they were also the Pennsylvania Railroad's main research facility for everything from parts to chemicals to foods (for passenger trains). At

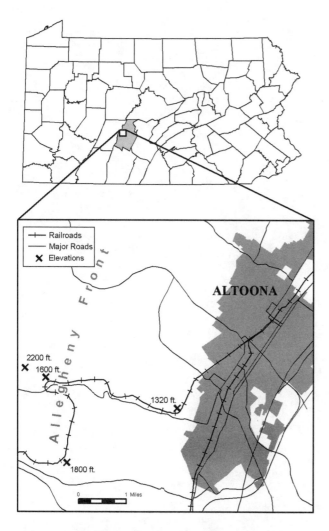

Figure 2.8 **Altoona and the Horseshoe Curve:** Altoona sits at the base of the Allegheny Front in Blair County. The spot elevations show a climb of 480 feet in about 5.3 miles, a 2 percent grade that is still a challenge for many trains. The roads climbing the Front follow streams that cascade down the steep hillside. Why does the road go straight where the curve makes its famous U-turn?

its peak, the Pennsylvania Railroad's Altoona works employed over 17,000 workers (Trainweb 2004).

Altoona's economy crashed when the Pennsylvania Railroad went into severe economic decline after the middle 1900s. The Pennsylvania Railroad was merged into the Penn Central Railroad, which was merged into Conrail. Recently Conrail's operations were split and bought by two other railroads, the Norfolk-Southern and CSX. Early in this process of mergers and buyouts, Altoona's operations were seen as luxuries that the new owners could not afford. Today virtually all of the original railroad-related operations in Altoona are closed.

That does not mean that railroads are dead in Altoona. Railroad operations are so ingrained in the culture of the city that, in their effort to create a new economic base, residents of Altoona are again turning to the railroad, this time as a tourist attraction. They failed in a bid to host the Railroad Museum of Pennsylvania (a state operation), which was built in Lancaster County near the town of Strasburg. This irks Altoonans because Strasburg's claim to railroad history is as one of the first very small operations, still active as the Strasburg Railroad (a very popular tourist attraction); Strasburg never had major rail facilities, and certainly never had any industrial railroad activity. The Railroaders Memorial Museum has been opened as a privately owned museum to capitalize on the area's rich railroad heritage.

Conclusions

Landforms are the foundation on which landscapes rest. We will see how they influence everything from the weather to the economy of an area. As we examine these other topics, it will be important to compare maps of new phenomena to the map of landform regions presented in this chapter. In Pennsylvania, the most important contrasts between landform regions will be between the Appalachian Plateau and all of the areas to its southeast. Keep in mind that these are differences in topography (the relatively flat plateau cut by rivers versus the strong hilliness of the southeast), elevation, latitude and longitude, and distance from the ocean, as well as differences in population, accessibility, economy and other human factors.

Bibliography

Barnes, John H. and William D. Sevon 2002. The Geological Story of Pennsylvania (3rd edition. Harrisburg, Department of Conservation and Natural Resources, Bureau of Topographic and Geologic Survey, Pennsylvania Geological Survey. 4th series, Educational Series 4.

Briggs, Reginald P. 1999. "Appalachian Plateaus Province and the Eastern Lake Section of the Central Lowland Province." Chapter 30 in Shultz, Charles H. (ed.) The Geology of Pennsylvania. pp. 362-377.

Marsh, Ben and Peirce Lewis 1995. "Landforms and Human Habitat." Chapter 2 in Miller, E. Willard (ed.) A Geography of Pennsylvania. pp. 17-43.

Potter, Noel, Jr. 1999. "Southeast of Blue Mountain." Chapter 28 in Shultz, Charles H. (ed.) The Geology of Pennsylvania. pp. 344-351.

Thompson, Glenn H. and J. Peter Wilshusen 1999. "Geological Influences on Pennsylvania's History and Scenery." Chapter 57 in Shultz, Charles H. (ed.) The Geology of Pennsylvania. pp. 810-819.

Trainweb 2004. Horseshoe Curve: Background. Trainweb Internet site: <http://www.trainweb.org/ horseshoecurve-nrhs/Guide.htm> visited 8/26/04.

USGS 1999a. Digital Ortho Quarter Quad (aerial photograph): lititz_pa_nw. Downloaded in TIFF format from http://www.pasda.psu.edu/access/ doq99list.cgi, 8/30/04. US Geological Survey, Washington, DC.

USGS 1999b. Digital Ortho Quarter Quad (aerial photograph): halifax_pa_se. Downloaded in TIFF format from http://www.pasda.psu.edu/access/ doq99list.cgi, 8/30/04. US Geological Survey, Washington, DC.

USGS 1999c. Digital Ortho Quarter Quad (aerial photograph): honesdale_pa_ne. Downloaded in TIFF format from http://www.pasda.psu.edu/ access/doq99list.cgi, 8/30/04. US Geological Survey, Washington, DC.

Chapter 3

The Geological Story

Pennsylvania has not always looked as it does today. We have no photographic or artistic renderings of landscapes, even from the earliest Native Americans who most likely arrived as early as 14,000 years ago (Richter 2002, 5). What we do have is the geological record: samples of rock that can be dated based both on their relative vertical position and on their chemical and atomic composition. This geological history presents evidence of fascinating differences and changes, some very gradual and subtle, others rapid and cataclysmic.

Why is it helpful to learn about events of hundreds of millions of years ago? For one thing they demonstrate the slow rate of change that characterizes most natural environmental processes. Understanding time and rates of natural change in this manner can help us to appreciate the significance of many changes that we impose on the landscape. A second reason is that it helps to show us why the landscape looks as it does today. This is not just an aesthetic consideration, because the presence of hills or ridges or plateaus in some areas and flat plains or relatively low-lying valleys in others, has greatly influenced the historical movement of people into and through the state, and can help to explain the current population's distribution. The third, and probably most important, reason is that it helps in future explorations for such geological resources as coal, petroleum and even water.

In order to follow the discussion that comes next, it helps to have learned some basic geology or physical geography. Two sets of geological concepts are particularly important. One of these sets of concepts is the twin processes of erosion and deposition. Water has been present from very early times in the Earth's existence, and is the principle medium for these processes. Basically, any land that is above sea level will gradually and eventually erode. Greater elevation differences will mean steeper slopes for the water to descend, resulting in faster flows and consequently faster erosion. If given enough time before tectonic forces kick in and raise elevations, all of that land will be worn down to a flat plain at sea level. All of the sediment eroded from land areas (above sea level) ends up being deposited at lower elevations, and will eventually wind up in the oceans.

The second set of geological concepts, whose purpose is to explain those tectonic forces, is known as the theories of plate tectonics, or continental drift. These theories say that the earth's crust is not a single solid shell enveloping the Earth, but is actually broken into pieces known as plates along lines that have been identified and that are being watched carefully. These plate edges are under constant stress, and occasionally release that stress via earthquakes or volcanoes. This set of theories is accepted by the vast majority of geologists and other earth scientists because it fits so thoroughly and convincingly with the evidence encountered. It has enabled the construction of a three-dimensional representation of the Earth showing changes over its 4.5 billion years of existence. Pennsylvania's place in that story is fascinating, and sometimes surprising.

Pennsylvania as Part of the North American Plate

Pennsylvania is located on the eastern edge of the North American continent, but nearer the middle of the continental plate. The North American plate is a development that followed from other partitions and arrangements of the Earth's crust. Initially, the formation and solidifying of the planet was followed by over a billion years of bombardment by meteors. The impacts of the meteors contributed extraterrestrial minerals and also the energy to heat and deform the rocks. Microplates formed and gradually coalesced into larger continental plates. These rocks still exist at the base of each continent. At one point in this early stage of Earth's geological history, nearly all of the microplates joined to form one vast continent, which has been given the name Rodinia. This continent grew because the crust was (and is) relatively thin and fragile, virtually floating over a fluid layer of molten magma (the upper layer of the mantle), and the currents of movement in the magma were directed toward Rodinia (Barnes and Sevon 2002, 8-9).

About 725 million years ago the currents of motion in the mantle changed, and began moving away from the center of Rodinia. One portion of Rodinia separated to form Laurentia, the base of modern-day North America. Beneath Pennsylvania's surface lies one of the last microplates joined onto the Laurentia portion of the earlier continent. By the time Laurentia had formed, the portion that was to become Pennsylvania was positioned south of the equator, with the Pittsburgh and Erie areas at its northernmost end, and the Philadelphia and Allentown areas located in the southern hemisphere's mid-latitudes (Barnes and Sevon 2002, 9-10).

Life existed by the time Laurentia had formed, but it was nothing more than bacteria, algae, jellyfish and worms. It was enough to form the beginnings of a more oxygen-rich atmosphere, though. Life was represented by simple organisms which evolved into more complex and better adapted creatures. The entire geologic "era" from 570 million to 250 million years ago is known as the Paleozoic, which means "ancient animals." Most of the Earth's living organisms inhabited the ocean. Over the first 150 million years of the Paleozoic Era, the oceans remained the dominant areas where life could be found (Barnes and Sevon 2002, 10).

Mountains, Mountains Everywhere

To the east (from today's point of view) of Laurentia lay an ocean which has been named the Iapetus. By the beginning of the Paleozoic Era, 570 million years ago, the portion of the crust which held Pennsylvania had sunk relative to the rest of Laurentia, and was under the Iapetus Ocean. Because it was along the margins of the continent, it was receiving sediments from land areas to what is today our west. These sediments eventually solidified, and form the layers of rock that lie immediately above those that were the original microplates (Barnes and Sevon 2002, 11).

Over the next 300 million years, during three different periods, there were vast mountain ranges to the east (again, from today's perspective) of Pennsylvania and parallel to today's coast. The first finished developing about 490 million years ago, when a line of volcanic mountains grew up out of the Iapetus Ocean. The mountains probably took nearly 50 million years to form. Some of their volcanic ash collected in central to southeastern Pennsylvania, and eventually became an additional layer of carbonate rock. From the time they erupted, and well into the next geological period of 26 million years, they were eroded down to sea level, some of that sediment also forming another layer of Pennsylvania's bedrock (Barnes and Sevon 2002, 12).

The second mountain range formed beginning around 410 million years ago, this time due to a collision between North America (by this time the geologists stop referring to Laurentia) and continents to our east (again, today). During this event, what is today eastern Pennsylvania was covered to a point where it became dry land at the feet of the mountains, while northern and western Pennsylvania was under water. At this time we were still in an equatorial climatic zone, with much more abundant fish, amphibians and even land plants. Some of the deposits therefore included organisms found in coastal ocean (salt water) areas, while others show evidence of land-based environments such as fresh-water swamps. Once again, the mountains eventually wore completely down to sea level, and the deposited sediments eventually hardened into additional layers of rock, but this time they contained fossilized salt water or fresh water organisms. Because of our equatorial location, these environments were tremendously rich in life. The rich organic content of those buried layers of rock gradually formed into our oil and coal resources (Barnes and Sevon 2002, 16-7).

The third set of mountains began forming about 290 million years ago, once again due a great collision between the North American plate and another continental plate to our east, this time Africa's. This mountain range was much higher (over 13,000 feet) and wider (150 miles) than the earlier ones, and even extended into eastern Pennsylvania. The collision this time was so violent that layers of rock to our east were sheared nearly horizontally and thrust toward central Pennsylvania. If Philadelphia and Harrisburg had existed back then, in fact, they would have been pushed some 50 miles closer together. The effects under the Atlantic Coastal Plain and Piedmont landform regions are evident today as metamorphic rock types in highly contorted layers. The effects along the arch of the Ridge and Valley landform region were a little less violent, producing the undulations in the bedrock which led to the differences in exposure at the surface, and resulting in the ridges and valleys we see today once erosion processes set to work (Barnes and Sevon 2002, 22).

By the end of the Paleozoic Era, much of this great mountain range had already been worn away, but the process continued into the Mesozoic Era (248 million years ago to 65 million years ago). The Mesozoic is well known as the time of the dinosaurs, and Pennsylvania has produced many of their fossils, but it meant other major changes as well. First, it was a period of drier climate, as eastern North America (due to its collision with Africa) became much more landlocked. In addition, Pennsylvania had already begun its journey north (Barnes and Sevon 2002, 23-5).

Secondly, a rebound effect of the great collision that created the last mountain range, and a shift in mantle currents, began the separation of North America from Europe and Africa, and the opening of the Atlantic Ocean. That separation continues to this day, as the Atlantic continues to widen by several inches per year. The stresses that have pulled these continents apart also created some weaknesses in the crust that allowed molten material from the mantle to intrude along the northwestern edge of today's Piedmont landform region. (Barnes and Sevon 2002, 25)

Thirdly, as in the previous periods, this entire mountain range wore almost completely away, leaving the rolling hills of the piedmont, new layers of sediment in the coastal plain, and a greatly deformed crust as the most lasting evidence of its existence.

The Final Uplift and Erosion

The current period of the geological record is called the Cenozoic Era, which began about 67 million years ago. During this time span the landscape was given its current appearance. Much of this sculpting was relatively subtle, compared to previous eras. It involved the final erosion of the landscape to current elevations. In the southeastern part of the state, that meant removing almost all of the overlying layers of rock to reach the original microplate. These are the oldest rocks in Pennsylvania (Barnes and Sevon 2002, 28). In parts of the southeastern piedmont region, the magma that intruded during the Mesozoic Era was exposed.

In the rest of the state there has been a gradual uplift of land to higher elevations. This raising of the land is nothing like the previous mountain-building periods, and yet western and northern Pennsylvania is still referred to as "the mountains" by southeasterners. Before this period, the landscape of almost the entire state was worn down to a flat plain. Then in three stages the land was raised and rivers went to work eroding it back down. Each time they were unable to complete their job before the next uplift occurred. This has resulted in the ridge and valley pattern of that landform region, in the terraces along some of the major rivers, such as the Susquehanna and the Schuylkill, and in the now deeper valleys of the streams and rivers throughout the state (Marsh and Lewis 1995, 24-25).

The final significant events have occurred within the last 800,000 years (Barnes and Sevon 2002, 29-30). Beginning then and in at least three separate events, glaciers emerged from northern Canada and entered Pennsylvania's northwest and northeast corners (see Figure 3.1). Pennsylvania had reached its mid-latitude location (after journeying from its equatorial origins) and the Earth experienced prolonged periods of colder temperatures. Snow accumulated in northern Canada, and packed underlying layers into near ice conditions, until the entire accumulation reached a height at which it could not support itself. The central areas of accumulation pushed the marginal areas out in all directions, some of them heading south towards Pennsylvania. These

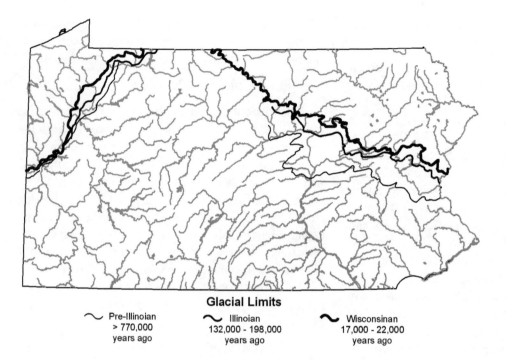

Figure 3.1 **Glaciers in Pennsylvania:** The three glacial periods had some of their greatest impacts on Pennsylvania's rivers. In the northeast the glaciers flooded larger rivers such as the Delaware and Susquehanna with greater amounts of erosive water and with sediment that ended up far downstream. In the northwest, the glaciers helped alter the directions that many smaller rivers flowed (Sevon and Fleeger 1999).

Glacial Limits

Pre-Illinoian
> 770,000
years ago

Illinoian
132,000 - 198,000
years ago

Wisconsinan
17,000 - 22,000
years ago

walls of ice acted like both sandpaper and bulldozer, scraping off northern surface materials and pushing them into Pennsylvania. The debris was left at the farthest reaches of the glaciers, and sometimes in pockets beneath the glaciers, the points at which temperatures were warm enough to melt the ice as fast as it arrived.

The three "ice ages" that we know the most about, and that impacted Pennsylvania, are the Wisconsinan, Illinoian, and Pre-Illinoian. Figure 3.1 shows their dates of activity and their furthest advances into Pennsylvania, at least as nearly as geologists can tell today (Sevon and Fleeger 1999, 15). In each case, the glacial ice sheet in Pennsylvania was likely 2000 to 3000 feet thick (half a mile high!), compared to 10,000 feet or more at its source around Hudson Bay in Canada (Sevon and Fleeger 1999, 11). We know the most about the latest, Wisconsinan, ice sheet because it obliterated features left by the earlier glaciers in many areas.

The ice and, for a long time after the ice disappeared, the debris blocked the former paths of many rivers, including the Allegheny in northwestern Pennsylvania. Rivers which once flowed northward were aided in carving new valleys, supplied with extra surges of water from the melting glaciers. Even rivers whose courses did not change significantly, such as the Susquehanna and the Delaware, still received so

much additional water and sediment as to change their nature.

After-effects of glaciers include swamps and randomly scattered lakes in areas where the surface was scoured or where the debris was left, disrupting former drainage patterns. The debris itself is also a valuable resource today, and is often excavated for its sand and gravel contents. The Pocono region of Pennsylvania is the best example of such an irregular and uneven glacial landscape. In the northwestern corner of Pennsylvania the glaciers had a different effect: that of taking a relatively rough landscape and making it smoother by filling in the uneven terrain carved by flowing rivers (Marsh and Lewis 1995, 40-41).

The Grand Canyon of Pennsylvania

Pennsylvania hosts many impressive displays of natural scenery. Most are preserved in some way, by either the federal or state government or even by private or non-profit land owners (see Chapter 31). Each site represents an outcome of natural processes that fit the geological history explained in this chapter.

An excellent example is Pine Creek Gorge, also known as "The Grand Canyon of Pennsylvania." Located on the Appalachian Plateau in Tioga County (see Figure 3.2), it is a broad deep valley carved by glacial and river erosion. While it doesn't rival Arizona's Grand Canyon in dimensions

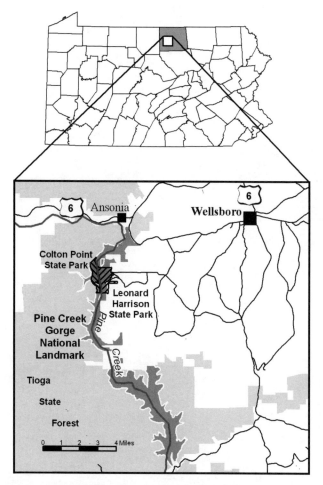

Figure 3.2 **The Grand Canyon of Pennsylvania - Pine Creek Gorge:** State Park and State Forest lands make up much of the "Grand Canyon" area today. The area was once an actively-traveled valley with a thriving lumber industry.

(Pennsylvania's is 47 miles long, up to 1450 feet deep and less than a mile wide, while Arizona's is hundreds of miles long, nearly a mile deep and over ten miles wide in places), it features a profile that is much more dramatic than other valleys in the region, lookout locations for viewing the tree covered valley from the rim, and several impressive rock outcrops (PA DCNR 2004).

An earlier valley formed during the Mesozoic erosional era by streams that usually flowed toward the north. During the last ice age, around 20,000 years ago, the glacier that entered this part of Pennsylvania covered the valley and later receded northward. It paused during that process, probably for many years, near the north end of today's gorge and left a pile of glacial debris that dammed the valley near present-day Ansonia, and created a huge lake filled with

meltwater from the glacial ice. The only escape for the water was to overflow the natural dam southward. The tremendous volume of water gave that overflow strong erosional power, and helped to carve the gorge, on its way to a new exit via the Susquehanna River. Pine Creek returned to its more modest size after the glacier completely disappeared, but has continued to carve its channel over the last 20,000 years.

The area is part of Tioga State Forest. It was once privately owned and deforested in the late 1800s, but has since reverted to state ownership (see Chapter 22). Along the river in the valley bottom was a path heavily used by Native Americans for travel in its earlier years, which became a railroad route in the 1880s that operated until 1988, and is now a hiking trail. One side of the rim, with wonderful overlook sites, was donated to Pennsylvania for preservation in 1922 by landowner Leonard Harrison, and is a state park today. The opposite rim is preserved as Colton Point State Park (PA DCNR 2004).

Conclusions

Each of the periods of our geological past, whether it was primarily a time of land building or erosion, has left some legacy in today's landscape. Earth scientists before 1900, studying maps of the Ridge and Valley and attempting to explain its cause, founded many principles of modern geology and physical geography. Modern geologists have put this story together by examining evidence, from mine explorations and drilling test cores, as well as from electronically generated ground-penetrating electromagnetic signals and seismography. Continued investigations will refine their dating and sequencing of past events.

As solid as the ground feels beneath our feet, it is always changing. Pennsylvania even has a history of earthquake activity, with modestly active areas in the northwestern and southeastern corners of the state and a more active area in lowland parts of the Piedmont through Lancaster, Berks and Lehigh Counties (Scharnberger 2003, 8).

On a geological time scale change is inevitable. There are forces constantly at work, so slowly that we have only relatively recently learned to measure them. The bottom line is that every part of Pennsylvania has been hundreds or even thousands of feet higher or lower in its geological past, and probably will be again in the future. As a result, there are many dramatic landscape scenes in Pennsylvania.

Bibliography

Barnes, John H. and William D. Sevon 2002. The Geological Story of Pennsylvania (3rd edition). 4th series, Educational Series 4. Harrisburg, Pennsylvania Geological Survey.

Marsh, Ben and Peirce Lewis 1995. "Landforms and Human Habitat." Chapter 2 in Miller, E. Willard (ed.) A Geography of Pennsylvania. pp. 17-43.

PA DCNR 2004. "The Grand Canyon of Pennsylvania." Pennsylvania Department of Conservation and Natural Resources State Parks, Harrisburg, PA. Internet site: <http:www.dcnr.state.pa.us/stateparks/prks/leonardharrison.aspx>, visited 8/29/04.

Richter, Daniel K. 2002. "The First Pennsylvanians." Chapter 1 in Miller, Randall M. and William Pencak (eds.) Pennsylvania: A History of the Commonwealth. pp. 3-46.

Scharnberger, Charles K. 2003. Earthquake Hazard in Pennsylvania. 4th series, Educational Series 10. Harrisburg, Pennsylvania Geological Survey.

Sevon, William D. and Gary M. Fleeger 1999. Pennsylvania and the Ice Age (2nd edition). 4th series, Educational Series 6. Harrisburg, Pennsylvania Geological Survey.

Chapter 4

Mineral Resources and Hazards

Beneath our feet lies much of the wealth that has served as the foundation for Pennsylvania's past success. If we think of it as a bank account containing a large sum of money, how and when we have spent that mineral wealth reflects to a great extent how we have perceived our individual and collective futures.

The minerals that have played the biggest roles in Pennsylvania's past accomplishments are iron, coal and oil, among a variety of others. We have seldom had the best or the most of these resources (the biggest exception being our anthracite coal deposits), but we have generally been the first in America to discover and exploit them. Being first may bring some advantages such as name recognition and longer experience, but it also brings some disadvantages such as being the one who starts out knowing the least about all your alternatives and being the one to make the mistakes.

In order to understand our use of the mineral resources (in later chapters) it is necessary to see why the minerals are there in the first place. What are the characteristics of these minerals that determine their qualities? Where are they found, and why there? This discussion connects our understanding of the natural processes that work to form the Earth, with that of the locations and qualities of these resources.

Sure it's a Mineral, But Is It a Resource?

Let's say you are walking along a stream and see a beautifully rounded rock about a foot long. You'd been looking for something to help edge a garden bed at home, and this rock looks perfect. All of a sudden this rock has value to you, but would you pay money for it? The question of when something like this decorative rock becomes a resource depends not on one consumer, but on whether an entrepreneur (or an existing business) sees many potential consumers. Is it worth investing time and buying equipment and supplies to collect many such rocks, and offering them for sale at a price that would earn a profit? If so, then such rocks become a resource. The real proof that the decorative rocks (or any other mineral) represent a resource in Pennsylvania is when the state's Department of Revenue requires the seller to charge sales tax, the Department of Conservation and Natural Resources requires paperwork to report how much is being removed from the environment, and the Department of Environmental Protection monitors the operation to see whether there are adverse effects on the stream.

The categories of minerals that are tracked as mineral resources are listed in Table 4.1 below. Notice that two of the minerals to be examined a little more closely in this chapter, iron and uranium, are not on the list. Iron, though never very abundant here, was very important historically, but is not actively mined in Pennsylvania today. Uranium has been produced in minimal quantities in Pennsylvania, but is much more important because of its association with a major geological hazard, that of radon gas.

Table 4.1 Mineral Resources of Pennsylvania

Coals	Non-Metals**
Anthracite	Dimension stone
Bituminous	Marble, Schist
	Diabase, Granite
Other Fuels	Sandstone, Bluestone
Petroleum	Serpentinite
Natural Gas	Slate
Uranium, Thorium	Aggregate
	Sand and gravel
Metals**	Crushed stone
Iron	Portland cement
Hematite, Magnetite	Bricks
Limonite	Clay, Shale
Siderite	Industrial minerals
Lead	Lime, Dolomite
Zinc	Slag
Chromite	Abrasive: corundum
Molybdenum	Graphite
Nickel	Agricultural minerals
Cobalt	Limestone, Dolomite
Copper	Slate
Gold *	Glass and ceramics
Silver *	Feldspar
Manganese	Kaolinite
	Magnesite (Epsom salts)

*   no longer mined   *
** Source: Barnes and
Smith 2001

Iron

Iron deposits have existed in almost every part of Pennsylvania, and made significant contributions to military and economic activity throughout the state and the US. The production and use of iron requires several steps (see Chapters 23 and 28). One clue to early locations of iron mining, purification (smelting), and manufacturing is in the names of the places where they occurred; look for names that include the word "Forge," "Furnace" or, of course, "Iron." Iron deposits located near abundant fuels and less expensive forms of transportation (rivers and railroads) were the most successful (Rodgers 1995, 285-287).

Interestingly, the deposits in these different landform regions of Pennsylvania occurred in very different geological conditions. Each of the different ores described occurs in a different type of rock (for example, metamorphic, sedimentary sandstone or sedimentary clays). They also vary in the percentage of that ore that is iron (from 15 percent to over 50

percent); even within one type of ore, different deposits will have different percentages.

The earliest deposits mined were in northern Bucks County and southern Lehigh and Berks Counties. The iron from these mines was used to make farm implements throughout colonial times, and weapons and ammunition during the French and Indian War. These ores, known as magnetites and hematites, formed in and on the margins of the Reading Prong area (the New England landform region) well over 800 million years ago (Inners 1999, 557-558).

Similar hematite ores occur in the Piedmont region, in York and Lancaster Counties, and were mined in the late 1800s (Barnes and Smith 2001, 17-20; Inners 1999 558-559). Even more famous is the Piedmont magnetite ore given the name "Cornwall-type iron" after the location in Lebanon County where it was mined from 1742 until 1973. These ores formed when magma intruded upward into the limestone-containing sedimentary deposits in the Piedmont Gettysburg-Newark Lowland areas. It took flooding caused by the heavy rains of Hurricane Agnes to close the last of these mines (Shultz 1999, 18; Gray 1999, 567-569).

In other Piedmont valleys, and in Cumberland and Lehigh Valleys (in the Great Valley area of the Ridge and Valley landform region), the ore was limonite, which occurred in sedimentary clay deposits (many of which started out as bogs over 500 million years ago) also high in limestone. It was generally important to have limestone available for the smelting of the ore, so these areas had an economic advantage for a while. In the middle to late 1800s many iron mines and related activities opened in the Ridge and Valley areas of Pennsylvania, most notably at Pine Grove Furnace in Cumberland County (now a State Park) and at similar operations in Franklin and Lehigh Counties (Inners 1999, 558-559).

Also in the middle to late 1800s numerous small iron mines opened in western areas of the state on the Appalachian Plateau. These iron ores, called siderite or iron carbonate, are also sedimentary and occur in more than a dozen of the bedrock layers that separate coal deposits. Siderite iron aided in the early stages of Pittsburgh's development as a major iron and steel manufacturing center. By the end of the nineteenth century, though, Pennsylvania's iron deposits were taking a back seat to richer deposits in Michigan and

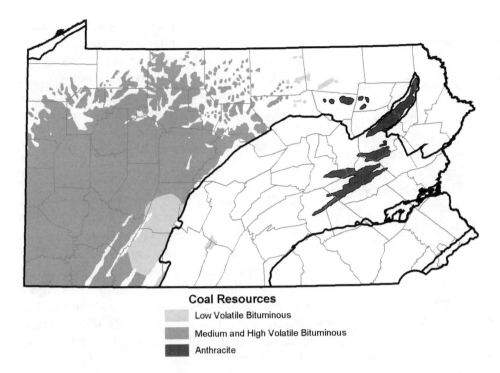

Figure 4.1 **Coal Resources:** The association of bituminous coal with the Appalachian Plateau and of anthracite coal withtthe Ridge and Valley is clear in the map. Only a few small patches seem contrary to that pattern.

Coal Resources

☐ Low Volatile Bituminous

▨ Medium and High Volatile Bituminous

■ Anthracite

Minnesota (Barnes and Smith 2001, 17-20; Inners 1999, 562-564; Rodgers 1995, 285-287).

Coal

Coal is found throughout nearly a third of Pennsylvania (see Figure 4.1). All coal has basically similar origins, as vegetation-rich freshwater swamps. In fact, the period of the Earth's geological history during which these swamps were repeatedly formed and submerged under new sediment, 290 to 330 million years ago, is referred to by geologists all over the US as the Pennsylvanian period (Edmunds 1999, 471).

The coal-rich areas generally have many layers of coal, stacked vertically and separated by non-coal-bearing sedimentary rocks. The sedimentary layers created the pressure required to turn the biological material, or biomass, into coal. The coal developed in that way is bituminous, while coal which endured the additional pressure during the later orogenies that created the Ridge and Valley landform region became anthracite coal. These differences show up in a way that has major implications for mining. Figure 4.2 shows the

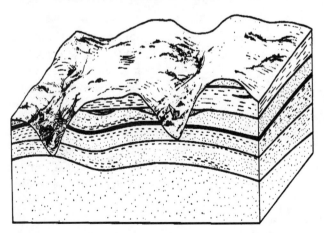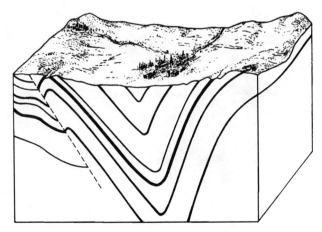

Figure 4.2 **Bituminous vs. Anthracite Bedding:** The diagrams below show the bedding of the coal in the two regions (Edmunds 2002, 10; illustration by Albert E. Van Olden used with permission). The plateau has been tectonically raised and lowered over its geological history, allowing sediments to pile up and be moderately compressed. Compression in the Ridge and Valley was horizontal as well as vertical, helping to further harden the coal into anthracite.

difference between the horizontal seams of bituminous coal and the contorted anthracite "bedding."

Bituminous coal occurs extensively around the minimally deformed Appalachian Plateau. Every layer in every region will have varying properties, such as the amount of carbon and its volatility, and impurities, such as the amount of trapped moisture and gases, intermixed mineral impurities, and sulfur (present as organic or mineral molecules). The sulfur content of Pennsylvania bituminous coals is usually high (up to 3% of its mass), presenting a challenge to its use, especially given today's stricter environmental regulations (see Chapter 35).

Approximately forty stacked seams of coal have been identified in the Main Bituminous Fields of western Pennsylvania, about 10 of which have been economically mined in some areas. To be considered for mining, a seam should be thick, accessible and have desirable mineral properties. While an estimated 11 billion tons of economically recoverable coal remain in seams at least three feet thick, far more exists in smaller or less accessible deposits. (Edmunds 1999, 479-481) However, how much of that coal can ever be removed depends on many other geographical factors (see Chapter 23).

Anthracite coal deposits are limited to several smaller fields in the northeastern end of the Ridge and Valley landform region in the eastern part of the state. Once again, the fields consist of multiple seams stacked vertically, but because of the tectonic forces that created the Ridge and Valley region the seams do not lie horizontally. Because the anthracite coal has endured more powerful forces, it has fewer impurities (less ash, and usually less than 1% sulfur) and somewhat greater heat value than bituminous. The anthracite coal has been important, especially because it is located closer to developed areas in the eastern part of the state, but there was less of it and it is harder to mine. Only 1.5 billion tons remain of anthracite likely to be economically recoverable. (Eggleston et al. 1999, 460-464)

Petroleum and Natural Gas

Oil may be the mineral for which we have the most colorful history of discovery and early use. This fluid mineral also formed in shallow waters teeming with life, but in this case they were arms of the oceans and were inhabited by saltwater species. The species were not vegetation, but fish, shellfish and algae that hugged the shore that stretched north to south and shifted westward and eastward at various times before, alternating with and after coal-forming episodes. In the best oil and gas deposits, the marine biomass was trapped in sandy sediment surrounded above and below by less porous layers of very fine sediment. After the hundreds of millions of years of pressure created liquid and gaseous hydrocarbons, the sandstone held the oil and natural gas and allowed them some mobility, while the impermeable layers trapped and even pressurized them.

Conditions at all stages of this process determine the quantity and qualities of the resulting resource. Just as crude oil can be refined into a variety of products (fuels, lubricants and tar, for example), in its crude state it is a mixture of different consistencies, including natural gas. Pennsylvania crude oil is relatively high in

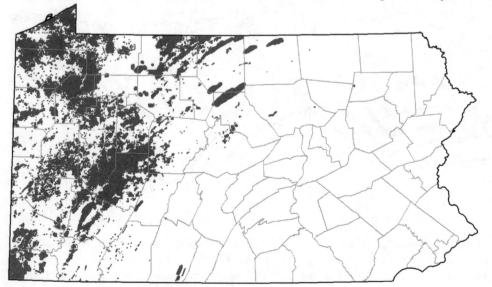

Figure 4.3 **Oil and Gas Fields:** These areas have produced oil or natural gas from at least one of many possible petroleum-producing bedrock layers. There are several Ridge and Valley and Piedmont areas which are under investigation as potential sources.

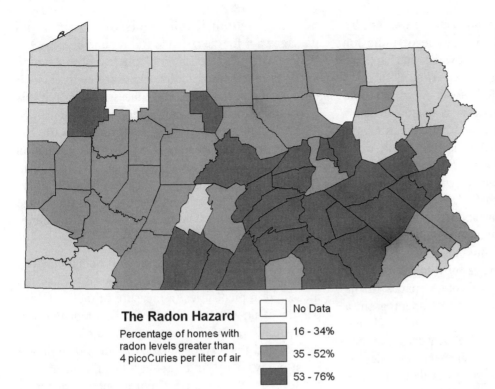

Figure 4.4: **Radon Levels in Pennsylvania Counties:** Radon levels are high virtually everywhere in Pennsylvania. Consider that even 16 percent represents one house in every six. Combine the percentages with the numbers of homes in the more densely settled lowlands of the Piedmont and Ridge and Valley regions to see the potential for harm.

The Radon Hazard

Percentage of homes with radon levels greater than 4 picoCuries per liter of air

☐ No Data
16 - 34%
35 - 52%
53 - 76%

paraffin, which makes it a high quality lubricating oil. Some deposits (see Figure 4.3) are very shallow, only hundreds of feet below the surface (Drake's original well struck oil at only 69.5 feet), while the most productive deposits are several thousand feet deep, and have been explored down to 20,000 feet and more. Most of the deepest wells produce much more natural gas than oil (IPAA 2004).

Even though Pennsylvania is no longer a major source for this petroleum- and natural gas-hungry world, many companies continue to drill exploratory wells and develop some into small producers, especially for natural gas.

Uranium and the Radon Hazard

Uranium is common as a trace element in many rocks, due to their tectonic origins. Occasionally, higher concentrations will result, though rarely will there be a uranium deposit as rich or thick as a coal deposit.

Uranium is important, of course, as the raw material for many atomic weapons, and as the fuel for most nuclear power facilities. The presence of uranium is also tied to the presence of other radioactive elements in our environment, and also to other elements, such as lead, that are not normally radioactive.

If you study the nature of radioactivity, you learn the reasons for its hazardousness. First, to call a material radioactive is really to say that the bonds that hold the pieces of its atom together are unstable. In the case of uranium, a very small proportion of the uranium atoms that occur in nature are extremely radioactive, while most of it is somewhat more stable. The more stable version, or isotope, of uranium, uranium-238, has 238 particles (protons and neutrons) in its nucleus, but the less stable isotope, uranium-235, has only 235 particles; three fewer neutrons are to be found among the particles that form its nucleus. The result of its instability is that all it takes is a relatively slow moving free neutron (that is, a bit of beta radiation) striking that nucleus, and it splits, or fissions, forming atoms of other elements, for example tin and molybdenum. Each fission also releases radiation (in the form of particles or electro-magnetic energy) and heat.

The less stable isotope of uranium, U-235, is known as fissile uranium, and is the real fuel of nuclear power plants. The "enriched" uranium that powers most of the nuclear power stations we have here in Pennsylvania is a mixture of U-235 and U-238. The enrichment refers to the fact that the naturally occurring ratio of 99.3% U-238 to 0.7% U-235 has been altered to a ratio of 97% U-238 to 3% U-235.

The transformation of a radioactive atom can also occur spontaneously, by the release of radiation (particles or energy). The rate at which that happens, and the type of radiation, varies from element to element, and from isotope to isotope. As with fission, the end result of radioactive decay is an atom of a new element, along with a release of radiation (particles or energy). Radium-226 (Ra-226) is an example of an element that decays in this way. It is the result of the fifth step in the decay chain for U-238, and Ra-226 then decays into another unstable isotope radon-220 (Rn-220). What is significant about this decay chain is that the first five steps result in solid elements, but the element radon is a gas. Thus, the radon is free to move if there are passageways for it to follow. If it follows those passageways through spaces in the bedrock and soil, through cracks or openings in a house's foundation, through the air in the house's living space, and into someone's lungs before it decays further, then the person is at great risk. When the radon decays into other elements which are not gases, those elements will remain in the lungs to decay further, releasing radiation (as particles or energy) into the vulnerable lung tissue.

This radium-to-radon-and-beyond decay chain had been observed before 1984, but came to public attention that year as a hazard when Stanley Watras was found to have high levels of radioactive radon gas in his Boyertown, Berks County home. It was discovered there because his house was built on ground whose bedrock contains high levels of uranium and passageways for the radon to follow, and because Mr. Watras worked at the Limerick nuclear power station and was tested for radiation exposure every day. It turns out that several areas in southeastern Pennsylvania have among the highest indoor radon levels in the US. They are most commonly associated with the tectonically young Reading Prong bedrock formation which comprises the New England landform region in Pennsylvania, and with other recent intrusions, and also with a swath of Pennsylvania that cuts across the Ridge and Valley and the Piedmont from Centre County toward the southeast (see Figure 4.4).

The radon hazard has been well documented to pose a serious lung cancer risk. Radon's presence is measured in picoCuries (a measure of radioactivity) per liter of air, with 0.1 pCi/l a typical outdoor air (safe) concentration, 1 pCi/l an average indoor level across the US, and 4 pCi/l (considered the equivalent of 400 chest X-rays per year, or about 8 cigarettes per day) the level at which the US Environmental Protection Agency recommends corrective action (US EPA 1986). The map in Figure 4.4 shows the seriousness of Pennsylvania's situation. The message is clear: every house should be tested (Rose 1999, 791; Geiger and Barnes 1994).

Limestone: A Resource and a Hazard

Pennsylvania has a number of well-known caves, and a map of their locations would correlate very strongly with a map of near-surface bedrock layers

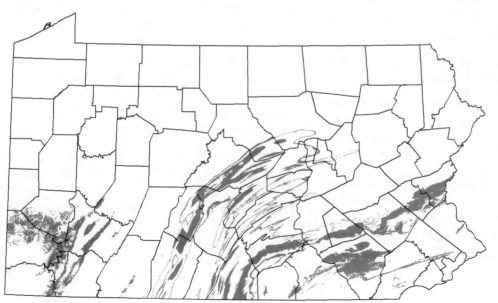

Figure 4.5 **Carbonate Bedrock:** Limestone and dolomite underlie lowland areas of the Piedmont, Ridge and Valley, and Pittsburgh Low (Appalachian) Plateau landform regions.

containing limestone and dolomite, otherwise known as carbonate rocks (see Figure 4.5). Carbonate rock is a known chemical "base" or "alkaline," which means that it counteracts the effect of acidic liquids by chemically reacting with them; in the process the carbonate dissolves into the liquid. Rain is always slightly acidic, a mild carbonic acid caused by the carbon dioxide in our atmosphere. When rainwater seeps into the soil, the water's acidity may increase due to the presence of decaying organic matter (see Chapter 9) or of minerals, such as sulfur, prone to increase the acidity of groundwater (see Chapter 36). If that water continues down into the groundwater system (see Chapter 8), it may come in contact with limestone or dolomite bedrock (Kochanov 2002, 5). Another situation that will bring similar results occurs where the carbonate bedrock is in direct contact with bedrock containing acid-inducing minerals (Dougherty 2004, 9). Both scenarios represent natural processes, but may be stimulated further by human activity.

Its chemical nature as a base gives lime some commercial value, especially as a soil additive. In fact, the soils that form naturally over carbonate bedrock are highly productive for agriculture, thanks to their natural alkalinity. Another valuable resource is created when impurities in a layer of limestone bedrock include the right amount of clay to make a natural Portland cement. Portland cement can be made almost anywhere by mixing the lime and clay ingredients. In the Lehigh Valley, especially around the town of Portland along the Delaware River in Northampton County, such mixing is not necessary; there are quite a few cement quarries and cement companies that operate in Lehigh and Northampton Counties (Dougherty 2004, 9). Additionally, limestone has value as crushed stone, or aggregate, most often used in roadbeds and concrete, and as a quarried dimension stone, cut into large (sometimes ornately carved) blocks for constructing monumental buildings.

Underground, however, the dissolving of the carbonate rock can potentially result in voids in the bedrock, which end up being natural collection points for groundwater, further continuing the process. If the void develops into a large cavern, the central part of it may become unable to support the rock and soil above it. The collapse of the surface into the cavern is called a sinkhole, and Pennsylvania has the second largest problem with sinkholes in the country, following only Florida. The problem is especially distressing in southeastern Pennsylvania due to the greater density of population and development (Dougherty 2004, 9). For the property owner whose home or business is suddenly impacted when a sinkhole develops on their property, it is often even more distressing. There have been many striking stories of huge buildings, roads, sidewalks, and parking lots disappearing overnight into newly-opened sinkholes. The greatest distress is usually the property owner's, when they learn that it is their responsibility (not the government's or anyone else's) to make the needed repairs.

Oil City and "the Oil Regions"

Many towns with economies based on local mineral resources have existed all around Pennsylvania. Depending on the resource (and other factors) some have seen continued success to the present day. Others had historically brief success with mining and related activities, only to suffer when these activities failed to survive. Oil City, like other successful survivors, has had to readapt its economy several times.

Oil City is located where Oil Creek flows into the Allegheny River in Venango County (see Figure 4.6). Oil Creek's name derives from the fact that even before the early 1600s, when its valley was settled by Seneca Indians, local Native Americans would collect oil that seeped into the river and floated to the river's edge, or dig shallow pits at oil seeps to collect it (Oil City Area Chamber of Commerce 2004). For a time it was known as "Seneca Oil," which was sometimes shortened to "snake oil," and sold at traveling shows and by traveling salesmen around the region to "cure" many ailments (Yergin 1991, 20).

In 1818, the Seneca Chief Cornplanter lost to white settlers the property that would become Oil City, a property granted to him in earlier negotiation with the Pennsylvania government. Soon after, the settlers built an iron "furnace," a mill and a foundry, which became the central businesses for a community based on supplying iron needs for the region. By 1840, the population was substantial enough that the town, by then named Cornplanter, was given a post office. In the 1850s the furnace went out of business and the town faced its first decline (Oil City Area Chamber of Commerce 2004).

In 1859, Colonel Edwin Drake began the nation's first "oil rush" when he successfully drilled for oil nearby (see Chapter 23). "Cornplanter" became Oil City and was granted a charter as a borough in 1862

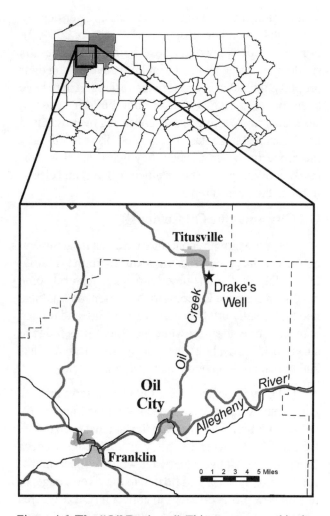

Figure 4.6 **The "Oil Regions:"** This term was used in the late 1800s when Pennsylvania was the only source area for oil in the US. Oil City and Titusville were the main centers for many oil-related activities.

(see Chapter 18). Initially, wagons driven by "teamsters" traveled the Oil Creek valley, overcharging the well owners to cart the barrels of oil to Oil City for loading onto barges bound for Pittsburgh and trains heading east. By the early 1870s, when Oil City was incorporated as a true city (Oil City Area Chamber of Commerce 2004), the oil was flowing to refineries and other industries and businesses in Oil City via local pipelines, and Oil City, nearby Titusville and other cities had "oil exchanges" for the same reasons we have stock exchanges today (Yergin 1991, 33).

The area around Oil City and Titusville, and smaller towns such as Pithole, Enterprise and Petroleum Center, became known as the "Oil Regions," and were nationally important for the remainder of the 1800s as the country's primary source of kerosene, and the only source of kerosene refined from petroleum (Yergin

1991, 29). Poor resource management (or the lack of any coordinated management) meant that high production levels in each newly discovered deposit were usually short lived. After 1900, attention shifted to larger and more easily refined sources in other parts of the country and the world. The area economy suffered a blow again.

When gasoline replaced kerosene as the primary byproduct of oil for the automobile market, local oil production had a new market and a new product: lubricating oil. Two of the largest producers of lubricants, Pennzoil and Quaker State, were headquartered in Oil City. The two merged, to form Pennzoil-Quaker State Company in 1998, and were bought by the Shell Oil Company in 2002 (Quaker State 2004). Today, Oil City is once again faced with reinventing its economy.

Conclusion

The minerals in the layers of bedrock provide resources and present hazards. Environmentalists are most concerned that the resources are finite and dwindling, and that their removal and use have left some significant hazards. Before mining began 200 years ago, for example, Pennsylvania had up to six times more coal than its current reserves. Air pollution from coal and oil combustion, water pollution from acid mine drainage, and even mine accidents, fires, and surface subsidence due to underground mine collapses (many intentional), are all examples that will be described further in later chapters.

At the same time, due to quality and accessibility issues, and especially due to competition from other parts of the US and the world, Pennsylvania's mineral resources are not as economical as they once were. Production levels are way down from peaks in the late 1800s or early 1900s. Even though reducing the production and use of our own mineral resources would no longer be as great a blow to the US economy as it once would have been, the decline in our resource productivity has had serious economic consequences, especially for many individual families and communities.

How do geologists find the resources and assess the hazards? They do it by making accurate three-dimensional maps of as many layers of bedrock as they can identify. The map in Figure 4.7 is highly generalized. The five age categories presented there are actually composites of nearly 200 more detailed

30

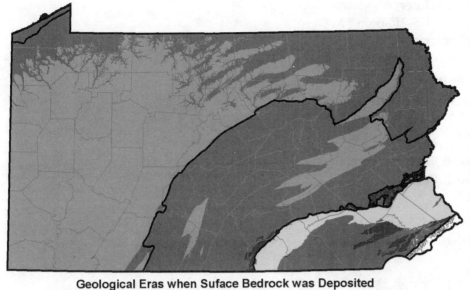

Figure 4.7 Age of Pennsylvania's Surface Bedrock: Geologists dig through the soil to find the topmost layer of solid bedrock, and then determine its type and its age. It is no accident or coincidence that the ages of those rock types correspond spatially with the landform regions (shown as black lines). The type of surface bedrock is a factor in how much or little the surface erodes over time. The older layers are usually also present under the youunger sourface rocks, a major clue to their locations of mineral resources and hazards.

Geological Eras when Suface Bedrock was Deposited

Numbers represent Years before Present

Cenozoic: 0 - 65 million

Mesozoic: 65 - 248 million

Later Paleozoic: 248 - 354 million

Earlier Paleozoic: 354 - 570 million

Pre-Cambrian: 570 million - 4.5 billion

identifying categories, and that is just for the uppermost layer of bedrock.

Over very extensive areas, younger layers of bedrock overlay the same layers of older rock, in the same sequence. Exceptions are clues to some of the major events of our geological past, as described in Chapter 3. Both the consistencies and the exceptions are also important clues to where certain types of minerals will be found, and at what depths. If we had known, two centuries ago, all that we know now about our geological underpinnings, our use of them would probably have occurred very differently.

Bibliography

Barnes, John H. and Robert C. Smith, II 2001. The Nonfuel Mineral Resources of Pennsylvania. 4th Series, Educational Series 12. Harrisburg, Pennsylvania Geological Survey.

Dougherty, Percy H. 2004. "Landform Regions of Eastern Pennsylvania and their Impact on the Cultural Landscape." In Dougherty, Percy H. (ed.) Geography of the Philadelphia Region: Cradle of Democracy. pp. 1-16.

Edmunds, William E. 1999. "Bituminous Coal." Chapter 37 in Shultz, Charles H. (ed.) The Geology of Pennsylvania. pp. 470-481.

Eggleston, Jane R., Thomas M. Kehn, and Gordon H. Wood, Jr. 1999. "Anthracite." Chapter 36 in Shultz, Charles H. (ed.) The Geology of Pennsylvania. pp. 458-469.

Geiger, Charles and Kent Barnes 1994. "Indoor Radon Hazard: A Geographical Assessment and Case Study." Applied Geography, 14(4): 350-371.

Gray, Carlyle 1999. "Cornwall-Type Iron Deposits." Chapter 40B in Shultz, Charles H. (ed.) The Geology of Pennsylvania. pp. 566-573.

Inners, Jon D. 1999. "Sedimentary and Metasedimentary Iron Deposits." Chapter 40A in Shultz, Charles H. (ed.) The Geology of Pennsylvania. pp. 556-565.

IPAA 2004. Oil and Gas in Your State: Pennsylvania. Independent Petroleum Association of America, Washington, DC. Internet site: <http://www.ipaa.org/info/In Your State/default.asp?State=Pennsylvania>, visited 8/10/04.

Kochanov, William E. 2002. Sinkholes in Pennsylvania. 4th series, Educational Series 11. Harrisburg, Pennsylvania Geological Survey.

Oil City Area Chamber of Commerce 2004. City of Oil City: Historical Overview. Oil City, PA. Internet site: <http://www.oilcitychamber.org/history.htm>, visited 9/1/04.

Quaker State 2004. Quaker State: A History of Industry Leadership. Shell Oil Products US, Houston TX. Internet site: <http://www.quakerstate.com/pages/about/history.asp>, visited 9/2/04.

Rodgers, Allan L. 1995. "The Rise and Decline of Pennsylvania's Steel Industry." Chapter 16 in Miller, E. Willard (ed.) A Geography of Pennsylvania. pp. 285-295.

Rose, Arthur W. 1999. "Radon." Chapter 55B in Shultz, Charles H. (ed.) The Geology of Pennsylvania. pp. 786-793.

Shultz, Charles H. 1999. "Overview." Chapter 2 in Shultz, Charles H. (ed.) The Geology of Pennsylvania. pp. 12-21.

Yergin, Daniel 1991. The Prize: The Epic Quest for Oil, Money and Power. New York, Simon and Schuster.

Chapter 5

Winter Weather

Pennsylvania's weather is driven by different forces in different seasons. Weather conditions are dynamic and are influenced by conditions in the atmosphere and also by many characteristics of the surface of the Earth. The term "climate" describes the typical or average weather conditions, along with the extreme deviations from the typical conditions, based on measurements over many years. Just as the earth scientists observe, measure and theorize about the causes of the landform regions, meteorologists and climatologists seek to improve their understanding of weather processes and patterns by analyzing a variety of atmospheric measurements.

Weather Factors

Put simply, the weather conditions in a given location are determined by the characteristics of the air mass currently over that location, or those conditions are the effect of a change from one air mass to the next: a weather "front." An air mass has relatively consistent properties of temperature, humidity (and therefore cloudiness) and turbulence. The movement of air masses is driven by differences in atmospheric pressure over much larger areas, such as the entire northeastern US and northern Atlantic Ocean. These pressure differences are primarily due to differences in temperature. The result is a tendency for air to move from higher pressures (cooler temperatures) toward lower pressures (warmer temperatures). Often, the lowest atmospheric pressure occurs at a junction between two or even three air masses, and becomes the focal point for very stormy conditions.

During the hottest parts of the summer, the land areas in the interior of the continent are dominated by lower air pressure than the ocean off the east coast of the United States. On the other hand, during the coldest part of the winter, the northern area of North America's central plains is dominated by higher air pressure than ocean areas. These patterns, coupled with the generally westerly (or eastward) flow of air across North America, contribute to the differences between summer and winter conditions in Pennsylvania.

Regions of similar climatic conditions generally correspond to our landform regions. Terrain factors that influence atmospheric conditions include the terrain's elevation, roughness and vegetation cover or lack of it, and the presence or absence of large bodies of water. Much also depends on the same factors in the region from which the air mass is coming. The land surface can change the temperature of the air by forcing it to rise to higher elevations, can influence wind and air mass speeds, or can contribute moisture to the air mass from surface waters.

A final major factor driving the variable weather patterns is the path of the "jet stream." The jet stream represents the strongest, most central winds, high up in the atmosphere, coming out of the west. It also represents the boundary between polar and mid-latitude air masses. If the jet stream shifts northward to a route over New York or beyond, Pennsylvania will mostly receive air masses from southern and southwestern source areas, and will usually experience warmer and more humid weather. If it shifts southward to a route across southern Pennsylvania or Maryland, most common in winter, then we are more likely to receive cooler drier air from the northwest. While it is shifting position over Pennsylvania, it operates as a "weather front," bringing these conflicting air masses together over us, often around a traveling center of low pressure; then, we experience more unsettled and often stormy weather (Yarnal 1995, 45).

Weather Data

The principle source for weather information is The National Weather Service, an office under the National Oceanic and Atmospheric Administration, which is in turn a branch of the US Department of Commerce. They maintain a network of reporting weather stations that record at least the temperature and precipitation. Many of the locations are airports or other locations that use the information themselves.

There are several issues concerning the locations of these reporting sites. For one thing, it is a well-known phenomenon that the air temperature in a city (especially downtown) is almost always a few degrees higher than the surrounding suburbs and countryside. If the airport is the regional weather station and is located outside of the city, the question is whether that is better due to not being artificially influenced, or worse since it is not measuring what that large urban population is experiencing. The second issue relates to cities such as those that are common in western Pennsylvania. The dominant air movement is across the plateau, but the cities tend to be located down in the valleys. Again the question is, "At which site should the weather station be located?" For purposes of tracking and modeling air mass movement, the plateau site is more true, but for reporting what residents are experiencing a valley location makes more sense.

Even though some of the numbers are questionable, it is still easy to see substantial temperature differences between northwestern and southeastern Pennsylvania (see Figure 5.1). The warmest area of the state, at over 56°, is at the point farthest downstream, and at sea level, along the Delaware River in Delaware County. There are two areas registering the coldest typical temperatures, less than 44°. One is in the high plateau area of the Appalachian Plateau landform region in McKean County in northwestern Pennsylvania. The other is in a similarly higher-elevation region of the Plateau in Wayne County in the northeastern corner of the state.

Areas in Figure 5.1 that show temperature isolines closest together represent significantly different temperature regions. On that map they correspond to the climb from the Ridge and Valley to the Appalachian Plateau in the southwestern and the northeastern ends of that landform transition.

Normal Winter Conditions

Pennsylvania's winter weather is most often driven by air flows from the northwest, with the jet stream shifted further to the south (see Figure 5.2). Northern central Canada is the source area (frequently a high pressure center) for many of the air masses that hit Pennsylvania during winter months. This is cold, dry arctic air which travels over Lake Erie en route to Pennsylvania. The more southern average path of the jet stream allows for greater penetration throughout the Appalachian Plateau, often into the Ridge and Valley landform region and periodically into the southeastern corner of the state.

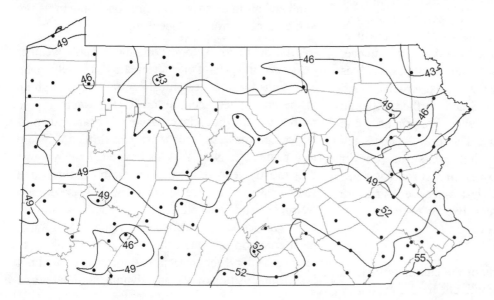

Figure 5.1 **Annual Median Temperatures around Pennsylvania:** NOAA maintains data from National Weather Service and other weather stations at these locations. Recorded data for 1971-2000 were used to create a median temperature at each station. Medians are usually preferred over means because means are easily affected by a single or few very high or very low values. Data source: NOAA 2002.

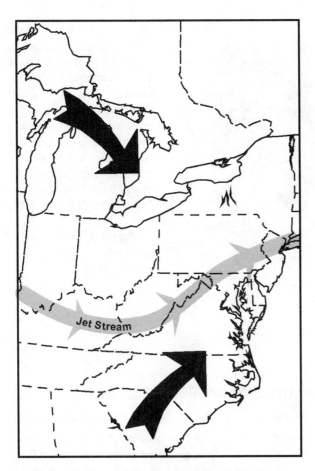

Figure 5.2 **Prevailing Winter Wind Patterns:** Arctic air pulls Great Lakes moisture into northwestern Pennsylvania to create Lake Effect snowfalls. Warmer and moister southern air moderates the southeast corner of the state.

Often, the air masses which enter Pennsylvania on this northwesterly track will pick up enough moisture over Lake Erie, and possibly the other Great Lakes, to hold the potential for precipitation. As the air mass climbs the slopes from Lake Erie to the Appalachian Plateau, orographic precipitation is induced. Figure 5.3 shows that Warren and McKean Counties are a prime location for snow in Pennsylvania, receiving over 90 inches per year (Yarnal 1995, 48 using US National Oceanic and Atmospheric Administration data from 1982). Erie and Crawford Counties are also part of this Snow Belt, feeling the impact of such Lake Effect snow storms (see Figure 5.3), with Edinboro University in southern Erie County recording an unofficial average of 137 inches from 1961 to 1990 (Stanitski-Martin and Williams 1999, 92) at the campus weather station.

Across southwestern Pennsylvania the Lake Effect will not be a factor. However, the other place which

receives the state's greatest amount of snow, about 90 inches per year on average, is located near the Maryland line in Somerset County. This precipitation is the result of different weather patterns and conditions. Moister air from the southwest travels across the lower southwestern Appalachian Plateau or from the south across the Piedmont and Ridge and Valley landform regions. In both cases the moist air must climb to the highest elevations on the Appalachian Plateau in the vicinity of Mount Davis and the Allegheny Front.

The northeastern corner of the state, again on the Appalachian Plateau, is another snowy region, with 80 or more inches a distinct possibility (Rossi 1999, 663, again based on US NOAA data, but from 1984). Here, though, the source of the precipitation is most often the southerly air that picks up moisture from the Atlantic Ocean and is guided north and east by the arcing ridges of the Ridge and Valley region. Scranton and Wilkes-Barre, upwind from this corner region, receive significantly less snow. Climbing onto the northeastern plateau, although not as dramatic a climb as the one over the Allegheny Front near the Maryland border, cools the air temperature and induces the orographic snowfall. Northwesterly air flows are also a factor, but they have to contend with the north-south orientation of many glacial features in western New York State.

Southeastern Pennsylvania, meanwhile, experiences winters with much less vigor: less frequent snowfall, and a somewhat greater likelihood that the precipitation will fall as rain, or will at least not remain frozen on the ground for very long. Snowfall totals of less than 36 inches, and even less than 24 inches in Philadelphia, are the norm. The southeast, however, is more subject to the effects of moist air off the Atlantic Ocean traveling up the east coast until it encounters cooler northern temperatures. When the jet stream is far enough south, these conditions can result in very large, though likely short-lived, snowfalls.

In addition to the precipitation effects of winter, homeowners also contend with the financial impact of winter heating bills. The National Weather Service records a weather statistic related to that impact. One Heating Degree Day is defined as one day during which the temperature stayed one degree less than 65° F, considered to be the highest outdoor temperature at which homeowners typically heat their houses. Ten

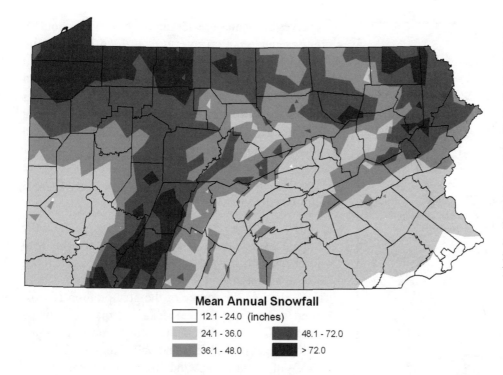

Figure 5.3 **Snowfall Totals:** Notice, first, the similarities between this map and Figure 5.1. Greater snowfall areas correspond to colder temperatures. The highest snowfall totals result from a combination of northern latitude, higher elevation, and sufficient distance from Lake Erie or the Atlantic Ocean as a moderating influence, but nearness to Lake Erie or the Atlantic Ocean as a source of moisture. (Map source: NOAA 2004)

Mean Annual Snowfall

12.1 - 24.0 (inches)	
24.1 - 36.0	48.1 - 72.0
36.1 - 48.0	> 72.0

heating degree days may result from the temperature staying 64° for ten days, or from the temperature dipping to 55° for one day, or various combinations of the two.

Figure 5.4 shows the distribution of heating degree days throughout Pennsylvania, on average. The lowest and highest values are located at the same weather stations referred to in the discussion of Figure 5.1: Appalachian Plateau locations for coldest conditions (due to elevation and latitude) and southeastern Pennsylvania for the warmest.

Extreme Winter Weather Events

During the winter, and even early spring, major snowstorms can strike in any part of Pennsylvania. In the southeast, a Low pressure center can be the focal point of a storm system which features contact between a cold arctic air mass and a warm moist air mass. In the northwestern part of the state, the lake effect snowstorms can similarly dump over a foot of snow within a day.

These blizzard conditions in the northwest can persist for weeks at a time, not necessarily snowing every day, but at least experiencing continuous gusting

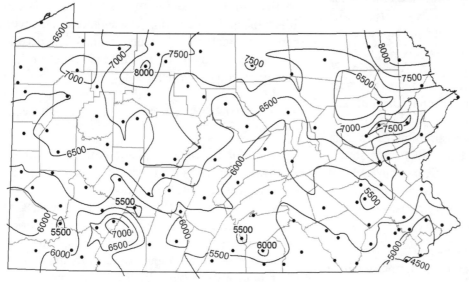

Figure 5.4 **Heating Degree Days around Pennsylvania:** This map portrays a relative measure of how hard your home's heating system has to work in any year. 6000 heating degree days represent a need to heat your home by 30 degrees Fahrenheit (F) for 200 days. (Data source: NOAA 2002)

wind from the Canadian plains. In the southeastern Piedmont and Atlantic Coastal Plain, you are more likely to hear "wait a few days and it will be gone."

The eastern part of the state has its own version of the blizzards emanating from the Canadian High: the Nor'easter. Nor'easters are known for delivering large amounts of rain or snow in New England. These storms result when a low pressure center, traveling along the jet stream, brings cold arctic air from the northwest into contact with warmer and moister air from the southeast. The intensity increases because the conflicting air masses strengthen the low much like they do in a hurricane; meanwhile a high pressure center located to the northeast of the low prevents the latter from moving out of the area. The result is a high-precipitation storm event that lasts many hours longer than a more typical storm. Winds can reach 50 miles per hour, and 24-inch snowfalls over large areas that extend into Pennsylvania are not uncommon.

Spring Transitions

As winter comes to an end, the sun reaches ever farther northward and the land areas warm faster than the ocean. When they do, the jet stream migrates more consistently to the north. In southeastern Pennsylvania, precipitation increases as the source of more "weather" once again includes more warm moist air masses from the Atlantic Ocean.

A big threat of flooding comes from a significant rainfall occurring while the ground is still frozen enough to prevent it absorbing very much of the water. In more northern areas, such as the Snow Belt around Lake Erie, an additional threat comes from a heavy rainfall during a warm spell which also melts a deep snow cover.

Warren's Winter Weather

Warren, in Warren County, is typical of Appalachian Plateau cities. Warren was founded in 1795 as a settlement for river- and forestry-related activity, and prospered also as a center for oil trade in the later 1800s. The Allegheny River valley, in which most of Warren is confined, provided natural avenues of transportation. Warren was served by three different major railroad companies simultaneously in the late 1800s and for much of the 1900s (PHMC 2004).

Transportation up to the plateau, however, has always been a challenge, due to both the natural landscape and the weather. The deep valley of the

Allegheny River protects the town from the worst winds and temperatures. However, the valley becomes a barrier to travel in the winter. Climbing the valley sides and dealing with the stronger wind and colder temperatures on the plateau are significant obstacles. Warren's elevation is about 1200 feet, while the plateau just outside town reaches over 1900 feet.

Table 5.1 shows a variety of weather data for Warren, broken down for each month of the year. Notice that its snowfall total is over 85 inches, close to the maximum for the region (City-data.com 2004).

Forest-related economic activity continues in Warren County, but little evidence of it exists within the city. There is still some remnant oil and gas industrial activity in the city. More significant today are industries focused on the molding of plastics and rubber, and on manufacturing specialty metal products.

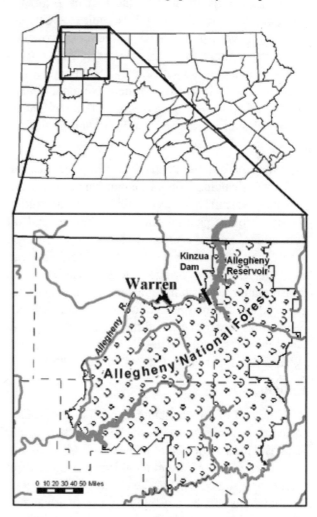

Figure 5.5 **Warren, of Warren County:** Warren is nestled in the Allegheny River valley, 700 feet below the surrounding plateau.

37

Table 5.1 Monthly Weather for Warren, Pennsylvania

	Jan	Feb	Mar	Apr	May	Jun	Jul	Aug	Sep	Oct	Nov	Dec
Average temp. (°F)*	24	26.2	34.8	45.9	56.7	65.6	69.9	68.7	61.7	50.6	39.7	29.4
High temperature (°F)*	31.3	34.8	44.6	57	69	77.3	81.3	79.6	72.1	60.3	46.9	35.8
Low temperature (°F)*	16.6	17.5	24.9	34.7	44.4	53.8	58.5	57.7	51.2	40.8	32.5	22.9
Precipitation (in)*	3	2.4	3.4	3.7	3.9	5.1	4	4.3	4.2	3.4	3.9	3.6
Snowfall (in)**	22.5	15.2	11.9	2.7	0	0	0	0	0	0.3	10.3	22.7

* Source: NOAA 2002.
** Source: City-data.com 2004.

The largest single employer, though, is a mail order clothing and housewares company.

In 1923 the federal government created Allegheny National Forest (see Chapter 10). In 1964, as part of efforts to control flooding in the Allegheny River valley, especially Pittsburgh, the Kinzua Dam was built, creating the Allegheny Reservoir (see Chapter 7). Both have had an impact on the local economy, attracting visitors to Warren as the largest city in Pennsylvania close to both sites. The city advertises itself as "The Gateway to Allegheny National Forest" and "The Capital of Kinzua Country" (City of Warren 2004). Winter recreation is an important part of the attraction.

Warren is another Pennsylvania city that has had to adapt to changing economic conditions. In this case, however, the shifts have been within the manufacturing sector. From a past dominated by lumber and oil companies, and to a present featuring specialized manufacturing operations, the city is looking ahead to greater activity featuring services such as improved transportation accessibility, housing and care for senior citizens, and tourism-related improvements such as a convention center, increased traffic to the Allegheny National Forest, and an Allegheny River "Musarium" (Warren Chamber 2004; City of Warren 2004).

Conclusions

One half of the challenge, as well as the beauty, of living in Pennsylvania is its winters. Parts of the state rival the snowiest areas in the country, but without the more extreme cold that the northern Rockies and Plains, or upstate New York, experience. This set of combinations may turn out to be quite an advantage as northern areas of the state are learning to market themselves as centers for winter recreation.

Bibliography

City-data.com 2004. Warren, Pennsylvania. Internet site: <http://www.city-data.com/city/Warren-Pennsylvania.html>, visited 9/14/04.

City of Warren 2004. City of Warren, Pennsylvania. Warren, PA, City of Warren. Internet site <http://www.cityofwarrenpa.org/>, visited 9/12/04.

NOAA 2002. Monthly Station Normals of Temperature, Precipitation, and Heating and Cooling Degree Days 1971-2000: Pennsylvania. Climatography of the United States series, Publication No. 81. Asheville, NC, National Oceanic and Atmospheric Administration: National Climatic Data Center.

NOAA 2004. SNOW1413.shp (digital map file). Internet site: <http://www5.ncdc.noaa.gov/cgi-bin/climaps/climaps.pl>, visited 7/13/04.

PHMC 2004. Warren County. Harrisburg, PA, Pennsylvania Historical and Museum Commission: Pennsylvania State Archives. Internet site: <http://www.phmc.state.pa.us/bah/dam/counties/browse.asp?catid=62>, visited 9/12/04.

Rossi, Theresa 1999. "Climate." Chapter 43 in Shultz, Charles H. (ed.) The Geology of Pennsylvania. pp. 658-665.

Stanitski-Martin, Diane and Kay R.S. Williams 1999. "A Climatological Summary of Pennsylvania's State System of Higher Education Universities." The Pennsylvania Geographer. Vol. 37, no. 2 (Fall/Winter), pp. 80-101.

Warren Chamber 2004. Largest Employers. Warren, PA, Warren County Chamber of Business and Industry. Internet site <http://www.wcda.com/>, visited 9/12/04.

Yarnal, Brent 1995. "Climate." Chapter 3 in Miller, E. Willard (ed.) A Geography of Pennsylvania. pp. 44-55.

Chapter 6

Summer Weather

Now that we have looked at the colder months and examined some of the processes at work in influencing Pennsylvania's weather, it is time to take a similar look at the warmer months. The same factors are at work during the summer, but there are some key differences.

Dominant westerly winds, landforms and the jet stream still play roles. The northward shift of the jet stream (see Figure 6.1) allows for air masses from the south and west to have a larger influence. The Ridge and Valley "mountains" and the Allegheny Front will guide them even more strongly than in winter. In addition, pressure-driven movements of air masses favor the sea-to-land breezes that are stronger in summer, whenever the westerly flows weaken.

One of the common principles of meteorology explains precipitation as the cooling of air that contains water vapor (water in a gaseous state). The cooled water vapor condenses into liquid droplets. Initially, the droplets form clouds, but they can also grow heavy enough to fall as rain or snow (or sleet or hail, etc.). The orographic Lake Effect snowfall presented in Chapter 5 is an example of this: the cooling was caused by the air being forced to climb hundreds of feet in elevation. Similar processes are at work in summer, with the temperature change induced either by elevation change (again, due to landforms), by altitude change (due to conflicting air masses) or by moving further north.

Typical Summer Weather

During the peak summer months, air movement over southeastern Pennsylvania frequently returns to

a situation of air from our south being pulled up the east coast (along the eastern front of the Appalachian highlands) toward New England. This air has

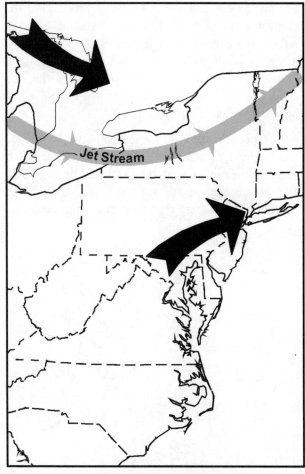

Figure 6.1 **Prevailing Summer Wind Patterns:** Warm moist air is guided to the northeast by the Ridge and Valley and Allegheny Front landforms. The Appalachian Plateau receives midwest air masses (after Yarnal 1995).

39

relatively high humidity because it has spent time near or over the ocean. The "pull" here is exerted by the northward shift in the Earth's tilt relative to the sun, and the corresponding northward shift in the jet stream.

Meanwhile, western and northern Pennsylvania in the summer are more commonly experiencing air masses carrying air from the west and southwest across the Appalachian Plateau on their way to the northeast. The air masses that come from Texas, for example, can be drier and susceptible to greater differences between daytime and nighttime temperatures. However, because these air masses are climbing in elevation from the Midwest toward Pennsylvania, they do contribute rain to the higher Plateau elevations.

These two patterns lead to the general trend of greater or lesser moisture influenced by local landscape features in Pennsylvania over the summer. Figure 6.2 shows that precipitation in the southeast decreases as we move from the Delaware River toward the northwest and the Allegheny Front, due to a combination of elevation and distance from the Atlantic Ocean. Similarly, precipitation tends to be lower in western Pennsylvania on the Appalachian Plateau, and to increase as the air moves farther east, except for a couple relative lows near the New York border (PA State Climatologist 2004a).

The combination of these two patterns produces the highest rainfall totals near the crest of the Allegheny Front close to the Maryland border. All in all, the summer season is more prone to turbulent conditions producing thunderstorms, and results in a few more total inches of precipitation than in the winter (PA State Climatologist 2004a).

The map in Figure 6.3 of Cooling Degree Days is similar in definition to the map of Heating Degree Days (Figure 5.4) in Chapter 5, but shows nearly the opposite pattern. In this case the calculation is based on the assumption that homeowners cool their houses when temperatures reach above 65 degrees (F). The highest cooling bills will be paid in the Philadelphia area, while the lowest bills will be in the coolest locations in northern Plateau areas.

Summer Extremes

While the typical summer weather is variable, the unusual occurrence is for the weather to get "stuck" in a prolonged pattern. Drought, so common in other parts of North America, does not generally result from the weather systems here in the east. Likewise, it is unusual for hot muggy conditions to persist without a break for weeks on end. Such extremes of summer weather have only occurred occasionally, with the drought of 2002 a recent example.

In addition to droughts and heat waves, other extreme summer weather conditions to develop in or affect parts of Pennsylvania are the hurricane (or similar tropical storm) and the tornado.

Summer is hurricane season in the tropical Atlantic Ocean and in the Caribbean Sea. Deeper summer heating of the ocean's surface waters contributes to strong updrafts. Turbulence between the mid-latitude westerlies and the tropical easterlies generate the cyclonic wind patterns. Most of the Atlantic storms

Figure 6.2 **Precipitation Totals around Pennsylvania:** Data recorded at teh same weather stations as Figure 5.1 show total annual rainfall and the liquid equivalents of frozen precipitation. The highest totals are in the southern Appalachian Plateau, Where the highest elevations are, and along the Delaware River. Other high totals correspond to areas that receive the highest snowfalls (see Figure 5.3). (Data source: NOAA 2002)

40

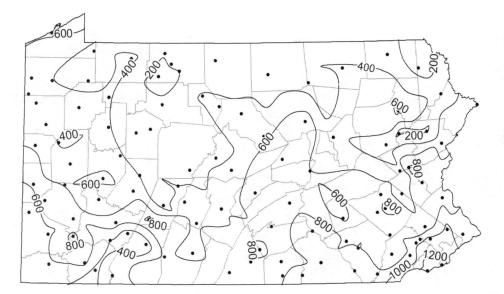

Figure 6.3 **Cooling Degree Days around Pennsylvania:** The opposite of heating degree days (see Figure 5.4), cooling degree days measure cooling demand on air conditioners and fans. Ten cooling degree days represent the need to cool the house by one degree F for ten days, or by ten degrees F for one day, or some combination of the two. (Data Source: NOAA 2002)

that reach this far north start in the open ocean and travel to the west or northwest, hitting the east coast but weakening rapidly as they reach inland. The impact of any storm depends mostly on the location of landfall and on the extent and strength of the storm. Southeastern Pennsylvania is the part of the state most likely to get heavy rainfall from the final remnants of such hurricanes.

In late June 1972, Hurricane Agnes came to Pennsylvania by a rather unusual route. It graduated on June 16th, to tropical storm status over Mexico's Yucatan Peninsula. The Caribbean storms are more likely to track toward Texas, but this storm came northward and then veered east toward the Florida panhandle. The storm gathered just enough strength to become a Category 1 hurricane, with maximum winds around 85 miles per hour, before striking the Florida coast on June 19th. It quickly lost strength to Tropical Storm and then to Tropical Depression status, but held together long enough to reach the Atlantic Ocean again after a two-day trip across Florida, Georgia, South Carolina and North Carolina. Back over the ocean again, it regained enough strength to reach Tropical Storm status again.

Agnes's most significant impact was the rainfall it generated as it traveled north, along the coast. It came ashore again around New York City on June 22nd, traveled perhaps a hundred miles, and reached the Catskill Mountains area of New York State's Appalachian Plateau landform region, just beyond Pennsylvania's northeast corner. The storm was broad,

hundreds of miles across, and part of it remained over the ocean collecting more moisture. The impact of the storm was felt mostly in New York and Pennsylvania as it stopped in the same area of New York State for over 24 hours while winding down. The town of Shamokin in the Ridge and Valley region received about 18 inches of rain in that period, while the rest of eastern Pennsylvania received 6-10 inches (MARFC 2004). This was a tremendous amount of rain, whose flooding impacts will be described in Chapter 7.

Another form of extreme weather is the formation of tornadoes. Although they are usually thought of as phenomena of America's Midwest, Pennsylvania has also experienced hundreds of tornadoes (since records were kept beginning in the late 1800s). Tornadoes are the result of atmospheric turbulence occurring when an advancing colder air mass forces its way under the edge of a more stationary warmer air mass. A single storm system can spin off several tornadoes. Such circumstances have developed many times in Pennsylvania. The worst of these was on May 31st, 1985, as part of a storm system moving from the Midwest across Ohio and into Pennsylvania. Twenty seven tornadoes touched down in the two states that day, causing a combined 75 deaths, 1025 injuries, and over $450 million in damages (NOAA 1992).

Autumn Transitions

Autumn is strongly influenced by the decreasing strength of the sun over continental North America, the corresponding decrease in the pressure differences

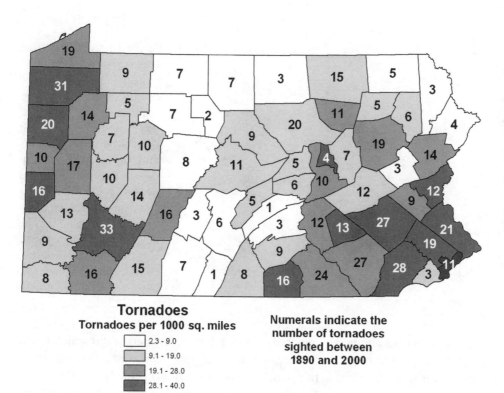

Figure 6.4 **Tornado Activity in Pennsylvania:** The shading is a better indicator of the relative hazard in each county, and shows that Philadelphia is at greatest risk. Since multiple tornadoes often appear in the same storm, few counties see them more than once in a decade. (Data source: PA State Climatologist 2004b)

Tornadoes
Tornadoes per 1000 sq. miles

- 2.3 - 9.0
- 9.1 - 19.0
- 19.1 - 28.0
- 28.1 - 40.0
- 40.1 - 76.7

Numerals indicate the number of tornadoes sighted between 1890 and 2000

between continent and ocean, and the resulting shift in position of the jet stream. By this time of year, even the southeastern part of the state is experiencing wind patterns and diurnal temperature ranges more like what the plateau areas have been experiencing all summer. The last of the strong tropical storms occasionally creates very stormy or unsettled conditions. The humid hot conditions are more frequently relieved by cooler, drier air traveling from the west or occasionally the northwest.

Gettysburg's Famous Summer

Gettysburg is best known for the Civil War battle fought there. Take away that battle, for the moment, and Gettysburg is still an important regional city. It is centrally located in the Gettysburg-Newark Lowland portion of the Piedmont, at around 500 feet above sea level. Its soils are not as productive as those of the counties to its east, but it supports a very strong orchard-based agriculture. Adams County, in which Gettysburg is located, is nicknamed Apple Capital,

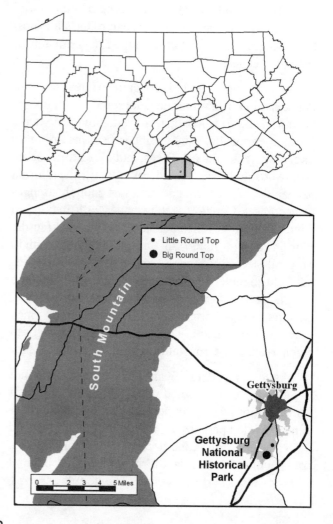

Figure 6.5 **Gettysburg:** Gettysburg's location east of South Mountain and in the Gettysburg-Newark Lowland played a significant role in Civil War strategy. They also factor significantly into local weather conditions.

42

Table 6.1 Monthly Weather for Gettysburg, Pennsylvania

	Jan	Feb	Mar	Apr	May	Jun	Jul	Aug	Sep	Oct	Nov	Dec
Average temp. (°F)*	29	31.4	40.6	50.4	60.5	69	74.2	72.5	65.3	54	43.4	34.5
High temperature (°F)*	38.7	42	51.4	62.2	72.2	80.7	85.8	84.5	77.3	66.9	54.9	44.4
Low temperature (°F)*	19.2	20.8	29.8	38.5	48.8	57.2	62.5	60.5	53.1	41	31.9	24.5
Precipitation (in)*	3.3	2.8	3.6	3.5	4.3	4.5	3.5	3.5	4.3	3.1	3.6	3.2
Snowfall (in)**	9.2	8.7	5.7	0.4	0	0	0	0	0	0	1.8	6.1

* Source: NOAA 2002
** Source: City-data.com 2004

USA and is at least the primary apple producing county in Pennsylvania.

Gettysburg's site made it a strategic location during the Civil War. It is just east of South Mountain, the first ridge of the Ridge and Valley landform region along the Maryland border. West of South Mountain is the Great Valley, connected to the Shenandoah Valley in Virginia. For Lee's Confederate armies, it represented an effective way for Confederate forces to penetrate the north beyond the focus of activities to the east in more developed areas of the Piedmont and Atlantic Coastal Plain. Gettysburg is located in the Gettysburg-Newark Lowland area of the Piedmont, which was intruded by volcanic rocks following the last collision between North America and Africa. In this area of the Piedmont, there are a number of relatively low hills and ridges, which have formed where the bedrock is locally more resistant to erosion. A number of these formations played significant roles in the famous Battle, such as Seminary Ridge, Round Top and Little Round Top (Brown 1962, 2-11).

Gettysburg's weather is more typical of southeastern Pennsylvania than of the other "mountainous" portions of the state (see Table 6.1). Even the following listing of weather conditions during the Civil War Battle of Gettysburg looks like conditions common to other parts of the Piedmont and of Philadelphia during the summer.

- Wednesday July 1, 1863: 72 to 76°. Cloudy with a slight breeze

- Wednesday July 2, 1863: 74 to 81°. Morning fog, then cloudy with a slight breeze

- Wednesday July 3, 1863: 73 to 87°. Humid and cloudy with a slight breeze (Brotherswar.com 2004).

Might the weather conditions have affected the battle? Which army would they have favored?

Conclusions

Summer weather in Pennsylvania is just as fraught with storm conditions and temperature extremes as its winters. From late spring through the summer, Pennsylvania receives slightly higher precipitation amounts than during the fall and winter. Compare Tables 5.1 and 6.1 to see both the typical patterns and the differences between northern and southern locations.

Pennsylvania's climate is defined by its wide ranging conditions. Winters in the northwestern part of the state are similar to those of states to its north, and summers in the southeast are similar to those of states to its south. To some of us it is the best of both worlds, but to others it is the worst.

Bibliography

Brotherswar.com 2004. The Battle of Gettysburg and the American Civil War. Internet site: <http://www.brotherswar.com>, visited 9/15/04.

Brown, Andrew 1962. Geology and the Gettysburg Campaign. Educational Series 5. Harrisburg, Pennsylvania Geological Survey.

City-data.com 2004. Gettysburg, Pennsylvania. Internet site: <http://www.city-data.com/city/Gettysburg-Pennsylvania.html>, visited 9/14/04.

MARFC 2004. The Life of Hurricane Agnes. State College, PA, Middle Atlantic River Forecast Center. Internet site: <http://www.erh.noaa.gov/marfc/Flood/agnes.html>, visited 2/12/03.

NOAA 1992. Tornadoes: Nature's Most Violent / Storms. Washington, DC, National Oceanic and Atmospheric Administration. Internet site: <http://www.nws.noaa.gov/om/brochures/tornado.htm>, visited 9/14/04.

NOAA 2002. Monthly Station Normals of Temperature, Precipitation, and Heating and Cooling Degree Days 1971-2000: Pennsylvania. Climatography of the United States series, Publication No. 81. Asheville, NC, National Oceanic and Atmospheric Administration: National Climatic Data Center.

PA State Climatologist 2004a. [No Title: a climatological description of Pennsylvania]. State College, PA, Office of the Pennsylvania State Climatologist. Internet site: <http://pasc.met.psu.edu/PA_Climatologist/state/index.html>, visited 7/6/04.

PA State Climatologist 2004b. Number of Tornadoes: 1881-Present. State College, PA, Office of the Pennsylvania State Climatologist. Internet site: <http://pasc.met.psu.edu/PA_Climatologist/state/misc/tornado00.jpg>, visited 7/6/04

Chapter 7

Surface Water Resources

All of the rain that falls in Pennsylvania heads in the direction of the Atlantic Ocean. However, it does so by a number of rather circuitous routes. There is an area of Potter County in which three of the largest drainage basins oriented to the Atlantic come together (Murphy and Murphy 1952, 21). Rain falling over much of the Appalachian Plateau area of western Pennsylvania exits the state via the Ohio River (or any number of its tributaries) and exits North America via the Mississippi River. Other streams flow directly toward Lake Erie or toward New York State's Genesee River, leaving the continent via the St. Lawrence River. From southeastern Potter County the water heads toward the southeast to the Susquehanna River, entering the Atlantic via the Chesapeake Bay.

In addition to traveling hundreds of miles in distance, the rain must descend from its point of origination in the atmosphere to as low an elevation as sea level. If the entire route is accomplished on the surface, the duration of the journey will be much shorter than if it descends into the soil and continues down into the geological layers beneath the surface (see Chapter 8). For the surface flow, the combination of vertical descent and geographical distance determine the slope of that flow, which in turn determines the rate of flow. From Potter County, water headed out the Chesapeake Bay gets there the fastest, while the Mississippi-bound water is the slowest.

The One-Way Network

The hydrologic cycle circulates water from the Earth's surface to the atmosphere and back. Upon returning to the ground, the rainwater (or the melted snow or other frozen precipitation) attempts to soak into the soil (see the next chapter) and, when that is saturated or if the "ground" it hits is an impermeable surface, then the precipitation becomes runoff. Once the runoff reaches a stream or other water body, it becomes part of the network of streams and rivers, with swamps, lakes and reservoirs acting as temporary delays in that flow.

Streams and rivers vary in their abilities to deliver water (and other river contents) downstream. The variability of the river's depth and discharge are the two biggest factors. Depth influences the navigability of any water course, determining the kinds of watercraft able to travel it and the load they can haul. Rivers were the first highways for natives and colonists, but Pennsylvania's rivers proved inadequate for getting goods very far inland. Our largest river, the Susquehanna, is very shallow and broad, and the Delaware River provides river access to the Port of Philadelphia only because its depth is maintained for today's larger ships by keeping a ship channel dredged.

A river's rate of flow is called its discharge, measured as a combination of volume and velocity in such units as cubic meters per second (m^3/s), cubic feet per second (cfs) or gallons per day (gpd). Its discharge is the most important factor in determining a river's ability to erode materials from the surface and carry them toward the ocean, and also determines water availability for human use. The discharge generally increases downstream after tributaries have entered the river, but can decrease downstream after significant withdrawals (for cities or factories) have taken place. The Susquehanna's discharge increases from 300 m³/s at Towanda, to 967 m³/s at Harrisburg and to 1171 m³/s at the Conowingo Dam just over the state boundary into Maryland (SAGE 2003).

The watershed, or drainage basin, of a particular stream or river is the area of ground from which all runoff exits through that stream or river. The watershed of a tributary stream or river is part of the watershed of its parent river; for example, the Schuylkill River watershed is a sub-basin of the Delaware River watershed. For these reasons, all issues involving water quantities or water quality first require a look upstream and uphill. For example, since a river's flow comes from its entire watershed, whenever a storm occurs runoff gradually gathers into the main river, increasing its discharge by both increasing its velocity (it flows faster) and its volume (its water level rises). Too much, of course, and flooding will occur.

Well, Sort of One-Way

The present courses of our rivers have been very persistent over the history of the earth since the last major mountain range east of Pennsylvania eroded (see Chapter 3). This is the main reason why the Susquehanna River appears to have plowed its way right through several ridges of the Ridge and Valley landform region. In fact, the ridges grew up around the river; all the river had to do was hold its course and keep eroding its channel as the magma currents in the mantle below pushed the land upward. During some periods, especially towards the ends of the ice ages, much larger quantities of water flowed with much

greater force and created a wider valley. With the extra debris in the upstream stretches of the river, more and larger sediment was carried downstream. Much of it ended up lining the river on both sides, creating level flood-plain terraces that are attractive for development but also subject to occasional floods (Marsh and Lewis 1995, 42).

While this natural engineering of the middle Susquehanna was impressive, the changes in the glaciated regions of the state were even greater. Before the ice ages parts of the Allegheny and other rivers flowed northward. The ice and debris blocked these rivers and filled them with water and sediment, all of which fed and empowered a southern flow connecting them with the Ohio River (Marsh and Lewis 1995, 41).

During and after the ice age a phenomenon similar to what formed the Susquehanna was at work on the Appalachian Plateau in western Pennsylvania. As the general elevation rose, the rivers held their courses and carved down into the plateau, exposing many of the lower layers of bedrock at these river cuts. These river valleys are relatively narrow and steep-sided, shaped more like a "V" (see Figure 2.6 in Chapter 2).

Dams and Their Uses

In the eyes of engineers the network of streams and rivers serves as a water delivery system. The movement of all this water also represents a valuable form of energy. A major difficulty with this natural system is its irregularity. The engineering solution is to build dams, which store water so that it can be

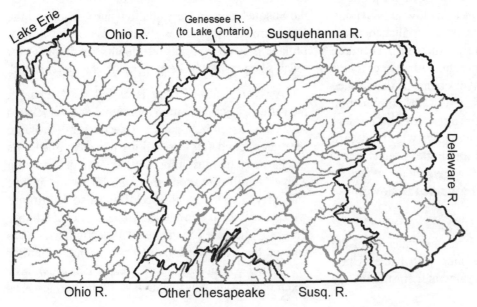

Figure 7.1 **Rivers and Watersheds:** These are just the major rivers of Pennsylvania! If every stream were shown, the map would be incredibly dense. The black lines delineate the major watersheds for the state: waters that leave the state via the different major rivers.

withdrawn or sent on its way when it suits people's needs to do so. Pennsylvania has about 3000 dams registered with its Department of Environmental Protection, ranging in height from seven feet to over 400 feet.

Their "V" configuration makes Appalachian Plateau river valleys ideally set up for "high" dams. The narrower valley reduces the amount of material needed to construct the dam. The deepness of the valley means that higher dams can be constructed. This is, of course, the principle behind the huge dams in the Colorado and Columbia River valleys in the western US, and was the reasoning behind the construction of many Pennsylvania dams (though on a somewhat smaller scale), such as the Kinzua Dam, just east of Warren in Warren County (see Figure 5.5 in Chapter 5). The Allegheny Reservoir created by the Kinzua Dam extends about twenty miles up river, well into New York State.

Most dams represent sources for municipal water supply. The water represents a stored supply "just in case," but ideally no more is withdrawn over any given period of time than is naturally replenished.

Dams are also used to create stable flows for hydroelectric facilities (see Chapter 26). The reservoir behind the dam is the utility company's way of assuring that they can generate electricity at any moment on very short notice. Many hydroelectric dams generate their electricity only for peak energy demand, because they are much easier to start and stop than other types of power stations. A key difference between hydroelectric facilities and other uses of the rivers'

natural flows is that the former is not a withdrawal; the flow continues uninterrupted and unpolluted. Not every hydroelectric generator requires a dam that completely captures a river. The York Haven Dam on the York County side of the Susquehanna River is an example. Its dam is angled upstream less than half way across the river, but that is enough to capture the water needed to run the dam at full capacity. Out of those 3000 dams in Pennsylvania, only nineteen are hydroelectric.

We have already said that a major function of dams is to store water for people who need to use it. Another function is to create the *capacity* to store water; this gives dams a flood protection role. In order to perform this function well, the water level in the reservoir has to be low. When the rains come the dam holds back the water, which can then be released slowly as conditions downstream permit after the storm. Many of the largest dams in Pennsylvania are called tailings dams, designed to hold back water flowing out of coal mining operations in the southwestern counties in order to prevent pollution and sedimentation downstream (see Chapter 36).

Some dams are built for the purpose of creating a recreational lake, or are at least maintained so that recreation can be carried on in addition to its primary purpose. Many hydroelectric and flood protection dams get adapted in this way. The owners, whether private or public, create swimming beaches, build boat docks, stock the reservoir with fish, and add amenities to the surrounding land areas as well, in order to enhance their company's or government agency's public image. Some may charge visitors in order to cover costs or create an additional source of income.

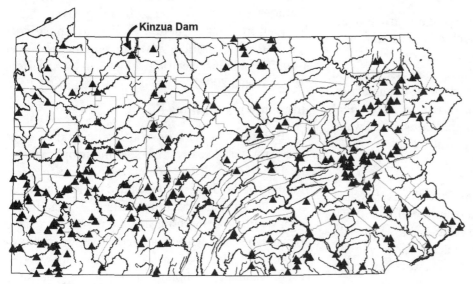

Figure 7.2 **Just Some of Pennsylvania's 3000 Dams:** This map shows nearly 300 dams (some with several separate sections). Most were built and are maintained by the US Army Corps of Engineers. Notice the clusters of dams around Pittsburgh and in the eastern Ridge and Valley region. Most of these play some flood control function for downstream areas.

The Kinzua Dam was created as a flood control dam on the Allegheny River. The need for this dam arose from repeated flood damage to areas as far away as Pittsburgh: the Kinzua is one of at least sixteen dams on the Allegheny River or its major tributaries protecting Pittsburgh. Engineering projects with such a large scope are most often the responsibility of the US Army Corps of Engineers, the engineering arm of the military in particular and the federal government in general. The process begins with site selection (for the dam and its reservoir) and preparation. The site chosen lies largely within Pennsylvania's only national forest, Allegheny National Forest. However, it also included land set aside by a 1794 treaty for the Seneca Indians. The Seneca Nation fought the project for several years in US courts, but ultimately lost.

The land projected to be under water was cleared and construction began in 1960. By 1966 the dam rose 179 feet above the Allegheny River. It was not designed as a hydroelectric dam, but a subsequent project created another type of hydroelectric facility: a pump storage system. A pump storage system is designed to use cheap or free electricity at low demand times to pump water uphill to a higher-elevation storage reservoir; then, when electricity is in high demand, the water can be released back downhill through turbines to generate electricity. The dam's main reservoir holds 573,000 acre-feet of water (one acre-foot is equal to a one foot deep layer of water covering an acre of land), with levels maintained in order to support an extensive water- and outdoors-based recreation facility. Its pump-storage reservoir adds 6720 more acre-feet of water storage (US ACE 1999).

The impacts of building dams are not all beneficial, though. First, much otherwise attractive land area is flooded when the reservoir is filled. Remember, river valleys have long been prized locations for everything from factories to communities. In addition, dams represent artificial ecological systems. Everything from fish species to altered water quality can have impacts far downstream.

Flooding and Its Impacts

Now that we have a sense of how the river system works, it is also important to be aware of how it can break down. Because river discharges vary so much, communities built along rivers adapt to the varying velocities and levels of the river. High water levels that occur regularly (annually, perhaps, or even every fifty years) define the floodplain. Floods that reach significant levels above normal are described by the amount of time that has passed or is expected to pass, on average, between other floods reaching that same level. A "hundred-year flood" is an extremely high flood, since it is expected only once every hundred years. Riverside development must be mindful of the cost of potential damage to anything built, and how often the owner can afford to repair or replace such damages. Most commonly, the landowners locate storage and transportation facilities within the 50-year flood zone because they can afford to sustain such damages. For many years, the federal government subsidized insurance companies who sold flood insurance to such developers.

Hurricane Agnes, whose meteorological prowess was described in Chapter 6, created some of the most extensive flooding in Pennsylvania history. Because the rainfall was greatest in the northeast corner of the state, its effects were felt in the Susquehanna and Delaware River valleys. Before the storm even hit, the river was a few feet above its normal water level, due to recent rains. By the end of the day that the eye of the storm reached New York's coast, June 22, 1972, the storm's rains had already raised the water level of the Susquehanna in Wilkes-Barre to 25 feet above normal. It kept rising the next day as the storm gradually broke down, and did not crest until 7:00 pm on June 24th at 40.6 feet above normal. Wilkes-Barre was the town that experienced the worst flood impacts: they had the largest share of the 72,000 people evacuated from homes during the flood, and lost electricity for up to thirteen days. Altogether, 150 factories were closed due to flooding; over 65,000 homes were damaged or destroyed; roads and bridges required $300 million worth of repairs; and over 14,000 people lost electricity (Agnes in Northeastern Pennsylvania 2004).

Johnstown's Famous Flood

Before 1889 Johnstown's flooding experiences were familiar to many river-side industrial cities of the late 1800s. Their adaptations to the frequent flooding of the flood plain included the fact that most of the heavy industrial development, especially US Steel's mill, and the train depot were there. Since Johnstown was virtually a one-industry town, the low-income mill laborers mostly lived in cheaper wooden dwellings along the floodplain provided by the steel

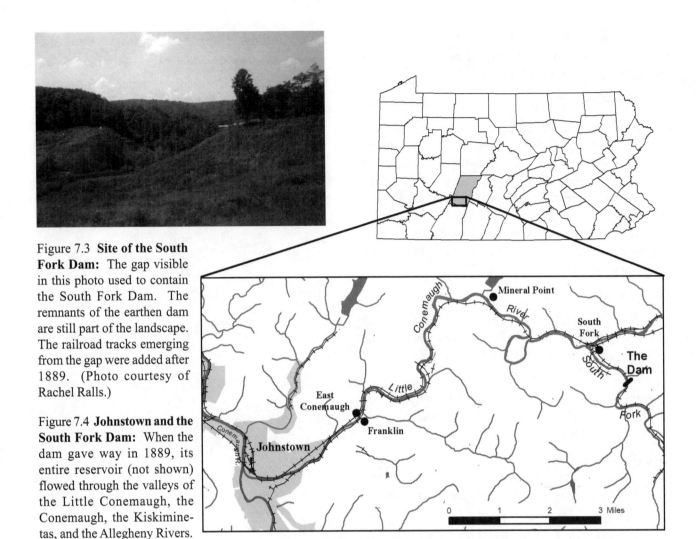

Figure 7.3 **Site of the South Fork Dam:** The gap visible in this photo used to contain the South Fork Dam. The remnants of the earthen dam are still part of the landscape. The railroad tracks emerging from the gap were added after 1889. (Photo courtesy of Rachel Ralls.)

Figure 7.4 **Johnstown and the South Fork Dam:** When the dam gave way in 1889, its entire reservoir (not shown) flowed through the valleys of the Little Conemaugh, the Conemaugh, the Kiskiminetas, and the Allegheny Rivers.

company. Flooding was a frequent enough occurrence that most people, given sufficient warning, simply moved their valuable possessions and furniture upstairs, waited out the flood, cleaned up the mud and debris, and moved things back downstairs.

Fourteen miles up the Little Conemaugh River from Johnstown was privately owned Lake Conemaugh. The lake had been created many years earlier by damming the South Fork, a small tributary of the Little Conemaugh River (see Figure 7.3). Its original purpose was to hold water needed to operate a canal paralleling the river, but the canal had been put out of business by the growth of the railroads (see Chapter 24). The land had been sold to the South Fork Fishing and Hunting Club, who had not maintained the earthen dam properly, and had even reduced its flood-worthiness in favor of maximizing its reservoir's fish-holding ability. The potential collapse

of the dam had been suspected before (Lavine 1990, 68).

The spring of 1889 had seen frequent rains, including the several days leading up to the fateful date of May 31, so the ground was saturated and there was already some minor flooding. On this occasion the entire dam—a triangular wall of earth at the downstream end of the lake—was so saturated that it simply slid out of position, and the entire lake emptied in little more than half an hour. The burst dam released a wall of water up to forty feet high that traveled downstream at speeds close to forty miles per hour. It devastated the smaller communities of South Fork, Mineral Point, Franklin and East Conemaugh, picking up debris (trees, houses, railroad cars, etc.) along the way. Forty-five minutes after the dam broke, the water hit the floodplain of Johnstown with devastating force. Factories and low-income housing were demolished. Brick buildings of the commercial district survived, but wooden ones did not fare so well. Higher up on the slopes of the valley, the homes of the more affluent were unscathed (US NPS 1998).

Conclusions

Pennsylvania generally has sufficient surface water for its needs. The major issues involve balancing management of those resources between different purposes which are sometimes at odds with each other. For example, are swimming and boating acceptable in a reservoir used for drinking water? Water quality issues such as this will be taken up in Chapter 37. It is important to understand how this very important set of natural systems works.

Bibliography

Agnes in Northeastern Pennsylvania 2004. Flood Facts. Internet site: <http://www.agnesinnepa.org/modules.php?op=modload&name=News&file=article&sid=3>, visited 2/4/04.

Lavine, Mary P. 1990. "The Legacy of the Johnstown Flood." The Pennsylvania Geographer. Vol. 28, no. 2, pp. 68-80.

Marsh, Ben and Peirce Lewis 1995. "Landforms and Human Habitat." Chapter 2 in Miller, E. Willard (ed.) A Geography of Pennsylvania. pp. 17-43.

Murphy, Raymond E. and Marion F. Murphy 1952. Pennsylvania Landscapes: A Geography of the Commonwealth (2nd edition). State College, PA, Penn's Valley Publishers.

PA DEP 2004. DSAW Dam Graphs. Harrisburg, PA, Pennsylvania Department of Environmental Protection: Bureau of Waterways Engineering. Internet site: <http://www.dep.state.pa.us/deputate/watermgt/we/damprogram/ndsad/main/graphs.htm>, visited 6/28/04.

SAGE 2003. River Discharge Database. Madison, WI, University of Wisconsin-Madison: Gaylord Nelson Institute for Environmental Studies: Center for Sustainability and the Global Environment. Internet site: <http://www.sage.wisc.edu/riverdata/scripts/keysearch.php?numfiles=50&startnum=2000>, visited 7/8/04.

US ACE 1999. National Inventory of Dams (1999 update). Alexandria, VA, US Army Corps of Engineers: Army Topographic Engineering Center. Internet site: <http://crunch.tec.army.mil/nid/webpages/nid.cfm>, visited 9/17/04.

US NPS 1998. A Roar Like Thunder. US National Park Service. Internet site: <http://www.cr.nps.gov/nr/twhp/wwwlps/lessons/5johnstown/5facts1.htm>, visited 10/8/02.

Chapter 8

Groundwater Resources

While surface water is a very important source of fresh water for everyday municipal use, it is more subject to weather variability, pollution and competing uses such as swimming, boating and fishing. Groundwater, literally water in the ground, has fewer uses, is less subject to short-term fluctuation (especially deeper sources), and there is generally much more of it (Fleeger 1999, 2).

Even though it appears obvious to contrast surface water and groundwater, any water readily flows between the two spheres. Not only must water pass through the surface on its way into the ground, but groundwater frequently re-emerges back to the surface. Natural mechanisms for this include springs and every body of surface water in Pennsylvania. Humans also accomplish this by digging wells, of which there are about a million in Pennsylvania (Fleeger 1999, 2).

There is generally an adequate supply of groundwater, but different landscapes present different challenges. The questions to be addressed in this chapter are, "What properties make groundwater a valuable resource?" and "What uses are made of groundwater in Pennsylvania?"

Where Is the Groundwater?

Much water is trapped in the soil layer and in many of the layers of bedrock. The water's origin may be as recent as this morning's rain, or as ancient as hundreds of millions of years old, when the rock particles making up a bedrock layer were newly deposited surface sediment. Most groundwater is fresh water, especially in Pennsylvania, but it can be salt water, too. For example, coal's list of impurities included water (fresh water in this case), which may well date back to the deposit's origin as a swamp.

Likewise, oil and natural gas deposits are often found in or near bedrock that contains salt water, or brine.

The mineral content of some layers of rock or soil is composed of finely sized, tightly packed or fused particles. This rock or soil may contain open spaces or pores that tend to trap water and hold it in place, or may prevent water from moving. In other rock or soil layers, soils and certain sedimentary rocks especially, the particles and the pore spaces between them are larger, and the water moves more freely. Groundwater flows downward (both vertically and following sloping layers of rock; see Figure 8.1) from space to space until a) it is blocked by impervious rock or soil, b) it is blocked because the area underground is already saturated, or c) it emerges back to the earth's surface. Almost all groundwater flows in such restricted conditions, though large crevices and caverns do occur in unusual circumstances. Because of those restricted

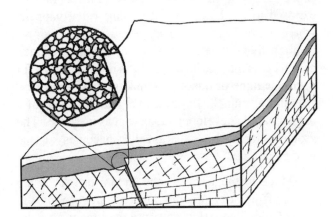

Figure 8.1 **Groundwater Flow:** Groundwater flows *with* the slope of the land. It travels in the spaces between particles that make up soil and certain layers of bedrock. Its speed depends on the size and connectedness of those spaces. If the groundwater encounters a more vertical fissure it may drain downward into lower layers of bedrock.

conditions, groundwater usually moves much more slowly than surface water.

In the vertical direction, the question of "Where is the groundwater?" becomes a question of how far down to drill or dig. The destination will be either the water table or an aquifer. The water table is the top of the layer of groundwater that sits above the first impervious layer of bedrock. It consists of water that has soaked through the soil and the soil's parent material (bedrock that is fractured or weathered to an extent that allows water to penetrate). An aquifer is a layer of sufficiently porous bedrock sandwiched between layers of impervious bedrock. The water could have been trapped there millions of years ago, or the aquifer, due to erosion or tectonic folding or faulting, could be exposed at some uphill site and receiving fresh inputs of water ("recharging").

Geographically, the question of "Where is the groundwater?" depends on understanding the porosity and slope of all the underground layers of bedrock. The surface is often the best clue, especially in the Appalachian Plateau and some valleys of the Ridge and Valley region. However, the bedrock underlying the Piedmont and especially the Atlantic Coastal Plain have been so contorted by tectonic forces that prediction is much more difficult; two wells drilled a short distance apart can produce in very different layers and at very different rates of flow.

The Groundwater Returns

The layers of ground or bedrock containing a downward-sloping water table or aquifer may be exposed to the land's surface. This often happens where a stream or river has carved its valley down through these layers. It may occur where human activity cuts into the layers, such as in excavations for a surface mine or quarry, an underground mine, or a road cut or tunnel. Exposure may be the result of a large sinkhole development, as well as its cause. The groundwater in this situation then becomes surface water again.

Much of the water in any stream or lake is the result of the streambed or lakebed intersecting the water table. Water flows out into the stream from the water table on both sides of the stream (see Figure 8.2). In Pennsylvania this seepage will often form a large portion of the stream's base flow.

The ideal situation for a homeowner or even a community planning to tap into a groundwater source

Figure 8.2 **Groundwater Re-emerges:** Groundwater feeds back into streams, rivers, and lakes where the land dips below the water table. The water table elevation helps to determine the water level in the stream or lake.

for their domestic water uses is to find a system of groundwater that is at a low point of a sloping aquifer with a large active recharge surface area. A well's capacity, or yield, is measured in terms of the number of gallons of water per minute that can be sustained, especially at the driest time of year, which is usually the summer. This capacity must be compared to the amount needed to satisfy all of the users and uses for that water.

The yields that can be achieved in Pennsylvania wells vary greatly depending on the source layer and the area topography. The most productive aquifers are in sand and gravel deposits such as those left by glaciers and floodplain deposition, and in carbonate rocks. Sand and gravel tends to be very porous, while carbonate rocks are likely to contain pathways caused by dissolving due to water acidity and by fracturing. Yields of 3000 and 8000 gallons per minute from such sources are possible. Other source materials yield as little as 20 gallons per minute (Becher 1999, 670).

Water Quality in the Ground

Groundwater is susceptible to pollution, though not to the degree that surface water is. First, it must be noted that naturally occurring water (in rain or on the Earth's surface) is never pure. The naturally occurring impurities include gases, other liquids and dissolved particulates (see Chapter 36). The process of flowing downward through the soil and upper layers of bedrock filters out many impurities, especially larger particulates. However, it may also introduce impurities into the groundwater, since the water is one of the agents helping to break soil's parent materials down into a more integral part of the soil.

There are also chemical reactions that can take place between the water (and its contents) and the minerals it contacts. Some of these reactions neutralize the pollutants (such as when an acid and a base come together), or remove the pollutants by attaching them to the minerals. Other impurities are not readily filtered. For example, farming areas whose farmers use more fertilizers than necessary for their crops frequently have problems with nitrates in the water table. Residential areas whose homeowners and golf courses over-fertilize their gardens and lawns can be even worse. The nitrates are not easily removed, and if the water then re-emerges into surface water bodies, the problems are compounded with surface runoff from those same fields. Nitrates are a problem for human water use and for the plants and animals living in that water downstream (see Chapter 36).

Pennsylvania's Use of Groundwater

Pennsylvania uses significant amounts of both surface water and groundwater. Each has advantages and disadvantages. Surface water is easier (that is, cheaper) to tap into, but more pre-use treatment is likely to be needed. Depending on the distance of the community from the river or reservoir, there may be a significant length of water mains to build and maintain. Groundwater requires the drilling and maintenance of wells, but they can often be drilled on or near an entrance point into the community system. Fewer pollutants are likely to be found in the water, but they may be more costly to remove.

The map in Figure 8.3 shows the distribution of groundwater use. One factor related to higher usage of groundwater appears to be distance from a large river. Notice that Philadelphia has a very low usage of groundwater, while neighboring Montgomery County has one of the highest quantities and rates of use (see below).

Souderton and Montgomery County

Figure 8.3 shows Montgomery County using the highest amount of groundwater and falling in the category for the highest percentage of water consumed coming from groundwater sources. Souderton (see Figure 8.4) is a small borough of 6,730 within Montgomery County, which has over 750,000 people (US Census, 2002). Souderton is a typical small-town middle-class community: it became a borough in the late 1800s, grew to its present area by the early 1900s, and attracted a largely white population living in single family homes, working and shopping in the local community. It grew again after World War II with migration out from Philadelphia. It benefited from its location near Pennsylvania Route 309 a commuter road, and north of the Pennsylvania Turnpike.

Souderton's drinking water is provided by the North Penn Water Authority, based in Lansdale, Montgomery County. Contrary to the map information, North Penn states that 75 percent of its water is taken as surface water from the Delaware River, and only 25 percent comes from water wells.

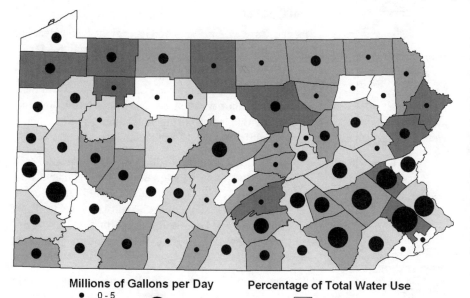

Figure 8.3 **Groundwater Consumption in Pennsylvania Counties:** The circles show the quantity of groundwater consumed in each county, which is influenced by population, groundwater potential, and the local cost of obtaining groundwater and perception of its suitability. The county shades show what percentage of that county's total water use is provided from groundwater sources. (Data source: Fleeger 1999)

Millions of Gallons per Day
- 0 - 5
- 6 - 20
- 21 - 40
- 41 - 63
- 148

Percentage of Total Water Use
- 1 - 20 %
- 21 - 55
- 56 - 85
- 86 - 98

They declare this mix to be ideal because, firstly, the groundwater is considered "hard" water because of the dissolved minerals it contains, but is effectively diluted by the "softer" surface water in the system. The surface water passes through a modern water treatment plant in Chalfont, Bucks County. Secondly, the groundwater temperature is constantly cool, which again is an effective mixer with the more seasonably variable surface water temperatures. Finally, the wells, located in the communities served by North Penn, provide an effective reserve during times of reduced surface flow, and serve as input points for chlorination (NPWA 2003).

Souderton's and the North Penn Water Authority's major concern these days is that at least six groundwater contamination sites have been located in their area by the US Environmental Protection Agency. Sites of significant contamination are targeted for cleanup, according to their position on the National Priority List. The first site to be identified, investigated and confirmed is within the borough of Souderton, where North Penn had a municipal well. They

discovered tetrachloroethane (also known chemically as PCE) in the well in 1979, notified the EPA, and shut down the well. PCEs are used as solvents (they dissolve organic chemicals) in several types of facilities. The EPA's investigation revealed a dry cleaner that had experienced a spill of dry cleaning chemicals years before, one hosiery knitting mill still in operation in 1992, another knitting mill closed in 1991, a mechanical engineering firm that uses solvents, and an apartment building that had several former uses and contained on-site underground storage tanks for fuel-like chemicals. The investigation has not concluded that there are specific health risks to the population, but that potential exists and caution and eventual cleanup are required (US CDC 1992).

Conclusion

It is always interesting and relevant, and even important, to know where your home's water comes from. Local governments were the first to assume the responsibility for providing domestic water, although many have signed on with larger regional providers. It is required that each source be tested frequently, and the results of those tests are distributed to the public through various means. Be informed: find out who your provider is, what their water source is, and how well their water quality measures up.

Bibliography

Becher, Albert E. 1999. "Groundwater." Chapter 44 in Shultz, Charles H. (ed.) The Geology of Pennsylvania. pp. 666-677.

Fleeger, Gary M. 1999. The Geology of Pennsylvania's Groundwater. 4th series, Educational Series 3. Harrisburg, Pennsylvania Geological Survey.

NPWA 2003. Water Currents. Lansdale, PA, North Penn Water Authority. January 2003 newsletter. Internet site: <http://www.northpennwater.org/pdf/newsletter 0103.pdf>, visited 12/30/04.

US CDC c. 1992. Public Health Assessment: North Penn-Area 1, Souderton, Montgomery County, Pennsylvania. Atlanta, GA, US Department of Health and Human Services, Centers for Disease Control, Agency for Toxic Substances and Disease Registry. Internet site: <http://www.atsdr.cdc.gov/HAC/PHA/penn/npa_p1.html>, visited 12/31/04.

US Census 2002. Census 2000 Summary File 1 (SF1) 100-Percent Data. Washington, DC, US Department of Commerce, Bureau of the Census, Population Division. Internet site: <http://www.census.gov>, multiple visits.

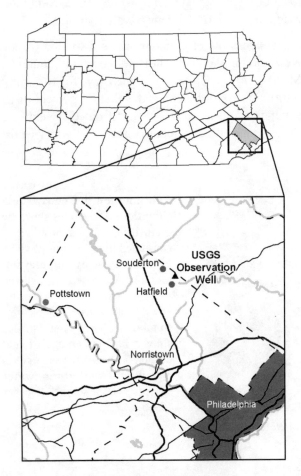

Figure 8.4 **Souderton, Montgomery County**

Chapter 9

Soils

Soil was one of the prime resource discoveries for early settlers in the Pennsylvania region of North America. It wasn't what they came here for, but they knew what to do with it when they realized what they had. The news they sent back to Europe encouraged many new settlers to make the journey. As the new settlers ventured further and further into the colony's interior, the soils in many of these new areas turned out to be different. By the time they realized that the soils of the Appalachian Plateau were not nearly as productive as those of the southeast, other highly productive economic activities had taken the place of agriculture.

High quality soils are common enough in Pennsylvania to make this state a prime agricultural producer even today. However, land featuring prime soils is usually under the greatest pressure for development into suburban residential neighborhoods and shopping centers. One of the dilemmas facing land owners and planning officials is how best to manage the land containing these soils. Those issues will be treated in the chapters on urbanization (Chapter 17) and agriculture (Chapter 21).

What are the characteristics that make Pennsylvania's soil so productive? To answer that, we will need to understand a little about what goes into making soil, and a lot about how its quality can be maintained.

Soil Ingredients

Soil's components are, generally, mineral particles, biological matter, living organisms, air and water. The minerals are either weathered (broken down) from the topmost layer of bedrock underlying the soil, or previously reduced particles washed in from uphill or upstream. The mineral particles are classified by soil scientists according to their sizes. The finest are "clay"-sized. The largest in a true soil (not counting pebbles and other stones) are "sand"-sized, while the particle sizes in between these extremes are referred to as "silt."

The mineral particle sizes are important because they define how water behaves within the soil. Surface movement of soil, better known as erosion, will be described in Chapter 11. The deeper mineral particles, if they can be separated from their neighbors, can be transported downward by water moving through the soil. At the same time, the smaller of those particles will pack more tightly together, and will therefore slow down the rate of water movement. Different mineral compositions result in different rates of movement. Soils with higher proportions of sand-sized mineral particles allow greater freedom of water movement, and more easily lose clay- and silt-sized particles from upper levels. Since Pennsylvania has a relatively wet environment, this downward vertical movement is almost always happening; soil is always changing or developing.

The varying mineral contents give different soils several identifying characteristics. One is a soil's texture, the mixture of sand-, silt- and clay-sized mineral particles, which can be detected by touch as well as by laboratory analyses. Another characteristic is the soil's color, which again reflects the minerals present, such as redness where iron is present and yellowness where aluminum is more abundant. A third soil characteristic results from the mineral's influence on the acidity or alkalinity of water passing through it (see Chapter 36).

Soil's biological components include living plants, dead and decaying plants, and insects, earthworms and other "critters." The living plant life includes grasses, plants and trees. In addition to being the source of the dead vegetation, the plants help to break down many of the minerals brought in with water through their roots. The roots also work to force apart soil particles and create space for water and air to flow. The dead vegetation is breaking down by decaying into rich brown organic material. It provides organic nutrients that the plants need, and helps to give topsoil its loose texture. Finally, the living animal and insect and other non-plant organisms in the soil also aid in both the weathering and decaying processes.

The water is an agent in the mineral weathering process as well as a transportation mechanism and a biological requirement for the plant, animal and insect organisms. Weathering may be the result of complex chemical processes or of simple freezing (expanding) and thawing (contracting) actions. Air is another biological necessity for many of the soil organisms, and it also provides the oxygen needed for the decay process to occur.

In addition to those material ingredients, two other factors influence the ongoing soil formation process. The first of these is temperature. During different seasons, especially here in Pennsylvania where we do experience widely ranging extremes, the temperatures have their greatest influence on the water content of the soil, not only freezing and thawing at the particle level, but also expanding and contracting the upper layers.

The second non-material factor is time. All of the soil-forming processes take time to do their work. Depending on local conditions such as slope and bedrock type, some soils continue to accumulate components over time, while others may be experiencing long term erosion. In addition to determining the balance between erosion and accumulation of soil, time plays a factor in how the material components of the soil are arranged. Soils formed over greater lengths of time (on the order of hundreds or even thousands of years) develop a more organized series of different layers. On the other hand, the material components in more recently developed

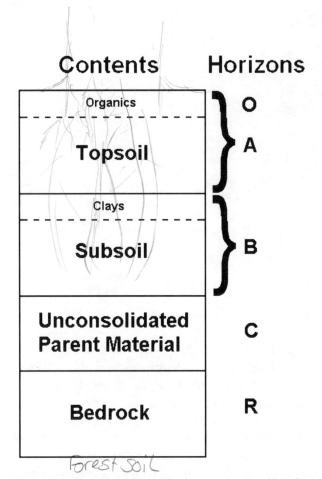

Figure 9.1 **Soil Horizons:** The layers of soil, which form naturally over time, exhibit distinctive characteristics. The letter designations are used in Soil Survey publications and by soil scientists. The depths and thickness of each layer vary widely among soil types.

soils tend to be thoroughly mixed throughout their entire depths.

Soil Structure

The end result of the ingredients and factors at work in any location is that the soil components get sorted into several sub-layers with distinctive characteristics, called horizons (see figure 9.1). The top-most "A" horizon is richest in living and decaying organic material and creatures. The top couple of inches, often called the "O" horizon even though it is really a part of the A horizon, are essentially pure organics. These components keep a rich soil full of organic nutrients that plants of all sizes need for growing. Plants also need minerals, which they primarily get from the dissolved inorganic mineral particles. Water, of course, is also taken in by plants.

The "B" horizon, or the subsoil is where particles removed from the A horizon accumulate. In some soils this process is so well developed that the top of the B horizon is a very dense layer of clay sized particles, much more difficult to dig through than the rest of the soil. The B horizon would not function as a soil itself, since it has no organic content and usually a different chemical content than the A horizon. However, it does influence the behavior of the topsoil, especially its water content.

Below the topsoil and subsoil are two additional named horizons. The "C" horizon has a much higher proportion of stones and boulders, parent material in the process of weathering into smaller and smaller particle sizes, while the "R" horizon is solid, or all-but-solid, bedrock. Soils forming over bedrock that is resistant to weathering may have those stones and boulders throughout the soil horizons.

As the last example suggests, the type of bedrock and its situation can lead to varying thicknesses and consistencies of the layers. For another example, in Pennsylvania, soils that form in glaciated areas are different in character than those that form over non-glaciated landscapes. The horizons of glacial soils are poorly developed and the soil layer overall is frequently not deep even though it is well drained. In other areas, if the bedrock is shallow and impermeable to water movement, then the soil may be permanently wet and the landscape dominated by swamps and frequent lakes. Even in areas with the normal arrangement of horizons and no unusual wetness conditions, the horizons may form distinctive sub-

horizons that vary in texture, color, or other characteristics.

Soil Types

All of these elements mean that no two soils are exactly alike. Soil science is the study of such differences. Soil scientists have collectively examined soil differences and developed an organized system, called the Soil Taxonomy, for naming and classifying soils.

With so many factors influencing soil formation, the challenge of organizing soil types was that of deciding which criteria should identify the most distinctive differences; the types identified at this level are the soil "orders." New criteria differentiate subsets of each soil order, and so on down through many levels of identification. Each level of categories has a general name, and each soil type at that level is given a unique name. To further complicate the process, each the categories within one soil order are not differentiated using the same criteria as those in another order. The next levels below the soil order are: the "suborders," the "great groups," the "subgroups," the "families" and, finally, the "series." Soil scientists further differentiate soils within a series, depending primarily on the land's slope. Ultimately each type is represented in many "mapping units."

Four of the twelve soil orders occur extensively in Pennsylvania: alfisols, ultisols, entisols and

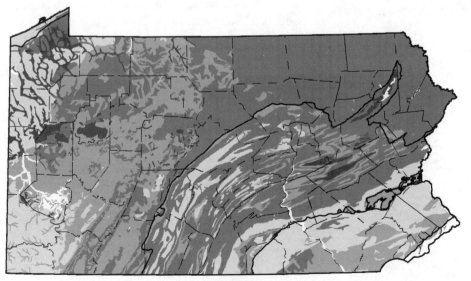

Figure 9.2 **The Soil Orders of Pennsylvania:** The soil orders' names and their properties are described in the text. White areas represent either areas of extensive urban development (paving and large buildings, for example), or areas under water. Both are areas in which there is no true soil because the surface is covered or highly modified. Notice also the correspondence of some patterns of soil orders with the landform regions.

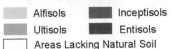

Soil Orders

　Alfisols　　　　Inceptisols
　Ultisols　　　　Entisols
　Areas Lacking Natural Soil

inceptisols (see Figure 9.2). Both the alfisols and ultisols are well developed and potentially productive soils, while the entisols and inceptisols are usually less developed and less productive. That productivity refers to its agricultural potential only. Almost all of the soils in Pennsylvania are fertile enough to support (and have supported) slower-growing forests.

Alfisols are the most productive soils, formed over carbonate rocks at lower elevations in the Piedmont and Ridge and Valley landform regions, and over limestone and other bedrock in some areas of the Central Lowland and Appalachian Plateau. They have higher quantities of clay-sized particles, which improves water retention, and greater alkalinity. Ultisols, on the other hand, are less productive unless fertility is added, and occur over non-carbonate sedimentary rocks as well as over igneous rocks. Find them in the upland areas of the Piedmont, some parts of the Ridge and Valley, covering the New England region, and widely present throughout the Appalachian Plateau. They are similar in structure to the alfisols, but beneficial minerals have been more thoroughly dissolved and removed from the topsoil (Miller 1995, 72).

Entisols are very poorly formed soils because they are developing over recent deposits of sediment in river valley floodplains. Since floods can occur at any time, these soils do not get a chance to mature and develop horizons. Adding fertility can make them productive, but must be repeated for every crop. Inceptisols are similar to entisols in that they are forming over material that does not represent native bedrock or over bedrock that does not easily break down. Even where the soils are deep, they develop little fertility in the A horizon.

The best place to learn about the soils of your area is to obtain (for free!) the county Soil Survey, a publication of the US Department of Agriculture Natural Resources Conservation Service (formerly the Soil Conservation Service) and available from your county's USDA "agent." The Soil Survey will show soil types identified to the series and map unit level, and will contain many pages of explanatory text. That text will classify such characteristics of each soil as its horizons, and water holding characteristics, its suitability for specific crops, its appropriateness for varying engineering applications such as house foundations, septic fields and road building, and many other uses. Property owners can get as much useful information from Soil Surveys as farmers and other professionals.

Hazleton: The State Soil of Pennsylvania

You may know that Pennsylvania has a state flower (the Mountain Laurel), a state animal (the Whitetail Deer), and a state bird (the Ruffed Grouse), but did you know that we also have a state soil? Out of more than 1,000 soil series found across Pennsylvania, state soil scientists were asked which soil best characterizes the state. They chose the Hazleton soil series, so we'll use it as an example to illustrate how much the study of a soil can tell us. The Hazleton soils are found in thirty three (about half) of the state's counties, in the Ridge and Valley and Appalachian Plateau landform

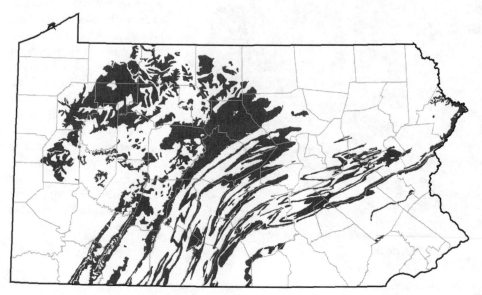

Figure 9.3 **Hazleton Soil in Pennsylvania:** The Hazleton soil series occurs over a variety of landform regions but represents some of the poorest agricultural soil in each area. Contrast carefully to the areas of alfisol soils (Figure 9.2) and carbonate bedrocks (figured 4.5).

58

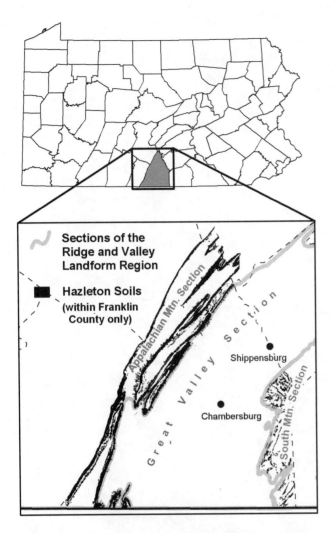

Figure 9.4 **Hazleton Soils in Franklin County:** Hazleton soils form on steep slopes as this map illustrates. The broad Great Valley and the smaller vallerys and flat areas of its bounding ridges have none, but they occupy most of the ridge slopes.

to brown, its texture is sandy loam, and it is considered stony to "channery." Hazleton soils have distinctive A and B horizons, identifiable more by color than by texture, and its parent bedrock (R horizon) is sandstone (PA APSS 2004).

The less technical interpretation is that it has formed on the tops and sides of ridges, in the Ridge and Valley region and also in the Appalachian Plateau, that are underlain by relatively young, weathering-resistant sandstone bedrock. Stones, small and large, are common throughout the entire depth of the soil right down to bedrock, which is a fairly deep 60 inches below the surface.

Most Hazleton soils are covered by woods, which have been cleared in some areas for pasture or crops (not very productive). This vegetation cover gives the soil a darker brown color for the first couple of inches; the next eight inches of that topsoil is a gray color. The subsoil adds another ten inches to the soil depth and is generally dark reddish brown with very little clay-sized minerals. These soils drain quickly after rain, partly because of the dominance of larger particles and stones, and partly because of the usually steeper slopes. The sandstone parent material and lack of alkaline content give the soil a higher level of acidity than other soils (PA APSS 2004).

This type of description can be written for any soil at any level of classification. Any area's soils have been analyzed and described in this way. Again, a county soil survey is the best resource.

Conclusion

Soils are critical elements of our environment in many ways, some commonly appreciated and others rarely considered. Pennsylvania's productive soils will be noted in later chapters on agriculture. This chapter has given us a chance to take note of what soil is and why it is not all alike. Its relation to earlier chapters on landforms, weather and water resources should be apparent. Its role in vegetation resources and in erosion issues will also be shown.

Bibliography

Miller, E. Willard 1995. "Soil Resources." Chapter 5 in Miller, E. Willard (ed.) A Geography of Pennsylvania. pp. 67-73.

PA APSS 2004. Hazleton: Pennsylvania State Soil. Harrisburg, PA, Pennsylvania Association of

regions, and on more than 5% of the state's land area: 1.5 million acres of land (PA APSS 2004). The map in Figure 9.3 shows the extent of Hazleton soils across Pennsylvania, while the map in Figure 9.4 shows its relationship to the different land features of the Ridge and Valley region within Franklin County.

The Hazleton soil series belongs to the order Inceptisols, and to the suborder and great group (combined in one word "Dystrochrepts" (USDA 1975, 111) or "Dystrudrepts" (PA APSS 2004). Is subgroup is termed "Typic," and the criteria used to define its family are "Loamy-skeletal, mixed, mesic." Hazleton soils are usually found adjacent to Dekalb soils, which share some if these qualities. Hazleton's color is gray

Professional Soil Scientists. Internet site: <http://www.papss.org/hazleton.htm>, visited 7/19/04.

USDA 1975. Soil Survey of Franklin County, Pennsylvania. US Department of Agriculture: Soil Conservation Service in cooperation with Pennsylvania State University: College of Agriculture and with Pennsylvania Department of Environmental Resources: State Conservation Commission.

Chapter 10

Forests

If people were completely removed from Pennsylvania today, then in a few hundred years the landscape would be almost completely forested, as it was before Europeans arrived. The exceptions would be the huge areas of land, especially in cities, that have been covered in buildings, pavement or any other impermeable surface. It might take thousands or tens of thousands of years, but they too, without upkeep, would eventually decay (think of the oldest ruins that have been found). Almost any area would develop a layer of soil, just as the areas scraped and covered by glaciers in the last ice age have done, and the areas where coal mining wastes were dumped and abandoned within the last hundred years are doing. Eventually that soil will support new vegetation and eventually that vegetation will be forests.

Of course, no one is planning to evacuate the entire state any time soon, but smaller parcels of land are frequently abandoned and that process of natural succession will occur on any of them. Forests are the final stages of that succession process. The most mature mix of trees, which would then persist until tectonic or climatic changes brought a new environment, is its climax vegetation. The geographical questions are "What types of climax forest would grow in each area?" and "Why?"

The same process describes Pennsylvania's past forests. Forests, and any other vegetation, were able to adapt to changes in the surface landforms as they developed, and to grow anew following the ice ages, because biological adaptation happens faster than geological change. Thus, colonists encountered forests that had been evolving for millions of years, with trees that were hundreds of years old. On the other hand,

those same forests that developed subject to only natural disturbances have undergone tremendous changes in the last one to three centuries.

Mid-Latitude Forests

We have seen that Pennsylvania's climate provides moderate average temperatures that vary both seasonally and regionally. All areas of the state get plenty of rainfall. Its soils are weathered from parent material that varies widely in its potential. Some, along ridges and valley slopes, drain quickly, hile others in the valleys and other lowlands, are more continuously moist.

The state's forest vegetation is well adapted to its mid-latitude location. The most northern areas of Pennsylvania, up on the Appalachian Plateau, produce trees that are able to withstand extended periods of freezing in winter, just like areas to the north in New York and Canada. The biggest challenges for trees in the southern-most areas (especially the southeastern Piedmont and Atlantic Coastal Plain landform regions) are the intense heat and humidity for short spells in the summer. Trees of central areas of the state reflect transitions between these characteristics.

Within a landscape area, species and their locations will also vary. Relatively high ridge tops and plateau uplands feature different species than river valleys. Stream banks and floodplains are more likely to feature willows and elms, while higher and drier land would not. Each species grows where it has: a source of seeds, suitable soil, a tolerable range of temperatures, adequate sunlight during its early years, sufficient rainfall or other water, and enough space. While a tree must compete with the other trees in the

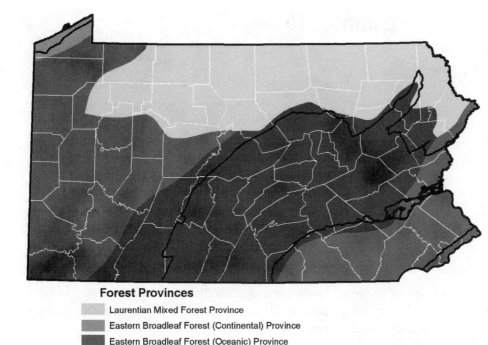

Figure 10.1 **Forest Regions of Pennsylvania:** These regions are actually subdivisions of larger US-wide forest types. Each in turn can be divided into more specific combinations of tree species. Note their correspondence with the landform regions, although in this case factors such as climate and soils come into play.

Forest Provinces

Laurentian Mixed Forest Province

Eastern Broadleaf Forest (Continental) Province

Eastern Broadleaf Forest (Oceanic) Province

Central Appalachian Broadleaf Forest-Coniferous Forest-Meadow Province

forest, more serious challenges against its survival to old age come from diseases, pests and especially humans.

In Pennsylvania, the species that have thrived are a combination of softwood/evergreen and hardwood/deciduous species. Several species each of hemlock, spruce and pine dominate the evergreen category. White pine, in particular, has been an important species economically. Add to that such hardwoods as cherry, oak, maple, hickory and the virtually extinct chestnut. Many other deciduous species, such as ash, aspen, beech, birch and elm, round out the forest populations.

Forests and Ecosystems

Forest types are mostly a matter of what combinations of tree species grow together, and account for variations in landforms, soils and climate. Smaller forest tracts can be further described according to their tree density, crown height, canopy closure, or their lumber potential (measured in board feet), among other qualities. All of the tree species listed above represent the largest and tallest trees in the forest, but the forest ecosystem also includes the other vegetation (such as grasses, bushes and smaller trees), animals, insects and other species. The US Forest Service has classified and mapped these larger all-inclusive combinations as Ecoregions (Bailey 1995, McNab and Avers 1994). The discussion here will focus on the forest vegetation of the ecoregions.

The Forest Service's ecoregions are hierarchically identified, like the soil types were in the last chapter. The top level of the hierarchy are Domains, which are divided into Divisions. The divisions are further broken into Provinces, some of which are specially grouped as Mountain Provinces within their division; within mountain provinces strong variations show up according to altitude. At the lowest level, each province is subdivided into Sections.

Four broad types of forests are shown on the map in Figure 10.1, representing portions of two different US Forest Service divisions all within the Humid Temperate domain. The Laurentian Mixed Forest province is the major portion of the Warm Continental division. The continental and oceanic Eastern Broadleaf Forest provinces are part of the Hot Continental division, and the Central Appalachian Broadleaf Forest-Coniferous Forest-Meadow province is a Mountain Province within that division (Bailey 1995).

The Laurentian Mixed Forest is the dominant province across northern Pennsylvania. It is primarily represented by mixtures of trees known as "northern hardwoods and Appalachian oak." Hemlocks populate moist sites, while beech and maple are more common on better-drained areas (McNab and Avers 1994). Coniferous trees, especially white pine and other pines, occupy poorer soils (Bailey 1995, 21). These natural vegetation combinations have been strongly modified by natural and human impacts including fire, tornadoes and a variety of diseases and pests (McNab and Avers 1994).

Along Lake Erie lies Pennsylvania's portion of the continental Eastern Broadleaf Forest province. The smaller forests within this region include various hardwoods, beech, maple, elm and ash (McNab and Avers 1994). The dominant hardwoods are oak and hickory species, and this province is the only one in which those two are featured in great number (Bailey 1995, 27). Forest clearing, land drainage and urban, suburban and agricultural development affect these forests (McNab and Avers 1994).

The oceanic Eastern Broadleaf Forest province dominates in two regions: the first is western areas of the Appalachian Plateau, and the second is the New England, Piedmont and Atlantic Coastal Plain landform regions and a small area of the Ridge and Valley along the Delaware River. The western region forests are more likely to include oak, beech, maple and pines, while in the eastern region the dominance shifts to oak and pine with other species in smaller numbers (McNab and Avers 1994), giving way to higher proportions of pine on the Atlantic Coastal Plain (Bailey 1995, 25). Fire and human development are or have been major stresses on these forests (McNab and Avers 1994).

The final forest province is characteristic of the central Appalachians, and combines broadleaf forests, coniferous forests and meadows. The Pennsylvania portion of this ecoregion includes the majority of the Ridge and Valley and part of the Appalachian Plateau near the Allegheny Front close to the Maryland border. Oaks, hickory and pine dominate, with the pines favoring steeper slopes and ridge tops and the oaks and other deciduous trees found lower on the slopes and in the valleys. Challenges to the natural forest here have included lumbering, fire (and now fire suppression), erosion and the Gypsy moth (McNab and Avers 1994, Bailey 1995, 29).

The Impacts of Humans

As we saw in Chapter 4, natural objects and phenomena become resources when they have economic value. Wood is a resource in great demand, and Pennsylvania's wood economy is among the largest in the country as we shall see in Chapters 22 and 27. However, extraction of a resource often creates impacts on the remainder of the resource, and this has been especially true for Pennsylvania's timber resources.

"Penn's Woods" enabled Pennsylvania's colonial settlement and growth, and had an even greater impact on Pennsylvania's industrial revolution in the middle 1800s. As settlers moved inland across the state, they turned many natural resources into viable economic activity. Wood was the most widespread of these. Some of the land was cleared of trees because settlers wanted to try to farm the land. They tended to be unsuccessful because of the thin Appalachian Plateau and Ridge and Valley soils. The wood proved to be so valuable, however, that by the early 1900s there were very few acres of Pennsylvania that had not been cleared of trees at least once.

That is the impact that is most mind-boggling of all. Most lumbering in the 1800s was "clearcutting" with no replanting, which meant that the forests had to completely regenerate themselves. Fire was a frequent occurrence in the clearcut regions, many occurring naturally, but indirectly the result of human practices. If you ever fly over or see aerial photos of a forested area of the state, imagine every tree cut down. Then multiply that scene to include the entire state. Then on top of that, realize that the majority of that clearing (using nothing but axes and saws), occurred within about a 70-year period, much less than the time needed for the forests to re-grow.

By the early 1900s forest management practices required that forests be replanted, for which the more valuable economic species received preference; in Pennsylvania, the black cherry is frequently mentioned because of its attractiveness for furniture and (earlier) the wood chemicals industries. In addition, fire suppression became the goal of forest management agencies. Some fires are natural, and serve to recycle nutrients and create more opportunities for some species than others (Mulhollem 2003).

Similarly, other devastating impacts that have included diseases and pests are indirectly human in origin. The chestnut, once one of the main hardwoods found throughout Pennsylvania, especially on the Ridge and Valley ridge tops has nearly disappeared from Pennsylvania. It is the victim of a disease known as "chestnut blight" which was accidentally introduced into New York City in 1904 and entered Pennsylvania as early as 1908 (PA Bureau of Forestry 1975). The gypsy moth is an introduced species that evolved in Europe and Asia, arrived in the US in the late 1860s near Boston, and entered Pennsylvania around 1932. Oaks, aspen and other hardwoods are the primary targets of the moth, which feeds on the foliage, ultimately killing the tree (Liebhold 2003, PA Bureau of Forestry 1975). Even white-tailed deer have an impact. As forests were cleared in the 1800s, the deer population declined. The initiation of licensed hunting led to a deer population boom (see Chapter 12), and their browsing (by favoring some tree species over others) can also cause changes in the forests (Mulhollem 2003).

Another impact may be due to human-caused air pollution. Acid precipitation is a well-understood pollution problem, but one that is expensive to prevent or correct (see Chapter 35). Acid rain may have some direct impact on the trees, but it is likely that its impacts on the forest soils are even more significant. Again, the changing balance toward a more acidic soil will favor some species over others (Mulhollem 2003).

The overall impact of these changes is showing up in the species composition of the forests. Just as chestnuts are disappearing due to the blight, the red oak also seems to be losing ground to the Red Maple. The oaks are tastier to young deer. The maples are acid-loving trees that used to grow mostly in swampy lowlands, and that used to be very vulnerable to fire. In addition to the expansion of the red maple, the small mammal, bird and even insect species that live in the forest are adjusting their populations.

Famous Forests in Pennsylvania

Pennsylvania is home to millions of acres of forests that are preserved from development, by both the federal and state governments. The US Forest Service, a branch of the US Department of Agriculture, created the concept of the preserved forest in 1891 (US Forest Service 2004b). Pennsylvania also has a system of State Forests, enabled by state legislation passed in 1897 and put into effect the following year with the first state purchases. The state has acquired much of its forest land by "tax sale," essentially taking over ownership from previous owners who had not paid their property taxes (PA Bureau of Forestry 1975).

Allegheny National Forest occupies about 513,000 acres in northwestern Pennsylvania (see Figure 5.5 in Chapter 5). Like most of the forest land throughout

the state, it had been completely cut over at the time. As privately owned land, it was "mostly Eastern hemlock and American beech, with white pine along river bottoms and oak on the slopes of river valleys" (US Forest Service 2004b). Since the US Forest Service took over management of the land in 1923, black cherry and maple have become dominant. The forest is managed for a variety of purposes, one of which is wood production. Black cherry is a preferred species because of its market value as a veneer in furniture making, and Allegheny National Forest produces one quarter of the US's black cherry sawtimber and one third of the world's black cherry furniture veneer (US Forest Service 2004a).

The management of the national forest also includes other purposes, including recreation, mineral extraction and fish and wildlife protection (although hunting and fishing are permitted). Since Kinzua Dam and the Allegheny Reservoir occupy so much of Allegheny National Forest, boating and other water-related recreation is a major attraction. Trails throughout the forest guide visitors to a variety of specific forest types and animal habitats, and also allow all-terrain vehicles, mountain biking, horseback riding and snowmobiling (US Forest Service 2004a).

Another important preserved forest in Pennsylvania is Cook Forest (see Figure 10.2), but it is preserved as a state park, not as a state forest. It too started out as a privately owned piece of Pennsylvania forest. Owned by John Cook in the 1820s and by several generations of his heirs, it was an important local lumbering operation with many water-powered sawmills. The land occupies over 6,600 acres in a valley of Tom's Run, a tributary of the Clarion River, which eventually reaches the Allegheny River (Frederick 2004).

Much of the Cook Forest was logged between 1828 and the early 1900s, but at the time it was purchased in 1927 for the state (using state money and a gift from the Cook Forest Association) it held one of very few tracts of virgin forest in the state. Many of those trees are still visible in an area of the park known as the Forest Cathedral. Unfortunately some have been lost, due to a fierce storm in 1956, and by a tornado that passed through the valley in 1976 (Frederick 2004). Such random natural events have always afflicted the forests of Pennsylvania (and everywhere else).

As a state park, Cook Forest has been modified and managed to promote recreation; it is not commercially logged. Instead, visitors can camp, swim, go horseback riding and cross-country skiing and snowmobiling, and see many of the original houses, sawmill buildings and workers' cabins, as well as exhibits of early logging tools (Frederick 2004).

Many other forest areas of Pennsylvania could also be presented here as unique or as the best representation of a certain forest type. These are both reasons for preserving forests. Some are also protected or managed in certain ways because they are important wildlife habitat areas (see Chapter 12) or for the health of our water and soil resources (see next chapter).

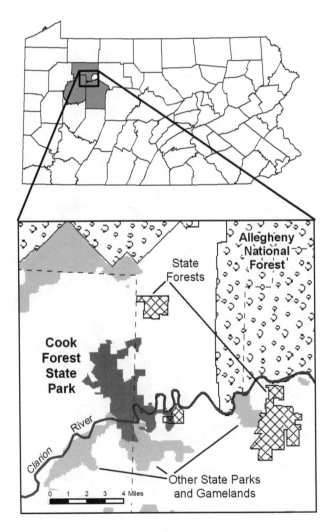

Figure 10.2 Cook Forest State Park: Covering areas within Clarion, Forest and Jefferson Counties, Cook Forest State Park holds some of the last stands of virgin timber in Pennsylvania.

Conclusions

Pennsylvania has a rich forest resource. It got us started industrially, and has continued to contribute to the state's economy. Questions about past exploitation of the forests and how to sustain them are being addressed both on federal and state forest lands and on private forest land. Forestry activity continues today (as we shall see in chapter 22), but with renewed respect for the nature of the resource.

Figure 10.3 shows the extent of land still forested in Pennsylvania. Even where they still dominate the landscape, they are not the same forests the Native Americans and the original settlers encountered. However, today, sustaining the publicly owned forest lands, at least, is state policy.

Bibliography

Bailey, Robert G. 1995. Descriptions of the Ecoregions of the United States, second edition. Washington, DC, US Department of Agriculture, US Forest Service. Miscellaneous Publication number 1391.

Frederick, Paul 2004. Take a Look at Cook Forest State Park. Allegheny-online. Internet site: <http://www.allegheny-online.com/cookforest.html>, visited 9/24/04.

Liebhold, Sandy 2003. Gypsy Moth in North America. Morgantown, WV, US Department of Agriculture: US Forest Service: Forest Service Northeastern Research Station. Internet site: <http://www.fs.fed.us/ne/morgantown/4557/gmoth/>, visited 9/29/04.

McNab, W. Henry and Peter E. Avers 1994. Ecological Subregions of the United States. Washington, DC, US Department of Agriculture, US Forest Service. Internet site: <http://www.fs.fed.us/land/pubs/ecoregions/index.html>, visited 8/10/04.

Mulhollem, Jeff 2003. Pennsylvania Forests Changing from Red Oak to Red Maple Dominated. University Park, PA, The Pennsylvania State University. News Release. Internet site: <http://aginfo.psu.edu/News.march03/forest.html>, site visited 3/26/04.

PA Bureau of Forestry 1975. A Chronology of Events in Pennsylvania Forestry Showing Things As They Happened to Penn's Woods. Harrisburg, PA, PA

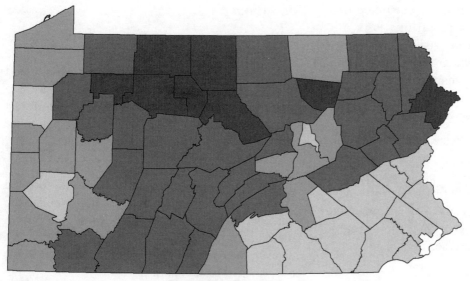

Figure 10.3 **Percentage of Each County Area That Is Forested:** Compare this map to many of the population and economy maps later in this text and you will see a strong similarity. Much of the forested land in central portion of the state, especially toward the northern border, is held by state and local governments. (Data source: US Forest Service 2004a)

Percentage Forested

None
11.3 - 40.0
40.1 - 60.0
60.1 - 80.0
80.1 - 93.9

Department of Environmental Resources (now the Department of Environmental Protection): Bureau of Forestry. Internet site: <http://www.dep.state.pa.us/dep/pa_env-her/historycalforestry.htm>, visited 9/29/04.

US Forest Service 2004a. Allegheny National Forest: Forest Facts. Milwaukee, WI, US Department of Agriculture: US Forest Service: Eastern Region – R9. Internet site: <http://www.fs.fed.us/r9/forests/allegheny/about/forest_facts/>, visited 9/29/04.

US Forest Service 2004b. Allegheny National Forest: History of the Allegheny National Forest. Milwaukee, WI, US Department of Agriculture: US Forest Service: Eastern Region – R9. Internet site: <http://www.fs.fed.us/r9/forests/allegheny/about/history/>, visited 9/29/04.

Chapter 11

Vegetation and Soil Erosion

Soil erosion is a threat to the sustainability of our agriculture. At the same time, soil erosion is inevitable in any landscape above sea level. How do we reconcile these opposite realities?

Considering the geological history of Pennsylvania, it might seem curious that we are concerned about soil erosion. After all, our landscape is the result of rock particles being deposited for hundreds of millions of years. At issue is the reality that the natural erosion process is altered by human development of the landscape. By clearing the forests and paving parking lots and roads, we increase the rate at which water, the main agent of erosion, flows across the landscape. Of course, the paving of any area also prevents any soil from ever eroding there.

In order to resolve these apparent contradictions, we will first learn a bit about how erosion occurs. Then we will review several programs aimed at reducing human induced erosion.

The Erosion Process

Erosion occurs because soil is made up of countless small particles. As a whole they create a rich, complex system for collecting, holding and delivering water and nutrients to plants. Individually, however, each particle is easily removed from the larger group. All it takes to dislodge one particle is a raindrop (or any solid object in motion) striking uncovered soil, a change of temperature which causes water to freeze (expanding) in a crevice of a rock fragment, or a chemical element carried in the groundwater reacting with another element in a rock fragment to remove bonds that tie particles to the surface of that rock fragment. Once detached, flowing water carries the particles away, so water is considered the primary agent of soil erosion.

Erosion is a concern in two parts of our water environment. The first is the removal of sediment from the majority of land affected primarily by overland runoff. The second is stream bank erosion. Both are accelerated during heavy rains and flooding.

Soil scientists and hydrologists have formulated an equation that predicts the amount of erosion that will occur from cleared agricultural land. Known as the Universal Soil-Loss Equation, it combines these factors: rainfall's erosive power, a soil's erodibility (due to the factors that define its soil series), slope length and steepness, and the presence or absence of crop management and erosion controls (Dunne and Leopold 1978, 523).

The map in Figure 11.1 presents the statewide measure of just the soil erodibility factor. Soil type influences the erosion process because lighter particles are more easily carried off by flowing water than heavier particles. Notice that the soils most susceptible to erosion in Figure 11.1 are the higher quality soils shown in Figure 9.2 in Chapter 9, including those higher in carbonate content (see Figure 4.5 in Chapter 4). Carbonates generally contribute clay-sized particles to the soil.

The rule of thumb for soil erosion by water is that the faster the speed of the water, the more easily it can pick up particles, the more it can carry, and the further it can carry them (because it will take longer for it to slow down). The particles already being carried in the flowing water can act as battering rams for dislodging more new particles. Two factors work to speed up the water. First, steeper slopes help accelerate the water, and longer slopes help to maintain or even increase that speed. Secondly, Soil that is already saturated with water will force all additional rainfall to run off over land. Floods, with their tremendous amounts of water, will almost always increase the rate at which erosion occurs. The extra volume of water means that there is greater capacity to hold eroded particles. Excess runoff over land or flood water in a stream always flows faster than normal water movement.

Natural Erosion Prevention

Two natural phenomena work to minimize erosion. The first is vegetation. Vegetation presents obstacles that make runoff slow down, and even reduces the force with which raindrops strike the ground. The roots of the plants, bushes or trees also help to hold soil in place. Different vegetative covers allow different rates of erosion, with forests providing the greatest protection.

The second phenomenon is a lack of slope: flat areas, or even still bodies of water such as lakes and swamps, will slow down the incoming water. The water then has a greater chance to infiltrate the ground, or move downstream more gradually.

Erosion across Pennsylvania averages greater than five tons of soil per acre per year on cultivated cropland, compared to a little more than one ton per

acre per year on non-cultivated land and just less than one ton per acre per year on pasture land (US NRCS 2000). The difference, of course, is the presence of permanent vegetation. Those extra four tons of topsoil are as little as a sixteenth of an inch, which doesn't sound like much. Over the course of a farmer's life, he may only see his land's elevation lower by a few inches—hardly noticeable. But multiply that by one million years and that is enough to reduce any part of Pennsylvania to sea level.

Human Factors That Increase Soil Erosion

Given our discussion about the components that make up soil, it should be clear that the factors which contribute to a soil's erodibility include: lighter and more easily detached particles in the soil, lack of protective vegetation, and anything which helps to speed up water overland or in streams.

An increase in development in any area will tend to increase erosion rates, too, unless precautions are taken. A hard surface, such as a roof or any pavement, especially one that is sloped, will allow the water to move unimpeded, and to pick up speed as it does move. It will create a problem if, at the bottom of that slope, the fast-moving water flows onto exposed soil. If that same piece of the Earth's surface had been covered with grass, bushes and trees, the water would have taken much longer to move, if it traveled at all before soaking into the ground.

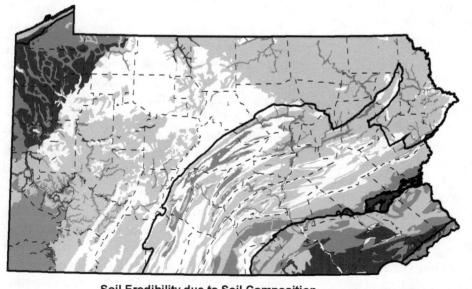

Soil Erodibility due to Soil Composition

☐ 0.1350 - 0.1900	▨ 0.2501 - 0.3100
▨ 0.1901 - 0.2500	■ 0.3101 - 0.3950

Figure 11.1 **Soil Erodibility Across Pennsylvania:** This is essentially a map of soil types in which each type has been classified according to the value recorded for its erodibility factor. These numbers are part of a larger equation that also includes total rainfall and slope measurements to calculate the number of tons per acre of soil eroded per year. Notice the strong differences between the ridges and valleys of the Ridge and Valley landform region, the subsections of the Piedmont shown in Figure 2.3 in Chapter 2, and the river valleys and upland areas of the Appalachian Plateau.

Even if the storm water is captured in a system of storm drains and sewer pipes before flowing into the nearby river, the water flow in that river increases higher and more quickly than it would naturally. The erosion danger for a faster moving river occurs along its banks.

People have also contributed to speeding up the erosion process with all of their forest clearing and agricultural and lawn management practices. Removing tree cover and replacing it with plants and grasses allows precipitation somewhat quicker and more direct access to the soil. Farming practices which remove or plow under the entire previous crop (plant and roots), even just for a short time in spring, also increase erosion totals. These differences mean additional tons of soil washing downhill and downstream over the course of a year.

Human Efforts to Prevent Erosion

Conscious efforts to plant and maintain vegetation cover are the main focus of many erosion prevention projects. Erosion problems have bee recognized and understood for well over a century, but became formally addressed with the formation of the US Department of Agriculture's Soil Conservation Service (now the Natural Resources Conservation Service) in 1935 (Helms 1998).

The challenge undertaken by the Natural Resources Conservation Service and its county agents is to reduce agricultural erosion to levels found on natural pasture or even forested lands. The reduction from five tons per acre per year to only one ton requires changes in many different farming practices. Many farmers are now using "conservation tillage," which leaves the stubble of last year's crop in the soil while this year's seeds are planted. The presence of the old roots and stalks reduces erosion. Another strategy is to align the crop rows perpendicular to the slopes, a practice called contour plowing, so that water cannot find easy downhill channels. A third practice, perhaps the most extreme, is to re-shape steeper slopes into a series of terraces, again aligned with the hillside's contours. A fourth approach is to leave a strip of natural grasses, bushes and even trees between any farm field and a nearby stream. This creates an area to catch and hold the eroded soil.

The challenge made a little more difficult by the fact that the federal and state governments have not asserted any authority over land management practices on privately owned land. Instead the approach has been to use farmers' dependence on financial assistance. Such assistance has been channeled through several programs, one of which is defined in the 1985 Food Security Act, and has been renewed and revised in the 1990, 1996, and 2002 Farm Bills. These are acts of the US Congress that have set the conditions under which farmers can receive federal money, even as a loan. They start by creating an official categorization of land, called "Highly Erodible" and identifying a soil loss tolerance level for each area of such land. Some of this land is being farmed already and some has not, before the farmer applies for the money. Previously farmed highly erodible land must be shown to have measures that reduce erosion at least 75%, to less than twice the tolerance level. Farmers breaking new ground ("sodbusting") must prove that they will not increase erosion by bringing the new land into production (US NRCS 2004).

Another approach has been applied to runoff concerns for housing and commercial developments. In many communities the runoff is collected in a storm sewer system, which directs the runoff through channels and underground pipes until released into a major stream, river or other water body. In others, the storm water flows through the pipes collecting sewage for the sewer system. In both situations, the biggest concern is heavy or long-duration rains directing more water into the system than it was designed to handle. One threat is that uncollected waters flood the community, becoming a public health hazard if sewage is not safely contained. Another is that flood waters from a number of communities get through the drainage basin too quickly and flood other communities downstream.

If you compare newer developments within your community with older developments, a prominent feature of the newer developments is their storm water retention basins, designed to collect water from the streets, parking areas and yards of the development, but slow down their flow into the community storm sewers. Even temporary situations at construction sites, are now required to prevent runoff. The policies controlling practices that create erosion flooding hazards were established by the federal Environmental Protection Agency as the National Pollutant Discharge Elimination System (NPDES) permit program, part of the latest version of the federal Clean Water Act.

This program required Pennsylvania to implement a means for policing stormwater runoff and discharges, which was accomplished in 2002 as the Comprehensive Stormwater Management Policy (PA DEP 2002).

Conclusion

The concerns about soil erosion revolve around the loss of this valuable component of our agricultural system, and the pollution it represents in such habitats and water resources as streams, rivers, lakes and the Delaware and Chesapeake Bays. It will take a more concerted effort to get to where the laws and other policies are intended to take us. We are not likely to reach a point of zero erosion, but to reduce the excessive human-caused erosion to something approaching natural background levels is realistic. Getting there will save both ecosystems and money.

Bibliography

Dunne, Thomas and Luna B. Leopold 1978. Water in Environmental Planning. New York, W.H. Freeman and Co.

Helms, Douglas 1998. Natural Resources Conservation Service Brief History. Washington, DC, US Department of Agriculture, Natural Resources Conservation Service. Internet site: <http://www.nrcs.usda.gov/about/history/articles/briefhistory.html>, visited 10/4/04.

PA DEP 2002. Comprehensive Stormwater Management Policy. Harrisburg, PA, PA Department of Environmental Protection. Document ID 392-0300-002.

US NRCS 2000. "Table 10 – Estimated Average Annual Sheet and Rill Erosion on Nonfederal Land, By State and Year," from Summary Report, 1997 National Resources Inventory (revised 2000). Washington, DC, US Department of Agriculture: Natural Resources Conservation Service. Internet site: <http://www.nrcs.usda.gov/technical/NRI/1997/summary_report/table10.html>, visited 10/1/04.

US NRCS 2004. Highly Erodible Land Conservation Compliance Provisions. Washington, DC, US Department of Agriculture: Natural Resources Conservation Service. Internet site: <http://www.nrcs.usda.gov/programs/helc/>, visited 10/1/04.

Chapter 12

Wildlife

Wildlife represents yet another part of the interactions that occur at the Earth's surface. It is challenging to define what is to be included in this discussion. The obvious reference is to animals, including mammals, reptiles, amphibians, fish and birds. However, there are also many vegetative species that grow wild in Pennsylvania's landscape. For many animals and vegetation, including the soil they depend on, insects and similar creatures are also an integral part of the environment, and are even capable of causing extensive change to those other categories of organisms. This broadest definition is reinforced by an extended look at endangered species lists.

At the same time, there is a large precedent for considering the initial narrow definition of animal species only. Wildlife management is generally aimed at animal species, and Pennsylvania has two agencies, the Game Commission and the Fish and Boat Commission, directly responsible for setting such policies. Vegetation species are the direct responsibilities of other state agencies: the Forestry Commission and the Department of Agriculture. Only the insects and arthropods are not protected by any legislation or agency, and are almost always seen as pests.

The public's awareness of, and encounters with, wild animal species within Pennsylvania are most commonly with a relatively few animals. The game species for which there are established hunting and fishing seasons, or the pest species which are routinely removed when encountered, will be the focus here. Other interesting and related topics, such as bird watching, butterfly watching, gardening methods which will promote wildlife, and caving and other encounters with bats, could also be considered, but will not be in this text.

Wildlife Ecology

The geographical study of a wildlife species emphasizes its range or territory, its potential for success or failure due to its ability to find food and shelter, and its interactions with other species including humans. Any single element is a potential limit to that species's ability to survive and reproduce. The entire population has to be large enough for there to be enough births to make up for all of the deaths that occur. If births exceed deaths, then the population will grow over time.

Ecologists take these concepts further by emphasizing the inter-relatedness of species in an area (an ecosystem). Each species depends on the vitality of all of the others. This collective perspective shows us that if the population of one species declines, then all of the other species are in danger unless they can adapt. On the other hand, if one species grows beyond normal proportions to other species, then it could either die back until it is back within its normal proportion or bring instability to the entire ecosystem.

These ecological principles can be said to apply no matter what the mix of species and no matter what biological kingdoms they belong to (animal, plant, insect, etc.). The two most unpredictable situations occur:

    a.   when a non-native species invades and disrupts the previous balance, and

71

b. (which is not necessarily different from a.) when one of the species in the ecosystem is humans.

White-tailed Deer

The White-tailed Deer is our state animal. It also has a reputation for causing trouble in suburban and agricultural areas. The white-tailed (or whitetail) deer is common throughout the US, and in fact its scientific name, *Odocoileus virginianus*, shows it to have been first identified in Virginia. It was economically important during the colonial and early American periods or our history, providing both meat and leather to settlers. William Penn and later colonial governors had allowed deer hunting within specified seasons, with fines levied against those who killed them out of season.

The deer prefer very young stands of seedlings and saplings where they can easily hide and forage for food. Edge areas between farm fields and forest are also desirable habitat because the food source is near their place of concealment. The rampant felling of forests in the late 1800s wiped out large areas of deer habitat, and left them exposed to widespread hunting. As a result, their numbers declined to a point where they became "rare" (Wildlife Information Center 1996). In 1896 the Pennsylvania Game Commission was created, with the protection of the deer as one of its initial tasks. By policies to regulate hunting, to conduct biological research and, in the 1920s, to purchase lands set aside as State Game Lands, the agency guided deer population back to a point where there are more deer in Pennsylvania today than when colonists first arrived (PA Game Commission 2004).

By the early 1900s hunting was permitted again, under a system which permitted only the males (bucks) to be taken. The goal was to preserve the female deer so that they could give birth to the typical two fawns and raise the previous year's young as well. The system proved so successful that populations have fully recovered, including in areas such as suburbs and around farms, where the animals are not as welcomed. For example, there is concern that browsing deer are altering the wooded areas of Valley Forge National Historical Park (Pomerantz and Welch 1996), that over-browsing of forests mean that the deer are out-competing other wildlife, including game species (Wallingford 2002), and that damage to farm crops and road encounters with vehicles have become severe problems.

The Game Commission now is attempting to change the hunting culture of Pennsylvania. They believe that part of the problem is that, because of past deer hunting policies which protected females and encouraged hunters to go after mature males only, the herd is unbalanced and lacking in mature male leaders. In order to control the deer population, the Commission is encouraging hunters to take more female deer, and is therefore limiting the number of male deer that may be taken. The challenge they face is that the hunting culture has developed an attitude that antlered deer are greater prizes and that antlerless deer are more like a consolation prize. Unfortunately, killing the larger-antlered buck will no longer be sufficient to control the deer population (Wallingford 2002).

The Elk of Elk (and Cameron) County

The elk is an antlered grazing animal, related to deer but much larger. A mature bull (male) elk is up to five feet tall at the shoulder and weighs 600 to 1000 pounds, and grows a new rack of antlers up to five feet long each year. The mature female is smaller, weighing in at 500 to 600 pounds, and can produce a calf annually. The life span is about 20 years. Because their food ranges from grasses to twigs and bark of trees and shrubs, elk prefer meadows, clear-cuts, stream valley bottoms and farm fields, though they are extremely shy of humans (Fergus 2003).

From the earliest times through the mid 1800s, the range of the Eastern elk covered most of the east coast of the US. However, hunting, first by increasing populations of Native Americans and later by colonists and Americans invading the elk's range, completely eliminated them by the 1870s. Within a few decades the Eastern elk was extinct throughout its range. The final areas of their range in Pennsylvania were in Elk and Cameron Counties (Fergus 2003, Kosack 2001).

This regional extinction preceded the creation of the Pennsylvania Game Commission by about 20 years. In its early years, the Game Commission focused on restoring the populations of turkey, quail and white-tailed deer. In 1912, they began discussing ways to revive elk because, at the national level, there was a need to relocate elk that were overpopulating areas in and around Yellowstone National Park. The Pennsylvania Game Commission agreed to take 50 elk in 1913 that were released in Clinton and Clearfield Counties, plus 22 more from a preserve in Monroe County (presumably a private operation that had also imported western elk) that were released on gamelands in Monroe and Centre Counties. In 1915 the Pennsylvania Game Commission brought an additional 95 elk from Yellowstone to Cameron, Carbon, Potter, Forest, Blair and Monroe Counties (Kosack 2001).

Trouble with farmers began soon after these releases, as herds ravaged crops, especially in Blair and Monroe Counties. The Game Commission requested additional money to compensate the farmers (to discourage the farmers from killing the elk), but the state legislature refused (Kosack 2001).

From 1923 to 1931 limited elk hunting was allowed. Over the entire period, 80 to 120 elk were legally killed (many additional elk were killed illegally). Despite the losses, there were still 200 elk in 1930, but their entire range was reduced to Cameron and Elk Counties. The Game Commission's interest in the elk waned until the late 1960s because the herd numbers declined, possibly to only 24 head at one point, and their range was reduced to the valleys of two streams in Elk County. Then, due to increased reports of crop damage from farmers outside those valleys, they came back onto the Game Commission's active radar (Fergus 2003, Kosack 2001).

One problem the Game Commission identified was that they had never studied how the elk live. A thorough ecological study, complete with annual censuses, began in 1970. Since then, both research and concerted efforts to improve the protection and habitat of the elk, including an elk management plan, have stabilized the herd. The first census reported about 65 animals, and through the 1980s it was consistently in the range of 120 to 150. Since the early 1990s it has grown to 500 and more. Additional rangeland has been bought for the state within Elk County, and new management areas have been started in Clinton County; the elk are also seen in Clearfield

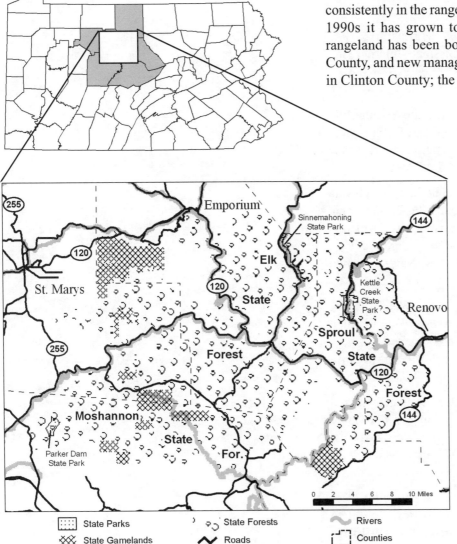

Figure 12.1 **Preserved Elk Habitat in North Central Pennsylvania:** A patchwork of State Forests, State Parks and State Gamelands creates the primary territory for elk in Pennsylvania. None of the roads are major ones. Much of the private land shown in the map and beyond is also forested or includes farm fields. The latter are a source of complaints by farmers and actions by state game officials interested in protecting both the elk and the crops.

73

and Potter Counties. Since 2001 limited hunting of elk has been re-instituted, based on a system in which a set number of hunting licenses are selected in a lottery, and the hunters selected may not re-apply for the lottery for five years, whether they kill an elk or not. (Kosack 2001)

Conclusions

While these wildlife management issues occupy a serious amount of governmental time and resources, they are not the only issues involving wildlife that do. Many environmental issues become publicized because of the harm done, or potentially done, to wildlife species as well as to humans. The same ecological principles that were so important in this chapter's discussion will be raised again in the last few chapters of this book. At that point we will again be reminded of the nature of the involvement of the human species.

Bibliography

Fergus, Chuck 2003. Elk. Harrisburg, PA, Pennsylvania Game Commission. Internet site: <http://sites.state.pa.us/PA_Exec/PGC/x_notes/elk.htm>, visited 5/28/04.

Kosack, Joe 2001. History of the Pennsylvania Elk. Harrisburg, PA, Pennsylvania Game Commission. Internet site: <http://sites.state.pa.us/PA_Exec/PGC/elk/history.htm>, visited 9/27/02.

PA Game Commission 2004. About the Pennsylvania Game Commission. Harrisburg, PA, Pennsylvania Game Commission. Internet site: <http://www.pgc.state.pa.us/pgc/cwp/view.asp?a=481&q=151287&pgcNav=|>, visited 10/4/04.

Pomerantz, Joanne T. and Joan M. Welch 1996. "Utilization of Woody Browse by White-Tailed Deer (Odocoileus virginianus) in Valley Forge National Historical Park." Pennsylvania Geographer. Vol. 34, no. 2, pp. 87-97.

Wallingford, Bret D. 2002. "Deer Management and the Concept of Change." Pennsylvania Game News, 73(7). Internet site: <http://sites.state.pa.us/PA_Exec/PGC/deer/GN0207.htm>, visited 9/27/02.

Wildlife Information Center 1996. White-tailed Deer in The Kittatinny Raptor Corridor. Wildlife Bulletin No. 17. Slatington, PA, Wildlife Information Center, Inc. Internet site: <http://www.wildlifeinfo.org/Bulletins/wildlifebulletin17.htm>, visited 10/4/04.

Section B: Chapters 13 to 19

Pennsylvania's Human Landscapes and Regions

The time scale of human activity in Pennsylvania is several orders of magnitude shorter than the Earth's time scale. Despite the difference, there are still complex systems at work, dramatic changes to describe, and plenty of evidence all around us.

Times and Changes

In most of the chapters in this section, and in the next section on our state's economy, you will recognize six major periods of our development (see Table B.1). The dates for these periods are not definitive. There is no single event that launches us in new directions, though the ends of three of these periods coincide with major wars. Social and cultural changes are often gradual, characterized by pioneers, followers and stragglers. Rather than focusing on the dates that begin and end each period, it makes much more sense to focus on what characterizes the dates near its middle in order to get a sense of what that period was all about.

Table B.1: Historic Time Periods

Time Period	Period Name	Activity
Before 1680	Pre-Colonial North America	Native American migration. Domestication of some native plants. Encounters with Europeans.
1680 - 1776	Colonial Pennsylvania	Establishment of Philadelphia and most boundaries. Treaties with native tribes extending European access. Development of agrarian life.
1776 - 1860	Early America	Settlement of the rest of Pennsylvania. Integration of Pennsylvania into US. Pennsylvania as US cultural hearth for agrarian living.
1860 - 1917	Industrial America	Aggressive extraction of natural resources. Railroads dominate transportation system. Growth of cities. Pennsylvania leads industrialization of US.
1917 - 1970	Modern America	Technological complexity increases. Peak growth and then decline of cities and industry. Establishment of the suburbs.
Since 1970	High-Tech America	Also referred to as the Information Age. Focus on computers and other modern technologies. Globalization of culture and economies.

Key events in the human history of Pennsylvania and North America begin with the initial migrations of natives to North America tens of thousands of years ago and then, a few thousands of years later, into what is now Pennsylvania. In fact, one of the earliest, reliably-dated archeological discoveries of native artifacts is at the Meadowcroft Rock Shelter in western Pennsylvania. It puts natives in Pennsylvania up to 20,000 years ago.

This was followed by the development of rudimentary aspects of native culture such as the domestication of certain plants. Ultimately, those same natives became organized into tribes and nations.

Native isolation ended with the "discovery" of North America by the Norse a thousand years ago and by Welsh sailors over 850 years ago. Major changes came with the initial discovery of North America by "civilized" European cultures a little more than 500 years ago, and their ultimate exploration and settlement of the area now known as Pennsylvania. Unlike the earlier European explorations of North America, these Europeans stayed permanently and the changes were tremendous.

The colonial era begins with the founding of the colony under William Penn, with its traditional date of 1680. However, even that date is a bit fuzzy, as is the date at which the colonial period truly ends. There were already European settlements here: Dutch, Swedish and English. Other colonies claimed parts of Pennsylvania's territory; that led to disputes, some of which were not resolved until the colonies became states. Even the process of turning colonial America into the United States of America as we know it took over a decade. So, not only were those dates fuzzy, but the lines were, too. Even so, Pennsylvania the colony was more independent of its neighbors.

Early America was focused on learning its geographical destiny and establishing successful practices for getting there, and Pennsylvania was a leader in doing so. In Chapter 1, Pennsylvania's role during this time was stated to be the training ground for much of later American culture. The early 1800s was the time when that relationship was established: Pennsylvania was a leading producer of agricultural goods and other natural resources.

By the time the Civil War ended Pennsylvania was poised to lead the country into industrialization. It had the resources, transportation facilities, and inventiveness, as well as the markets (or access to them) to define industrial America.

Modern America grew out of the aftermath of World War I, but also out of a redefinition of industry from many small companies to much larger and more integrated corporations. Some of Pennsylvania's participants in that industrial economy, such as US Steel and the Pennsylvania Railroad, were among the largest, but by the end of the period, they had mostly shrunk or even disappeared. This led to a rethinking of corporate purpose and organization, causing the shift of priority from "making" to "serving." Turning a population of blue-collar laborers into one of service providers has had profound implications for our culture.

It is debatable whether High-Tech is truly a new era compared with Modern America, but there are plenty of economic and cultural differences to support the claim. The factory jobs that characterized Industrial America and the service jobs of Modern America, though still important, are now joined by jobs and companies grounded in the age of computer, satellite and microscopic technologies. We communicate globally and have more in common with foreign cultures than ever before, from product brand names to retailers to foods and clothing styles.

The Geography of People

Studying the geography of any group of people, no matter how that group is defined, involves learning *where* they are, *who* they are (what are their *numbers* and their *characteristics*), how they are *changing* (in terms of their rates of *increase or decrease* and *migration*), and what impacts they have on humans and nature nearby. None of those elements is simple, because each can be analyzed at a variety of scales and for many types of groups and subgroups.

In the context of the entire US, there may be a sort of stereotypical Pennsylvanian, but if you have traveled at all within Pennsylvania, you have probably noticed that there are strong regional differences. Philadelphians speak differently than those from Scranton or Pittsburgh. At the same time there seem to be overlapping areas that follow the Phillies or the Pirates, the Eagles or the Steelers, and the Flyers or the Penguins. Those are "cultural" regions, and could be defined based on many different criteria. Like our early political boundaries, these cultural regions are almost always separated by fuzzy lines. In addition, human geography encompasses such things as types and amounts of economic activity, or any group's political preferences and behavior.

If we are really going to understand what makes Pennsylvania tick, we need to study the variety of Pennsylvanians, their similarities *and* differences. We have to ask the geographical questions, the ones that tell us where these similarities and differences occur, and why.

Chapter 13

People, By the Numbers

Pennsylvania's people have proven to be its most influential resource. With most resources, the more you have, the better your position is for the future. Does that hold true with people? The geographical perspective adds this qualifier: the people need to be in the right places at the right times.

The periods of greatest economic success have been those of greatest population growth. It may be a "chicken or egg" type of question as to which causes which, but the correlation is strong. Pennsylvania's population boomed early in its history, and then again at the height of the industrial revolution in America. During both periods, the state benefited greatly from better access for immigrants. As the US grew and spread out, and as the resources, labor and technologies became available elsewhere in the US, Pennsylvania's locational advantages disappeared.

Over time, population can increase or decrease. Only two factors influence these changes: the medical reality of birth or death, and the decision that is made by most people at some point in their life of whether to stay or leave. Both of these factors are strongly geographical in their nature and their effect. If we have a sense of how those factors are developing in all areas, can measure their quantities and record the changes in those quantities, and can relate those quantities and changes to measurements of related factors, then we have valuable tools for explaining the dynamics of population change. Since people are cultural beings, as we shall see in the next few chapters, understanding population change helps a great deal when we attempt to explain cultural differences and changes.

One who studies population patterns is a demographer. The main sources of information for any American demographer are the US Bureau of the Census and the National Center for Health Statistics (part of the Centers for Disease Control). There is a tremendous wealth of information published by both agencies which is freely available. The main difference between the two, beyond the specific types of data collected, is that the census data are presented for more detailed geographical areas than the NCHS data. The ability to examine the census characteristics of any number or arrangement of geographical areas is very valuable in our efforts to study people-related issues. The biggest challenge to these studies is that, because they tend to focus on endless streams of numbers, it can get mind-numbingly dry and complex. The use of tables, graphs and maps to present and examine those numbers will help in that regard.

The US Census collects data from each household, but only reports data for an organized hierarchy of areas. The smallest area is the census block, a collection of households. Each level of areas above the census block is a collection of the next lower level's areas, progressing like this: census block, census block group, census tract, municipality (township, borough, town or city), county, state and nation. There are other sets of areas that do not create a continuous coverage, such as metropolitan areas (See Chapter 15).

A key decision when examining population dynamics is the scale at which the examination will be conducted. We will start by looking at a few demographic descriptions of Pennsylvania as a whole. Greater insight can then be gained by comparing the counties of Pennsylvania demographically. We could continue that pattern by studying ever smaller divisions of area, but then the maps and numbers become more and more complex. The information to do so is available, and if you were a specialist in one county, such as Philadelphia, you would have a great deal of data available to work with. In fact, it is useful in studying any geographical area to look at it both as a whole and as a collection of its principle parts, as we will be doing here.

As a State

The last complete US census was conducted in 2000. Pennsylvania's population on April first of that year was 12,281,054. Knowing the state population demonstrates very little without other numbers to compare it to. There are four approaches we can take to develop the comparisons. One is to compare it to the US as a whole and the other 49 states. The second is to look at how that number has changed over time. The third approach is to break that total down into thematic subtotals, such as male vs. female, urban vs. rural, those whose houses have plumbing vs. those whose houses don't, or those who prefer each different genre of music. The census provides answers to the first three of those comparisons, but you will have to conduct your own research to calculate musical preferences. The fourth approach is to break Pennsylvania's total population down into geographical subtotals, which we will do presently.

Pennsylvania's total population gives us 4.4% of the total US population (281,421,906 in 2000). We are the sixth largest state, trailing California, Texas, New York, Florida and Illinois.

Historically, even including colonial times, Pennsylvania has been among the country's population leaders, ranked second in every census but one from the first one in 1790, to 1940. The graph of its totals over time, shown in figure 13.1, is a classic example of a population that grows exponentially until about 1920. From then until today, the population grows at a decreasing rate and appears to reach a maximum from 1970 to 1990 before increasing significantly again in the 2000 census.

Does our nearly level population of the last thirty years represent a physical limit to the number of people Pennsylvania can support? Why is Pennsylvania's position in our country's historical population growth declining? We have cities continuing to lose population to their suburbs, who are, in turn, trying to preserve the best farmland from development (see Chapter 21). Of course, even if all of the good farmland was preserved, there is still a tremendous amount of space to put people in this state.

The reasons for our declining population growth include a variety of factors, some best portrayed statistically (see Table 13.1) and others anecdotally. Our rate of natural increase is one of the lowest in the US, and people are leaving Pennsylvania in significant numbers. The birth rate for Pennsylvania is significantly less than the US birth rate, while our death rate is greater.

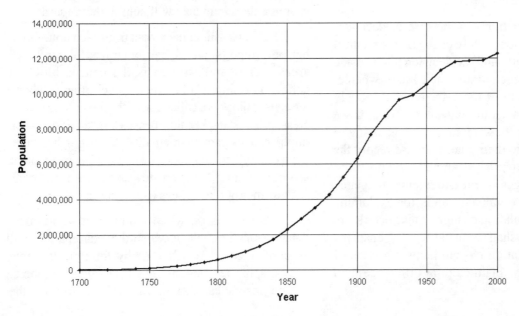

Figure 13.1 **Pennsylvania's Population Growth Since 1700:** The earliest numbers are estimates; the first US Census was conducted in 1790. Population is strongly linked with many other aspects of our nathion. For example, the dip in 1940's total population is due to factors related to the Depression of the 1930s, including changes in immigration rates (Data source: US Census Bureau 2002).

Table 13.1 Comparing US and Pennsylvania Population Characteristics

Data (year)	US	PA
Population (2000)*	281,421,906	12,281,054
Population (2003)**	290,788,976	12,406,292
Population (2004)**	293,655,404	12,370,761
Population Growth Rate (2003 to 2004)	0.99% 10 per 1000	0.29% 3 per 1000
Births (2003 to 2004)***	4,099,399	146,368
Birth Rate (2003 to 2004)	14 per 1000	12 per 1000
Deaths (2003 to 2004)***	2,453,984	130,322
Death Rate (2003 to 2004)	8 per 1000	11 per 1000
Natural Increase (2003 to 2004)	6 per 1000	1 per 1000
Net Immigration (2003 to 2004)***	1,221,013	19,188
Migration Rate (2003 to 2004)	4 per 1000	2 per 1000
Median Age (2000)*	35.3	38.0
% Age 65 or Older (2000)*	12.4	15.6
% Age 18 to 65 (2000)*	61.9	60.6
% Under Age 18 (2000)*	25.7	23.8

* Source: US Census Bureau 2002
** Source: US Census Bureau 2004a
*** July 1 to June 30; Source: US Census Bureau 2004b

The slowing growth visible on the graph was due more to emigration to other states or countries than it is to any change in birth or death rates. The difference between the population growth rate and the rate of Natural Increase is the net immigration (if positive) or emigration (if negative) rate. The reasons are pretty well documented: jobs, cost of living, climate, and environmental concerns or regulations.

Our population is older than the US as a whole, as represented by a higher median age and higher percentage of senior citizens, and by a lower percentage of people in younger age ranges. The aging of a population is something like the aging of an individual. Health care, mobility and income/wealth security are among their concerns. Many do depart for Arizona and Florida, but many, in a population that is living to older ages than ever before, are staying. The state and federal governments are committed to providing for their needs.

Counties

The questions raised by looking at totals for the entire state change when we look at distribution patterns within the state. The first pair of maps (figures 13.2 and 13.3) helps to show where population growth is occurring. The first map illustrates growth due only to the difference between the birth rate and death rate in each county. Generally, the eastern half of the state has an increasing population except in the area around Scranton, Wilkes-Barre and Hazleton, and the western part of the state shows a decline except in the area around Erie. In the areas of highest increase, either the birth rate is significantly higher than the state average, or the death rate is significantly lower than the state average. Speculate about what factors would lead to such differences. Are they ethnic, religious, environmental, economic or other?

The overall population change map shows a similar pattern, but with some interesting differences due to migration. First of all, most of the same eastern counties are showing population growth, though not all to the same relative degree. The biggest difference is Philadelphia, which has among the highest positive rates of natural increase, but is in the group with the highest rate of overall population decline. This is strong evidence of the number of people continuing to leave the city. Another example is Pike County. Despite having an average rate of natural increase, it has one of the highest rates of overall population growth. The difference is a recent trend of people from New York and New Jersey building vacation and retirement homes in the area so that they can stay or live outside the city but within an easy drive.

The other two maps represent ways to evaluate age distribution within the state. The area around Wilkes-Barre and Hazleton again shows up in the highest categories showing an older population. Similarly, the counties of the western plateau region tend to be in the higher categories on both maps. Both of those regions can be described as formerly more

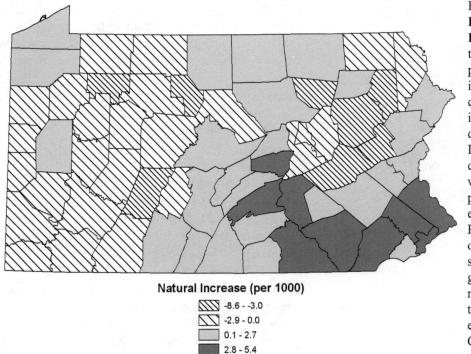

Natural Increase (per 1000)

▨	-8.6 - -3.0
▨	-2.9 - 0.0
▨	0.1 - 2.7
■	2.8 - 5.4

Figure 13.2 **Rate of Natural Increase Around Pennsylvania:** This map shows the the distribution of the rate of population change due to natural increase only, in 2000. The counties with a positive natural increase have more births than deaths each year, and vice versa. If this were the only population change factor, many counties would be losing population, particularly in the western and eastern areas of the Appalachian Plateau and the northeastern end of the Ridge and Valley. In the southeast, the counties are gaining population. Use this map together with the others in this chapter to think of explanations. (Data source: US Census Bureau 2002)

prosperous in earlier years of the iron and steel industry, but which are suffering somewhat today.

On the other hand, notice the odd situation of Centre County. It has a median age at least five years younger than the rest of the state, and is in the category with the fewest senior citizens. In this case, we are looking at an artifact of the census-taking process. The census records each person's residency by their location on April 1 of the census year (2000, in this case). Since this occurs during the academic year for universities, and since Pennsylvania State University draws students from all over the state and beyond (hence, relatively few students commuting from home), they are all considered resident in Centre

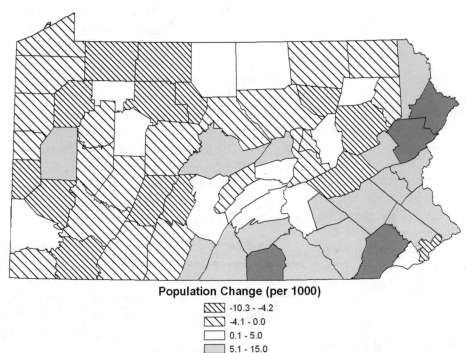

Population Change (per 1000)

▨	-10.3 - -4.2
▨	-4.1 - 0.0
☐	0.1 - 5.0
▨	5.1 - 15.0
■	15.1 - 36.6

Figure 13.3 **Rate of Population Change Around Pennsylvania:** This map shows the distribution of the rate of population change in 2000. Births and immigrants add to the population, while deaths and emigrants reduce it. Since the values mapped in Figure 13.2 are part of those mapped here, the first explanation to attempt is why some areas have opposing trends in natural increase vs. migration, while others are consistent. Philadelphia shows the most extreme inconsistency, while Chester, Cameron, Cambria, Luzerne, Schuylkill and Sullivan Counties are extremely consistent. (Data source: US Census Bureau 2000)

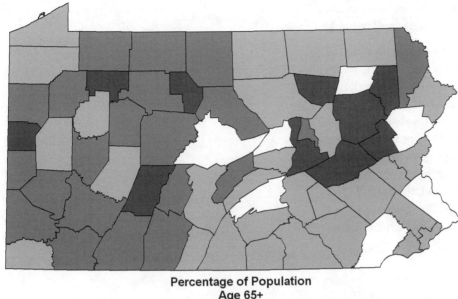

Figure 13.4 **The Percentage of the Population Aged 65 or Older Around Pennsylvania:** Pennsylvania's older citizens are not evenly distributed around the state. Note that this is not just a map of where the senior citizens are, but also a map of where the younger citizens "are not." Most of the counties that had the lowest rates of natural increase (Figure 13.2) have the highest percentages of seniors. (Data source: US Census Bureau 2002)

Percentage of Population Age 65+

	10 - 13
	14 - 16
	17 - 18
	19 - 22

County for census purposes. That does not describe Philadelphia, though. It has an average proportion of seniors, but a much lower median age.

Conclusions

The same process could be used to ask other population-related questions about Pennsylvania. Where do cancer-related deaths occur in greatest numbers, and what does that tell us about the most likely causes? Where are the greatest numbers of single head of households located, and what might be the pressures that lead to those situations? The process doesn't necessarily give you the answers, but it can stimulate speculation that leads to explanations.

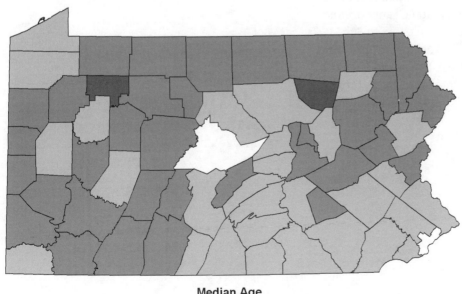

Figure 13.5 **The Median Age of the Population Around Pennsylvania:** Despite appearances, this map is generally consistent with Figure 13.4. No county shifts by more than one category. Forest, Cameron and Sullivan Counties have older populations, significantly more deaths than births per year, and are the only counties with populations less than 10,000. (Data source: US Census Bureau 2002)

Median Age

	29 - 35
	36 - 38
	39 - 42
	43 - 44

Sometimes, by comparing maps of related phenomena it is possible to reach less conventional explanations.

One question that studying the characteristics of population and their state-wide (or even nation-wide) distribution could also stimulate is whether population growth (or certain levels of it) is good or bad. Would we be better off holding still at one level and working to improve living conditions for the numbers that we have? Or, is it better to encourage growth, in order to increase the number of citizens/taxpayers contributing their money and energy for our mutual benefit? The simple graph in Figure 13.1 led to questions that demanded more data in order to provide explanations.

Bibliography

US Census Bureau 2002. Census 2000 Summary File 1 (SF 1) 100-Percent Data. Washington, DC, US Department of Commerce, Census Bureau, Population Division. Internet site: <http://www.census.gov>, multiple visits.

US Census Bureau 2004a. Table 1: Annual Estimates of the Population for the United States and States, and for Puerto Rico: April 1, 2000 to July 1, 2004 (NST-EST2004-01). Washington, DC, US Department of Commerce, Census Bureau, Population Division. Internet site: <http://www.census.gov/popest/states/NST-ann-est.html>, visited 1/6/05.

US Census Bureau 2004b. Table 5: Annual Estimates of the Components of Population Change for the United States and States: July 1, 2003 to July 1, 2004 (NST-EST2004-05). Washington, DC, US Department of Commerce,US Census Bureau, Population Division. Internet site: <http://www.census.gov/popest/states/NST-comp-chg.html>, visited 1/6/05.

Chapter 14

Ethnicity and Immigration

The vast majority of Pennsylvanians alive today were born here, but every one of us has an ancestor who was born outside of North America. In this chapter we examine our most direct links to our ethnic heritage.

The US Census attempts to record our ethnicity by asking a sample of the population, "What is this person's ancestry or ethnic origin?" (US Census Bureau 2002b). This becomes a difficult question for most of us, whose preceding generations have probably included many who married outside their ethnicity. It remains to be seen whether the question is useful for researchers, since as recently as 1950 the question was not asked. Ever since the earliest censuses, though, the questions have included something like "Where was this person born?" which presents the challenge that a German who immigrated in 1897, for example, was counted as a foreign-born German in every census from 1900 until he or she died. The census is very useful for tracking long-term trends, but less useful for current activity.

During any particular decade between censuses the leading countries of origin or birth may have shifted any number of times. Europe has contributed the vast majority of North American immigrants over the last 300 years. The period of maximum immigration to Pennsylvania, which peaked in the years before World War I and shows up in the 1910 Census, has long since passed (see Table 14.1). The subsequent decrease reflects national trends twenty years later related to the two World Wars of the 1900s, as well as the Depression that persisted through the 1930s. US immigration policies were changed, so overall US immigration totals and patterns changed also. Pennsylvania was also affected by the increased number of US states in the late 1800s and the shift of immigrant destinations to many of those newer states even up to the present day.

Many geographers, and other demographers, are interested in this dynamic larger picture, as it helps to describe the character of many communities at a variety of scales. You can relate to this because who you are as an individual is strongly connected to your ethnic origins.

Ethnicity

Your ethnicity is one way of distinguishing you from those around you. It is most often your national origin, such as German or Hispanic, or that of a recent or distant ancestor, traced back to a point where the individual in question was born in a foreign land. It can also be primarily racial, such as African-American, or religious, such as Jewish. In other words, where people have chosen to belong to a large group, or been lumped into such a group, which then becomes a primary way to identify the individual and the group, an ethnicity has been established.

Thus, your ethnicity may be beyond your control, but also may be your choice. It is most often determined by the ethnic identity of your parents. However, if your parents were from two different ethnic backgrounds, your ethnicity might be their choice, your choice, or some new blending of their ethnicities (as when we refer to the Scots-Irish below). For some, their ethnic identity is very important and given a great deal of significance in a wide range of decisions. For others, it is insignificant to their day-to-day lives.

Genealogy has become a very popular hobby as more previously out-of-print sources showing individual membership in various groups and activities are made available, many of them on-line. Tracking ancestors in order to learn more about their origins and other details about them has re-awakened interest in ethnic identification.

The Immigration Process

The decision to migrate is a complex one, especially when overseas, cross-cultural travel is involved. Think about it from the migrant's perspective. First, conditions in the country of origin have to be bad enough that migration holds the best hope. The migrant (with family) usually has to be well off enough that he or she can afford to move such a tremendous distance. Finally, the migrant has to have enough knowledge to make the decision of where (specifically) to go and how to get there.

Europe has been our largest source mostly because they had times of great political, religious and economic distress. These are referred to as "push" factors because they are reasons for starting the migration process. European countries also generally had the knowledge and technology (especially systems of communication and transportation) to travel this far, and in such numbers. Asian Americans and Latin Americans are immigrating in greater numbers today than Europeans, but the same push factors apply.

The destination decision of any migrating individual, family or group is most dependent on what knowledge they have received. Especially if they are selecting among several possible destinations, "pull" factors are at work guiding them to one destination above the others. For example, in the earliest years of Pennsylvania as a colony, William Penn himself made several trips back to travel around Europe canvassing for new settlers. Most of the later immigrants to a particular town or neighborhood came because someone from their foreign home town had already made the trip and sent back positive and welcoming news (which was not always completely accurate).

Thus, every immigrant has experienced a set of push and pull factors. Once here there are further challenges to surpass in order to be successful. These include language (for many), employment and finding a home. Modern times have not made this process significantly easier.

The Early Immigrants

Source areas for early settlers were limited to a relatively few parts of Europe. English, Welsh and many Germans came directly from their native countries. These were mostly members of religions out of favor with, and persecuted by, the governments and other established institutions there. In other cases,

Table 14.1 Historical Immigration to the US and Pennsylvania			
Year	US Foreign Born	PA Foreign Born	% of PA Population
1850	2,244,602	303,417	13.1
1860	4,138,697	430,505	14.8
1870	5,567,229	545,309	15.5
1880	6,679,943	587,829	13.7
1890	9,249,547	845,720	16.1
1900	10,341,276	985,250	15.6
1910	13,515,886	1,442,374	18.8
1920	13,920,692	1,392,557	16.0
1930	14,204,149	1,240,415	12.9
1940	11,594,896	976,573	9.9
1950	10,347,395	783,965	7.5
1960	9,738,143	603,490	5.3
1970	9,619,302	445,895	3.8
1980	14,079,906	401,016	3.4
1990	19,767,316	369,316	3.1
2000	31,107,889	508,291	4.1

Data sources (1850-1990): Gibson and Lennon 1999.
(2000): US Census Bureau 2002c.

the migrants were more opportunistic, looking for a chance to start afresh with more land or better soil.

Some of the immigrant groups had already migrated within Europe, for the same reasons. For example, many Scottish Protestants, Presbyterians, had moved to Ulster (now known as Northern Ireland), during the late 1600s. Life in Ireland was still a struggle because of economic hardship and religious persecution by Irish Catholics, so many came to the colonies. They were known in America as the Scots-Irish (Zelinsky 1995, 116). A better-known migration that involved Irish occurred in the middle 1800s. The potato, a plant native to America, had been grown successfully in Ireland for over 200 years, and had become a staple in the diet of many Irish and other Europeans. During the 1840s a disease called the potato blight emerged in Ireland, and they were poorly prepared to cope with the number who died or became sick from the diseased food. The reaction of many to the devastation surrounding them was to leave. Many went to a variety of destinations in Britain, Canada and the US. Most of these immigrant Irish were Catholics, in contrast to the earlier Scots-Irish, and once again many of them ended up in Pennsylvania.

The Later Immigrants

Once the US gained its independence, immigration was less a government policy and more a phenomenon that was already well-established. Philadelphia continued to be a major port of entry. Immigration really picked up and changed after the American Civil War. Clashing classes and periods of open war in Europe were two significant push factors. Ironically, the industrialization of Europe (and its changing economic and social systems) was another push factor there, while industrialization in America was a pull factor drawing them to this side of the Atlantic. Pennsylvania attracted many of the new immigrants because it was known to have the industrial opportunities, and because of news from its existing foreign-born population.

Immigrants in the late 1800s and early 1900s came from Italy, Poland and Austria as well as other parts of Europe. These newcomers headed for the big cities and for factory, mining and lumbering towns, unlike the earlier agricultural immigrants. The 1920 Census, for example, counted about 1.4 million Pennsylvanians born outside the United States, out of a total population around 8.7 million (Zelinsky 1995, 119).

Table 14.2 Ancestry of Pennsylvanians in the 2000 Census (US Census Bureau 2002c)

Area of Origin	No. of People	Percentage
England/ Wales	1,184,577	9.6%
Scotland/ N. Ireland	405,492	3.3%
Ireland	1,981,106	16.1%
Germany	3,246,341	26.4%
Austria/Hungary	190,186	1.5%
Other NW. Europe	554,185	4.5%
Poland	824,146	6.7%
Other E. Europe	746,954	6.1%
Italy	1,418,465	11.6%
Other S. Europe	176,986	1.4%
Latin America (mostly Caribbean)	76,778	0.6%

The Immigration and Naturalization Service was set up as a federal agency in 1891, with the responsibility to enforce new laws regarding the large numbers of immigrants. Congress handed criteria such as inability to read or to learn English, health problems, and prior criminal records to the agency as reasons for refusing admittance. In 1921 a quota system was established which limited the numbers and origins of most immigrants. Those numbers and origins were changed several times over the next thirty years. In the 1950s, immigration from Asia was permitted to increase, and in the 1960s the quota system was abolished. Since then, Asia and Latin America have been the largest sources of immigrants, though Pennsylvania's share of those immigrants is not consistent with its proportion of the national population (see Table 14.2).

As we saw in Chapter 13, the more interesting questions are stimulated by maps of where the immigrants settled within Pennsylvania. The maps of Pennsylvanians claiming English and German ancestry in the 2000 Census does not look much different than the overall population distribution. Notice the differences among the maps in Figures 14.1 and 14.2, which look at the same information presented for other ethnic groups, first, as a quantity and, next, as a percentage within each county. The absolute quantities of Figure 14.1 show that Philadelphia and Pittsburgh are consistently among the leading counties for any ethnic group, simply because of the size of their total populations. It also illustrates that the traditional nationality-based ethnicities, many of which go back five generations or more, are not as strongly claimed as those of the more recent immigrant groups: the

Pennsylvania Ethnic Ancestries: Quantities

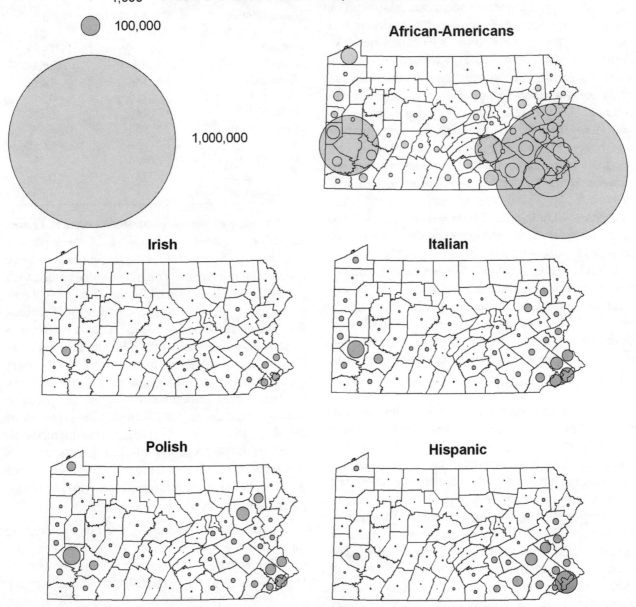

Figure 14.1 **The Number of People in Selected Ethnic Groups:** The size of each circle is proportional to the number of people claiming single or mixed ancestry in that map's ethnic group. About 20 percent of those responding do not claim any ethnicity, while others claim ethnicities that are difficult to interpret. (Data source: US Census Bureau 2002c)

African-Americans (early 1900s) and Hispanics (after 1950).

On the other hand, Philadelphia and Allegheny County (which includes Pittsburgh) are still very ethnically diverse, since they tend to fall in the higher percentage categories of Figure 14.2. The other counties around the state that include larger cities show the same tendency, with some interesting exceptions.

For example, the Irish population is one of the highest percentages in almost every county, and some 16 percent statewide; while Hispanics outnumber Irish in the southeastern counties, they comprise only 3.5 percent of the state population. It is also interesting to compare the urban diversity with the less urban counties. While many appear similarly diverse, that hides the fact that many smaller towns within their borders are much less diverse.

Pennsylvania Ethnic Ancestries: Percentages

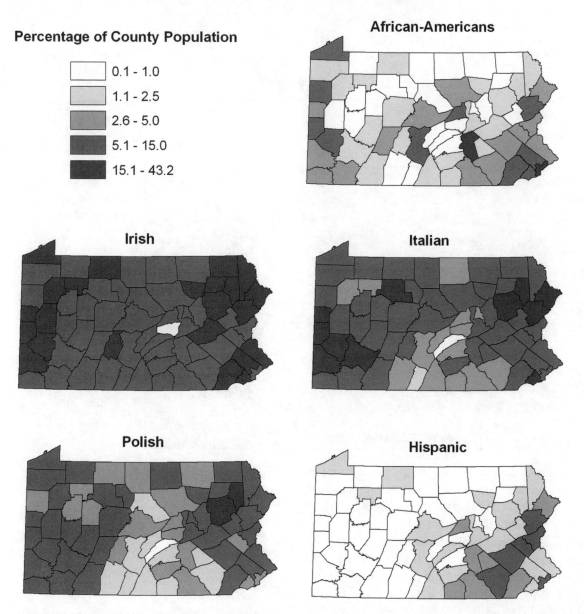

Percentage of County Population

- 0.1 - 1.0
- 1.1 - 2.5
- 2.6 - 5.0
- 5.1 - 15.0
- 15.1 - 43.2

African-Americans

Irish

Italian

Polish

Hispanic

Figure 14.2 **The Proportion of People in Selected Ethnic Groups:** More than 43 percent of Philadelphians are African-American, but no other ethnic group represents more than 30 percent of its county. (Data source: US Census Bureau 2002c)

The Welsh Tract, Today's Main Line

The first large immigrant group to Pennsylvania, in the first twenty years after Penn founded the colony, was not the English or the Germans. It was the Welsh, most of whom were Quakers. While Penn had been luring new colonists (especially those with money) by offering them plots in Philadelphia, the Welsh sought a larger area where whey could form their own

separate society. They wanted a single, contiguous area where they could preserve the Welsh language and have the authority to govern themselves. Penn set aside 40,000 acres (see Figure 14.3), which the Welsh called the Welsh Barony (Hartmann 1978, 43), though others referred to it as the Welsh Tract.

These earliest Welsh immigrants came mostly from the central areas of Wales (see Figure 14.4).

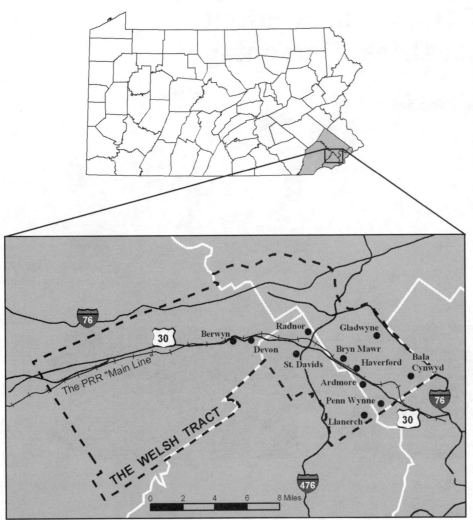

Many of them had owned or rented small farms in Wales and had been just barely comfortable; however, because of industrialization in Britain, changes in land policies and economic structure were squeezing them financially. America represented an opportunity for independence from English land policies, and Pennsylvania's land was known to be affordable (Pearson 1997, 44). In addition, Welsh Protestants of Quaker, Baptist and Presbyterian persuasions, like Protestants elsewhere, were feeling threatened by both Catholic and Anglican persecution (Hartmann 1978, 32, 40). The Quakers were the first group to migrate to Pennsylvania because of their more direct link to the English Quakers.

Because the early leaders were Quakers, their internal religious and governmental organization was based on "Meetings." They thought they had negotiated with Penn the right to self-governance within the Barony, but it turned out that Welsh separateness was short lived. By 1690, Penn and his administrators had established part of the Philadelphia-Chester County boundary through the Welsh Tract, had established several township governments to remove authority from those Welsh Quaker Meetings, and had sold some 10,000 acres within the original Welsh Tract (that were never paid for) to non-Quaker Welsh and to English and German settlers (Hartmann 1978, 44).

Other Welsh communities were established in Pennsylvania in those early years. For example, Gwynedd Township (also known as North Wales) in Montgomery County was another important early Welsh settlement, populated mostly by Welsh Anglicans. By 1730, additional settlements in Bucks, Berks and Lancaster Counties had been established. One of these areas straddling the Berks-Lancaster County boundary is known as the Welsh Mountains, and both counties have Caernarvon Townships and other similarly Welsh-named places in the area. However, after that date, the number of new Welsh

Figure 14.4 **Welsh Town Names Present in Pennsylvania:** Many of these places are relatively small towns. Among the place names that also appear in Figure 14.3 are two, Johnstown and Nantyglo, that are in Cambria County (itself a Welsh name).

immigrants fell to virtually none. Later waves of immigration included Welsh going directly to Cambria County starting in the late 1790s (Magda 1986, 2; Pearson 1997, 44), and Welsh starting in Pennsylvania but then moving on to Ohio in the 1830s (Pearson 1997, 44).

One of the fascinating aspects of the Welsh is the degree to which they integrated themselves among the English speakers of America over time. To those with no understanding of Welsh traditions, it is surprising to hear many names identified as Welsh in origin. Considering the unusual spellings for many place names, many Welsh family names seem to be more English than Welsh: Evans, Griffith(s), Hughes, Jones, Morgan and Owen are a few examples.

Most of the Welsh Tract was rather run down by the late 1800s. Then, the Pennsylvania Railroad was looking for an improved route westward out of Philadelphia. Most alternatives would have required them to take small properties and piece together a less direct route. Through the Welsh Tract the railroad could purchase large properties and keep their route more straight and efficient. Once the line, called the

Main Line because it continued to Harrisburg, Pittsburgh and the Midwest from there, was built, they could sell the unused sections of their properties for additional profit (Oser 2004). It turned out to be a very desirable area for wealthy Philadelphians looking to move out of the city but remain within an easy commute, because the train was the commuter's prime option back then.

Today the region is informally referred to as the Main Line; it is well known in the Philadelphia area for its large estates, now hemmed in by middle class housing developments and strip commercial development along US Route 30, which parallels the rail line. Figure 14.5 shows the variety of development now visible there, with US Route 30 clearly visible diagonally crossing the center of the photograph. To its north are the wealthier estates, and to its south is the later middle class urban-style development (see next chapter).

Figure 14.5 **Main Line Landscapes:** The railroad and US Route 30 separate lower-income higher-density urban housing (median household income about $55,000 in 2000) from higher-income lower-density estates (median household income about $112,000) to the north (US Census Bureau 2002c). (Image source: USGS 1999)

89

Reviving the area economically brought back attention to its cultural roots. In fact, at the time of the railroad's construction, the community located where Bryn Mawr is today was called Humphreysville. A railroad employee involved with acquiring the properties found a reference to Bryn Mawr on one of the early deeds and chose it to name the Pennsylvania Railroad stop and its surrounding development (Oser 2004).

Conclusions

The Main Line today holds very few links, beyond the community names, to its Welsh origins. That, too, is not unusual among ethnic communities. For example, poor neighborhoods of large cities are often referred to as "ghettoes." A ghetto, originally, was a Jewish neighborhood. Early in the existence and industrial expansion of major cities, when there was no country of Israel, most large cities in the world had Jewish communities that were generally ostracized and segregated by Christian majorities. In many of those cities the Jews eventually became financially successful and "moved up" to more affluent neighborhoods. Their original neighborhoods then became home for the newest poor immigrants, the ethnicity of which changed as each wave of immigrants settled, improved their economic situation sufficiently to afford to move out, and were replaced by the next wave.

Other communities have retained their ethnic heritage very strongly. This shows up in the forms of ethnic festivals, popular ethnic restaurants (that are not franchises of nation-wide chains), ethnic social organizations and other evidence of pride in that ethnicity. These strongly ethnic communities are often smaller towns or villages or neighborhoods within larger cities.

How much do you know about your ethnic roots? It can be a fascinating exercise to ask older family members if they know who the immigrants in your line of ancestors were, where they came from, where they came to (at all of their stops on their way to your home town), and when all of this occurred.

Bibliography

Gibson, Campbell J. and Emily Lennon 1999. "Table 13: Nativity of the Population, for Regions, Divisions, and States: 1850 to 1990." from: Historical Census Statistics on the Foreign-born Population of the United States: 1850-1990. Washington, DC, U.S. Department of Commerce, Census Bureau. Internet site: <http://www.census.gov/population/www/documentation/twps0029/tab13.html>, visited 7/14/04.

Hartmann, Edward George 1978. Americans from Wales. New York, NY, Octagon Books: Farrar, Strauss and Giroux.

Magda, Matthew S. 1986. Welsh in Pennsylvania. Harrisburg, PA, Pennsylvania Historical and Museum Commission. Internet site: <http://www.phmc.state.pa.us/ppet/welsh/page1.asp?secid=31>, visited 7/14/04.

Oser, David R. 2004. Bryn Mawr Historical Information. Bryn Mawr, PA, Main Line Real Estate/REMAX Executive Realty. Internet site: <http://www.mainlinerealestate.com/bryn_mawr_history.htm>, visited 7/14/04.

Pearson, Brooks C. 1997. "18th and 19th Century Welsh Migration to the United States." Pennsylvania Geographer. Vol. 35, no. 1, pp. 38-54.

US Census Bureau 2002a. Census 2000 Summary File 1 (SF 1) 100-Percent Data. Washington, DC, US Department of Commerce, Bureau of the Census. Internet site: <http://www.census.gov>, multiple visits.

US Census Bureau 2002b. Measuring America: Long Form Questionnaire. This longer publication has been separated into several documents for posting on the Census Bureau Web site. Washington, DC, US Department of Commerce, Census Bureau. Internet site: <www.census.gov/prod/2002pubs/pol02marv-pt4.pdf>, visited 8/14/04.

US Census Bureau 2002c. Census 2000 Summary File 3 (SF 3) 100-Percent Data. Washington, DC, US Department of Commerce, Census Bureau. Internet site: <http://www.census.gov>, multiple visits.

USGS 1999. Digital Ortho Quarter Quad (aerial photograph): norristown_pa_se. Washington, DC, US Geological Survey. Downloaded in TIFF format from <http://www.pasda.psu.edu/access/doq99list.cgi>, visited 1/4/05.

Zelinsky, Wilbur 1995. "Ethnic Geography." Chapter 8 in Miller, E. Willard (ed.) A Geography of Pennsylvania. pp. 113-131.

Chapter 15

Urbanization

A classic child's tale illustrates that people are either more comfortable in the city or in the countryside. To that observation, today, we should add the choice of the suburbs. Cities and suburbs represent two solutions to the challenge of housing ever-increasing populations. While cities and suburbs are not unique to Pennsylvania, their development here has been historically influential.

Population growth during the late 1800's created the great cities of today. It was a "chicken or egg" phenomenon: cities' populations and transportation accessibility grew faster when they had industrial or similar employment opportunities for newcomers, and industries or other employers were drawn to places that had abundant potential workers and transportation facilities. How these places have fared since the early twentieth century is a different matter.

Industrialization and urbanization went hand-in-hand. Would other phenomena, such as America's "melting pot" of cultures and fast pace of innovation, also have occurred without the concentration of people, transportation and communication that happened in cities in the 1800s? Cities in Europe evolved gradually over a long history. In the new world, and especially in Pennsylvania, they were intended from the very beginning. William Penn is known to have hired city experts to lay out a design for Philadelphia that would create an ordered but equally capable version of a successful European city. This is most evident in the original street plan for the city (see Figure 15.1) he saw as his base of operations and the focal point for the colony. You may well see urban features that you recognize in Pennsylvania cities that you know, because several features of the original plan for Philadelphia, especially its rectangular grid of streets and its centralized square (now occupied by City Hall), were widely copied in other Pennsylvania and US cities.

Cities represent two sets of processes. First, a city interacts with its surrounding area, which will have smaller towns, suburbs and rural land. The amount of this interaction that takes place depends on how far it is to the next similarly-sized (or larger) peer cities. The city is like a parent to these "children," and at the same time, interacts and competes with its peers. Geographers look at where cities of different sizes and types are located.

The second set of processes that a city represents is focused on its internal affairs. A city is a complex system of internal components. The components must be organized, and they must provide for all of the needs of the city's population and economy. Many cities specialize in certain products or functions because of their location and because of the talents of their populations.

Why are cities located where they are? Those larger location decisions were usually based on site characteristics, and on the relative locations of other cities, some of similar types and others which are different. What made some grow larger than others? What functions are common to all cities? What makes a city grow (or lose population) is really a matter of both scales of thinking: people move there (or move away) because of both where it is and what is in it. The geographer finds it interesting to try to explain such patterns.

City-ness

Not only is there a fundamental difference between the "countryside" and the city (including its suburbs),

91

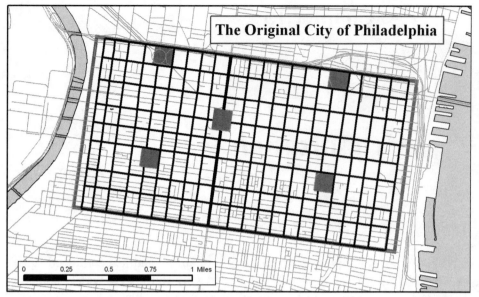

The Original City of Philadelphia

Figure 15.1 **The Original Philadelphia:** Philadelphia started as a town stretched between the Delaware and Schuylkill Rivers. In this depiction of a famous map, the roads laid out on the original plan are drawn in black, while today's roads are in gray. The darker gray squares were originally intended as parks, which three of them still are. The central park now houses City Hall, while the northwestern park is now an important traffic intersection, Logan Circle.

but there is also a mutual dependency. This is clearly illustrated in the development of America by the fact that we began as a primarily rural society in which over 90% of the households identified themselves as agricultural. An urban dweller cannot grow enough food to feed a large family, so he is dependent on the farmers to grow excess and sell it to him. The colonial cities and towns became market centers for agricultural goods as well as for the resources and products sent from America back to England. A number of cities and towns have institutionalized this aspect of their past by continuing to host farmers markets and market days, and many counties and other communities have their annual fairs including agricultural exhibits and rodeos.

Being a market center required an efficient transportation system, both internally and between the city and its surrounding countryside. That "system" may have started out as direct and passable roadways and navigable streams and rivers for individuals carrying their own loads, but eventually transportation itself became a form of economic activity. Over time, our large cities became hubs of extended networks of river, rail, highway and (later) air transportation. The same transportation system that brings food and other basic necessities into the city is also used to distribute higher cost goods and finished products to surrounding areas. These systems continue to be critical to the city's economic condition, as more money is spent within the city by commuting workers and by transient travelers and tourists.

Financial conditions are of vital importance in the life of a city. Just as city residents depend on rural

farmers for food and miners and engineers for energy, they also depend on each other for such necessities as water, food and fuel distribution, fire and police protection, waste removal, health care and private or public financial support. Since all of these necessities cost money, especially at the scale they are provided for in a city, taxes and disposable incomes must be sufficient to cover them. In good times, cities seethe with activity and possibilities, but in hard times the problems can quickly become compounded.

Patterns of Cities

There is a hierarchy of cities and towns. This can most readily be seen by sorting cities according to their sizes. There are no definitive limits to the categories of size we will follow, and in any event the categories will be different in the southeast than in, for example, the north central part of the state. The concept is based on the functional relationship between any central city and its surrounding suburbs and smaller cities and towns, at varying distances, in addition to the rural countryside (Erickson 1995a).

The smaller cities and towns outside the city (then or now) have usually been linked to each other, in addition to being linked to the larger city, in ways strongly related to their sizes (see Figure 15.2). The smallest of these communities house a limited range of goods and services, generally those needed frequently and costing relatively little. Examples include groceries, gasoline, small eating places, post offices and pharmacies. Stores that sell items required less frequently and that are more expensive, and businesses that choose to operate in larger quarters or

92

offer greater variety, need a larger number of nearby residents in order to maintain their critical mass of paying customers. Such enterprises will be attracted to larger towns and cities depending on that critical mass, and also reflecting how far customers are willing to travel. For example, while we may not like to drive more than five miles or so to buy groceries a few times a week, we will go to the nearest larger place to buy a car every few years. The larger places then frequently have a cluster of auto dealerships. Owners of a major professional sports franchise will build their facility within a very large city or within easy transportation access of an equivalently large population.

The sizes of these functional categories of cities vary. In southeastern Pennsylvania cities of 20,000 or more are likely to have a strip of highway packed with car dealers, and smaller towns will have something similar only if the larger city is not within a reasonable drive. In north central Pennsylvania, it might be a town of 4-5,000 that can be counted on for a selection of car dealers.

Another indicator of the importance of the relationship between a large city and its surrounding communities is the US Census's Metropolitan Area category for data reporting (see Figure 15.3). Every census Metropolitan Area is at least a county. The larger central cities have multi-county Metropolitan Areas; Philadelphia's even includes Burlington and Camden Counties, New Jersey. The metropolitan area represents an area larger than a city containing places that are integrally linked both economically and culturally.

Development of the Suburbs

The term "suburb" today usually connotes a particular style of residential development outside a city, with single-family homes surrounded by yards. Several other development types can also readily be classified as suburban: the townhouse residential development, the shopping plaza (or "center"), the shopping mall, and the industrial park. If these were to be used in a definition of "suburb" then its origins do not date back much further than World War II. Actually, though, there have been suburbs as long as there have been cities.

Urban expansion has always been associated with cities. In Pennsylvania, more than in the older colonies and certainly more than in Europe, many boroughs and cities were planned from their beginning. They

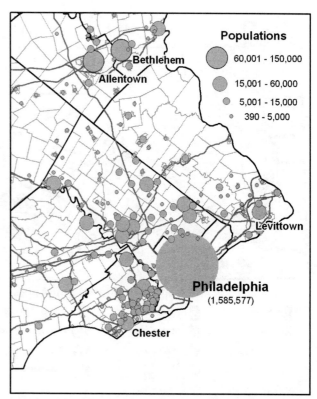

Figure 15.2 **The Relative Sizes of Cities and Towns in Southeastern Pennsylvania:** These circle sizes indicate the relative sizes of these urban places. The places within the larger ranges are likely to offer more services and a greater variety of goods than smaller places. People living in the smaller places will likely have to drive farther to find more expensive or more specialized goods and services. (Data source: US Census Bureau 2002)

usually started urban life as an unincorporated town, grew to become an incorporated borough, and grew to city status later. Once urban growth reached the original boundaries, smaller places were "annexed" (absorbed into the fabric and boundaries of the larger urban place). A state law in the late 1800s changed that practice by giving smaller communities the right to choose whether they wanted to be absorbed by the larger place, effectively creating the modern suburb.

The essential difference between pre-World War II and post-World War II suburbs is the central importance of the automobile in their planning. Before World War II the street pattern was usually a continuation of the existing urban street pattern of rectangular blocks. This pattern made it easy to extend such urban services as the water and sewer systems and public transportation routes. Early suburbs that accommodated autos were for the wealthy and featured larger properties. In some cases the developments

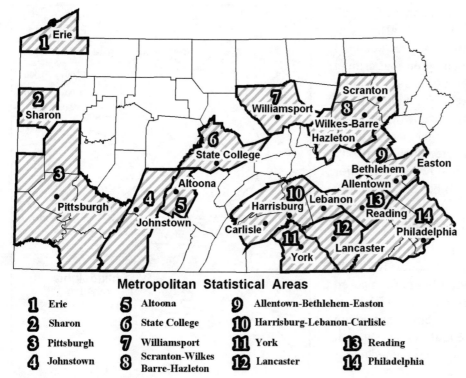

Figure 15.3 **Pennsylvania's Metropolitan Statistical Areas:** The metropolitan statistical area is the US Cenus Bureau's way of recognizing that major cities are integrally connected to their surrounding smaller cities, towns and rural areas. There are actually several categories of metropolitan statistical area (not shown here).

Metropolitan Statistical Areas

1. Erie
2. Sharon
3. Pittsburgh
4. Johnstown
5. Altoona
6. State College
7. Williamsport
8. Scranton-Wilkes Barre-Hazleton
9. Allentown-Bethlehem-Easton
10. Harrisburg-Lebanon-Carlisle
11. York
12. Lancaster
13. Reading
14. Philadelphia

were later annexed, but once the development had a choice, it didn't always agree to annexation.

After World War II, the automobile was affordable for the lower middle and the working classes. The 1950s suburbs added smaller mass-produced houses (see Levittown, below). These newer residential suburban developments usually adopted street patterns that were less interconnected with the urban pattern, and in fact became even less accessible

in more recent developments, and are one of the easiest ways to date suburban developments (see Figure 15.4). For other social reasons this process most often attracted whites, thus changing the ethnic character of the cities or towns also.

The effect of this was to burden those who remained in the city—by choice or by necessity—with all of the tasks and expenses shared by the formerly larger and wealthier population, as well as with the

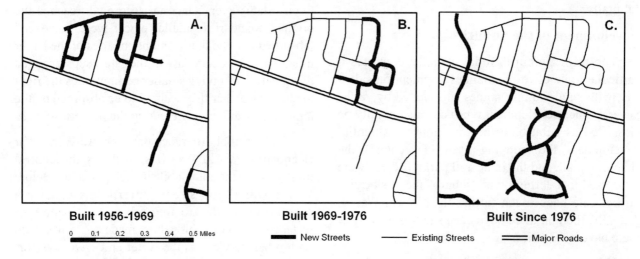

Figure 15.4 **Street Pattern Changes over the Years:** In the post-World War II era (map A), suburban developments were likely to approximate the urban grid as an efficient way to fill space. As time progressed (map B) the roads varied in shape and access in order to minimize the amount and speed of neighborhood traffic. The newest development in map C shows the latest adaptations, with even less access and several dead-end cul-de-sacs.

resulting vacancies and aging industrial properties and infrastructure. The implications include problems with traffic, services such as police and fire protection, and the city's eroding tax sources, especially along the urban-suburban edge.

Patterns within Cities

There are significant similarities to the general arrangement of functions in all urban places. First, consider the main categories of these functions: commercial and government, industrial, transportation (passenger and freight are sometimes separated), and residential. In larger towns and cities there are city districts which have a preponderance of establishments from one category only: an industrial district, a transportation center, or the central business district (CBD) where all of the major retailers and government offices and other institutions (museums, post offices, and convention centers) are located (Erickson 1995b). To some extent these functions may be intermixed: the grocery store within a residential neighborhood, for example, or the small town based on one industrial employer that is annexed by the larger city. South

Philadelphia is a prime example of this phenomenon, as were most of the riverside districts of Pittsburgh through the 1950s.

Several factors cause these patterns to emerge. Larger commercial businesses need customers above all else, and prefer to be as close to the center of the city as possible, equally accessible from all directions. There may be other such locations, desirable for their accessibility, and they are usually identifiable by taller buildings. Industrial activity is attracted to districts where services it requires—mainly transportation but also energy, cooling water, waste disposal options. Residential districts (neighborhoods) also tend to show uniformity of housing style and size, which was as much due to the needs perceived by city planners as by the developers' preferences.

Levittown

There are exceptions to every rule, and Levittown, in Bucks County, may appear to break most of the generalizations (see Figure 15.5). The key difference is that it was not seen as a classical city when it was

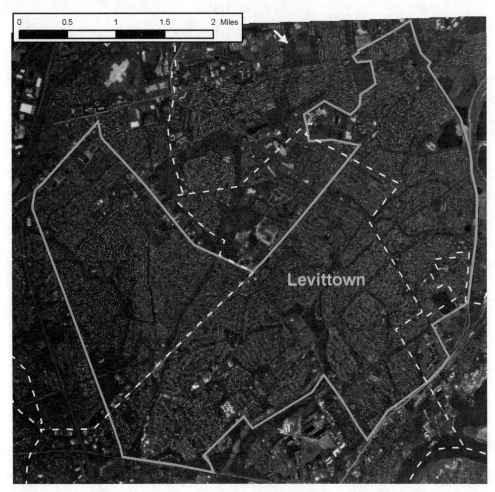

Figure 15.5 **Levittown, Pennsylvania:** The thicker gray outline surrounds the total Levittown development project. The thinner dashed white lines are the boundaries of the four municipalities Levittown spills into. the community "sections" are visible by their self-contained street patterns. Many of Levittown's original occupants worked at the Fairless Works steel mill. (Image source: USGS 1999a, b)

95

planned, and has not grown by evolution. In fact, Levittown represents a major turning point in the housing business. Although Levittown, New York was the first such mold-breaking project, the Levitts learned many lessons from it. When they set about building Levittown, Pennsylvania in 1951, the process was almost completely fine-tuned.

Levittown today is the equivalent of a city, Pennsylvania's ninth largest, in fact, that was built in less than six years. What is different about Levittown is that it consists almost entirely of single family dwellings (six styles to choose from) on individual lots. It was completely planned: shopping centers, community recreation facilities such as swimming pools and baseball fields, churches, and 17,311 homes each on a 7000-square-foot lot. The project brought to Pennsylvania the modern suburban development, beginning the "white flight" from the cities to the suburbs. The Levitts were catering to the middle and working classes. It succeeded partly because it coincided with US Steel's Fairless Works steel mill. These were low cost houses for wage-earning blue collar workers (PA State Museum 2003).

The Levitts, led by William Levitt, got the project started by secretly buying nearly 175 contiguous Atlantic Coastal Plain properties, mostly farmland, totaling 5,750 acres. They divided the project into 41 "sections" of 300-500 houses each, with construction progressing on the assembly line principle: one crew prepared each site in an area, another poured the concrete slab the house was built on, another crew assembled the house frame (each house's pre-cut materials arrived at the site as a kit), etc (PA State Museum 2003).

If the project had a failing, it was that so much effort had gone into making Levittown a self sufficient community, but that never became a reality. The land on which the project was constructed included parts of four municipalities. For a community to become a city, which would make it independent of those municipalities, it must "incorporate," which would require the approval of those municipalities. That process was never successful, so Levittown, Pennsylvania today remains a community without an officially recognized government (or, rather, with four governments shared with other communities) in suburban Philadelphia (Levittown 2002).

Conclusions

The development of cities and, later, suburbs changed the human landscapes of Pennsylvania and the US. They have implications for population distribution, economic activity and environmental problems. They even influence, as we shall see in Chapter 19, the political landscape. Cities, especially the larger ones, appear to be in a state of decline relative to their suburban fringes. The larger issue is whether cities and suburbs are and will be a necessary and permanent part of human existence, or are a temporary step toward a different, more efficient arrangement of people.

Bibliography

Erickson, Rodney A. 1995a. "The Location and Growth of Pennsylvania's Metropolitan Areas." Chapter 18 in Miller, E. Willard (ed.) A Geography of Pennsylvania. pp. 315-335.

Erickson, Rodney A. 1995b. "The Internal Spatial Structure of Pennsylvania's Metropolitan Areas." Chapter 19 in Miller, E. Willard (ed.) A Geography of Pennsylvania. pp. 336-355.

Levittown 2002. The History of Levittown. Levittown, PA. Internet site: <http://www.levittownpa.org/Levittown.html>, visited 10/19/04.

PA State Museum 2003. Levittown, PA: Building the Suburban Dream. Harrisburg, PA, The State Museum of Pennsylvania. Internet site: <http://server1.fandm.edu/levittown/>, visited 10/19/04.

US Census Bureau 2002. Census 2000 Summary File 1 (SF 1) 100-Percent Data. Washington, DC, US Department of Commerce, Census Bureau. Internet site: <http://www.census.gov>, multiple visits.

USGS 1999a. Digital Ortho Quarter Quad (aerial photograph): langhorne_pa_se. Downloaded in TIFF format from http://www.pasda.psu.edu/access/doq99list.cgi, 8/30/04. US Geological Survey, Washington, DC.

USGS 1999b. Digital Ortho Quarter Quad (aerial photograph): trenton_west_nj_sw. Downloaded in TIFF format from http://www.pasda.psu.edu/access/doq99list.cgi, 8/30/04. US Geological Survey, Washington, DC.

Chapter 16

Ethnicity and Cultures

Ethnicity and culture are two indicators of who we are. The differences between them are subtle, but thinking about them and seeing how discussions about them are organized help to bring attention to the truly human side of Pennsylvania. The mixing in of non-English newcomers with the English founders of the original colony made Pennsylvania into a cultural melting pot. The ethnic diversity of the early Pennsylvanians helped to create a unique new world culture, as we shall see in the next chapter.

Culture

The "culture" of a group of people is something like the personality of an individual. You, as an individual, are a unique combination of behaviors, learned from those around you or created with your own imagination. Examine the list of cultural elements in Table 16.1. There is tremendous variety in the ways each item in that list is expressed in this world. For each cultural element, if you look at the ways it is expressed, you will find that there are some that describe you, some that you recognize in others, and some that you've never seen or heard of. For example, if you look at a list of the names (or hear samples) of all of the different styles of music, you can readily distinguish your favorites from those you can tolerate and those you dislike or know nothing about. Many of our habitual behaviors and preferences reflect elements of our culture that call for our active participation. Your preferences and behaviors form your own personal expression of your culture.

You learn your behaviors and preferences by means such as family (very direct and strongly influential), community institutions such as school and church (very direct and secondarily influential), non-institutional community members such as friends and neighbors (more variably influential), and other media sources such as television, radio, magazines, newspapers and now, of course, the Internet. All of these are means for transferring culture from one generation to the next.

Notice that some of the means for transferring culture are potentially strong ethnic associations (family, church and neighborhood) while others are usually ethnically generic or diverse (school and news sources). In Pennsylvania, as in most of the US, a cultural group is most likely to be identified by its ethnicity, which refers to many cultural elements at once. Each named cultural group presents a combination of elements, from language and dialect to music, foods and clothing styles. Some of those elements become part of a stereotype image or representation of that culture group. Some culture groups have subsets (or, some cultures have subcultures), differentiated from the larger group by some key element, such as age. At some level, though, the culture's identity allows for individual variability.

Your culture group's characteristics are also influenced by environmental resources (for example,

---

Table 16.1 Elements of Culture

Language
Foods
Musical styles
Clothing styles
Belief systems
Ethical values
Architectural styles
Means of artistic expression
Entertainment choices
Solutions to economic challenges
Educational system

---

weather, water availability, food sources), employment and other economic factors and, increasingly, cross-cultural influences from other cultures. These will be discussed in the next chapter.

The Cultural Community

Some culture groups or subgroups are identifiable geographically as communities. They may be clearly defined by political boundaries, or they might be nothing more than a rural village or an urban neighborhood. Pennsylvania contains thousands of such communities.

Often, especially during colonial and early American times, an individual or a small group of investors bought a piece of land in order to create a town or city. In cases where profit was not the only motive for doing so, there was often some national or religious group that the new place was intended to accommodate. The Welsh Tract described in Chapter 14 was one example. The borough of Strasburg in Lancaster County is another. It was founded by a group of settlers from the Alsace-Loraine region in Europe, whose European home had changed hands several times between French and German rule, and who came to Pennsylvania seeking a settlement for French refugees from that area.

In a community, people interact, they share responsibility for educational and moral upbringing of all the community's children (and assert social control over those who violate the values taught), and community members contribute economic goods and services (whether volunteered or sold); in short, they contribute toward making the human environment within which everyone is born and spends their formative years. The members of that community share a common culture.

A community with strong ethnic or other cultural identity usually made that identity known at its founding by their choice of name for the place. Since that is the community's primary identity, it conveyed a lot of information very efficiently. Within the plans for the community would be named roads, named churches, and many other public parts of the site. These names were also often culturally consistent.

Cultural Diversity

The example of the Welsh Tract presented in the last chapter was fairly common in another sense as well: within the first few decades, people from other ethnic traditions began diluting any ethnic or cultural purity. This was usually welcome, when the new community members were bringing new wealth, knowledge and skills. The cultural changes, though, often took place with great reluctance and sometimes violence between older and newer community members. However, we now have such a mobile population that all communities are ethnically diverse, and many are also culturally diverse.

People who live together in a community must share at least some cultural characteristics (especially language), and have some common goals, in order to co-exist. If the community's members are not already part of the same culture group, then they will develop a community culture. The culture of the community will be the *typical or agreed-upon* choices for the cultural elements in Table 16.1 made by its members, which may differ from the cultures within individual families in the community.

Even where diversity altered the community fabric, the original place name usually remained. However, within that community, the new roads, taverns, farms or other elements of the landscape, located where the newer members came to live, were usually given names reflecting the new ethnicity. If a new neighborhood or district was added to an older community a hundred years later in order to house an ethnic minority different from the original founding group, its streets and churches and other parts of the community landscape might reflect the new group's ethnicity.

Pennsylvania's Cultural Regions

Some cultural elements allow the connection of one community to other nearby communities, enabling their members to interact socially, economically and politically within that larger region. At some distance, though, culture becomes a force that differentiates communities. Each cultural element in the Table 16.1 is a candidate for making those connections or distinctions.

Just as we can classify our type of physical environment based on climate, soil types, vegetation types, etc., it is possible and useful to classify cultures. The challenge in classifying cultures, as in classifying physical environments, is to decide which cultural element(s) should be the primary basis for grouping or distinguishing culture groups.

Ethnic identity is the most common basis for attempting to map culture groups, mainly because information about it is readily available from the US Census. The ethnic identity maps in the last chapter (Figures 14.1 and 14.2 in Chapter 14) are good examples. They show the distribution of those ethnicities within the larger area of Pennsylvania, although there is a great deal of detail in those distributions that is hidden because of the use of counties as the main mapping unit there.

At the same time that a community has one type of cultural connection with one group of neighbors, it may have another type of cultural connection with another very different group of neighbors. For example, geographers sometimes map religious affiliation to show one set of cultural boundaries. The challenge that they face is in deciding what level of dominance for that cultural element is enough, and at what point it gives way to another region with a different identity. The mapping may represent one cultural community as part of a larger Catholic region. That same community may be connected with neighboring Protestant-dominated communities based on common interests in agriculture which have led to adopting certain dietary preferences not available to the other Catholic communities.

Thus, it is much easier to map one cultural element than a fully complex set of elements to define culture groups. Even though the single-element map might represent an over-simplification of cultural identities, it may prove useful in uncovering current or past cultural connections. The first map in Figure 16.1 shows communities where the "pierogi" is advertised on restaurant menus. As a single food representative of Polish ethnic culture, it serves as a substitute for

mapping Polish communities. Similarly, the second map in Figure 16.1 is of communities with Irish origins. The communities on the second map are less likely to have as strong an Irish cultural identity today as the Polish identity of the communities on the first map. Thus, we must be careful about the conclusions we draw from mapping different cultural elements.

The Mummers: Which Culture?

Table 16.1 mentions 'music,' 'entertainment' and 'means of artistic expression,' which can include a wide variety of cultural displays. A New Years tradition in Philadelphia is the Mummers Parade, in which marching bands featuring banjos (and other instruments) and elaborate costumes decorated with feathers and sequins march to very up-beat music (see Figure 16.2). Parade units are organized into four categories: the Comics, the Fancies, the String Bands and the Fancy Brigades. The interesting puzzle is deciding which culture's traditions it reflects.

The term Mummers comes from an English tradition of troupes of actors who performed short impromptu comical and satirical plays, but is also tied to the Medieval German (Lukacs 1981, 27). And yet, different sources disagree on the source of the Mummer tradition in Philadelphia. One says that it arose from Finns and Swedes in the 1840s (Bauman 2002). Another says that the Mummers began strutting in the 1800s, and that "the spiritual home of the Mummers is still the Irish end of "South Philly" (Dubin 1996, 39, 21). Yet another says that the parade was first organized in 1901 when South Philadelphia was still most influentially inhabited by the English and "Scotch-Irish" [sic] working class (Lukacs 1981, 26). It appears that all of these culture groups enjoyed

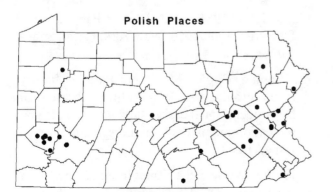
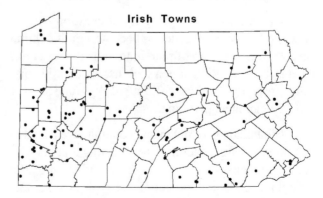

Figure 16.1 **Polish and Irish Strongholds in Pennsylvania:** The Polish map was created by, first, searching the Internet for restaurants advertising pierogies as a specialty or listing them on their menu. The Irish map shows all of the towns whose names start with "Mc..." In some cases the town's name has changed to a non-Irish one. (Data source: USGS 2004)

Figure 16.2 **The Mummers Parade Route, Philadelphia:**
The Mummers' march up Broad Street is an annual New
Years' event.

welcoming the New Year with noise-making, music
and costumes (Philadelphia 2005). All agree with
Dudden (1982, 592) that South Philadelphia and
Kensington were the focus of early Mummer activity.

The real answer to the cultural question may be
that the Mummers Parade has become a multicultural
event. It is made up of marching units that organize
themselves as clubs. Each club has its own history,
and probably its own cultural background. Over
10,000 marchers participate.

The Parade is as much an economic event as it is
a cultural event. In addition to attracting tourists who
buy from businesses along the parade route, the
marching units are competing for cash prizes. The
costs of the parade include the expensive costumes,
the extra traffic control and the police. The Mummers
Parade has seen better days now, though. Its peak

came in the early 1960s, when both the parades and
the crowds were much larger. The city even
experimented with parade routes along Market Street
instead of the traditional march through South
Philadelphia up Broad Street (Fischer 2002). Even
though the Mummers Parade is not likely to disappear,
it is likely to continue to change.

Conclusions

Your cultural identity may be strongly grounded
in your ethnicity, or it may be defined by other inputs
as we shall see in the next chapter. It may be relatively
pure, or very mixed up. It will probably change
somewhat with age and changes in social or economic
status. Even though your cultural identity does not
define your personality, it does to a significant extent
determine who others see when they look at you.

Bibliography

Bauman, John F. 2002. "Philadelphia (city,
     Pennsylvania)." in Microsoft Encarta
     Encyclopedia 2002. Microsoft Corporation.

Dubin, Murray 1996. South Philadelphia: Mummers,
     Memories, and the Melrose Diner. Philadelphia,
     Temple University Press.

Dudden, Arthur P. 1982. "The City Embraces
     'Normalcy' 1919-1929." in Weigly, Russell F.
     (ed.), Philadelphia: A 300-Year History. pp. 566-
     600.

Fischer, John 2002. 2002 Mummers Parade Part 1:
     Quaker City, Downtowners, Golden Sunrise and
     Murray Win… But at What Price? Philadelphia,
     About.com. Internet site: <http://
     philadelphia.about.com/library/gallery/
     aa010302a.htm>, visited 1/20/05.

Lukacs, John 1981. Philadelphia: Patricians and
     Philistines 1900-1950. New York, Farrar, Straus,
     Giroux.

Philadelphia 2005. Mummers Parade History.
     Philadelphia, PA, Philadelphia Recreation
     Department. Internet site: <http://www.phila.gov/
     recreation/mummers/mummers_history.html>,
     visited 1/20/05.

USGS 2004. Geographic Names Information System.
     Washington, DC, US Department of the Interior,
     Geological Survey. Internet site: <http://
     www.geonames.usgs.gov>, site visited 10/8/04.

100

Chapter 17

Pennsylvania Cultures

Cultures change and spread over time. The mere presence of America as an outlying piece of English culture is evidence of that. However, now that we are politically disconnected from the United Kingdom, our cultural attributes have taken on lives of their own as well. This phenomenon can lead us to two conclusions: that cultures start somewhere before they get anywhere else, and that the elements that identify a culture group today, no matter how well known the group is, are not likely to define it in the same way in, say, another generation. The idea that a culture is born and developed in one area before spreading is expressed by calling that area the "hearth" for that culture. The variations between, and even within, culture groups makes their identification tricky.

The ethnic identity used to label most culture groups is an artifact of our immigration-centered past. Is there an American culture, yet? If you went to a foreign country, what would give you away as being American? Certainly, there are enough global cultural differences that Americans would be readily identifiable in a foreign country. Is there anything within your American identity that would further identify you as Pennsylvanian? What if you were traveling around the US? In a similar manner, if you travel around Pennsylvania, you will encounter cultural differences, such as word pronunciations, that would alert people to your being an outsider.

Pennsylvania's Cultural Hearth

Southeastern Pennsylvanians do not usually detect an accent when listening to media reporters and announcers. We recognize New York City and Boston accents and the southern drawl, but much of the rest of the country seems to have our general accent. Wilbur Zelinsky and others attribute this lack of an accent to Pennsylvania's role as the origin of this generic American culture; its "hearth" was in Pennsylvania. Zelinsky and others call the area in Pennsylvania where these origins are most visible the Pennsylvania Culture Area or Region (Zelinsky 1977, 128; Lewis 1995, 2-9; Zelinsky 1995, 133-137), depicted in Figure 17.1. Notice that Zelinsky's Pennsylvania Culture Area extends outside the state via landscapes that are influenced by the natural landform regions.

Richard Pillsbury and Joseph Glass produced studies attempting to delineate the Pennsylvania Culture Area. Pillsbury worked by examining the street patterns of towns founded before 1815 across the entire state. He identified as Rectilinear one which carried a rectangular grid of streets characteristic of Penn's original layout for Philadelphia, usually featuring a "square" or "diamond" at what would be the most central intersection of the original street plan (Pillsbury 1970). Zelinsky added that they tend to feature greater population densities, the use of the "row home," trees planted along city streets, and a distinctive pattern of street names: numbered streets in one direction and streets named after trees in the other (Zelinsky 1977, 131-138). This street plan is superimposed on whatever natural landscape is there, resulting both in modifications of the land caused by adherence to the street pattern (diverting streams, moving earth) and in modifications of the plan caused by immovable natural features.

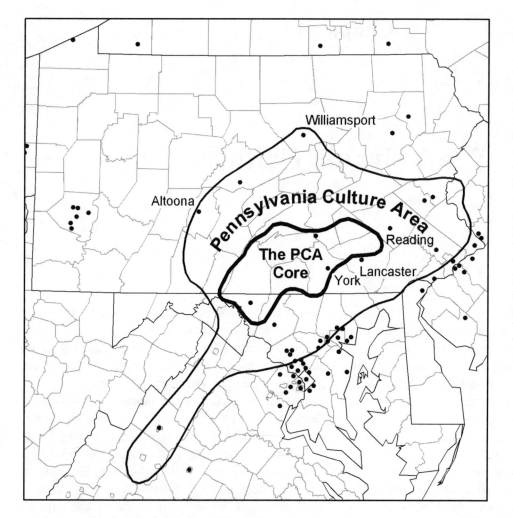

Figure 17.1 **The Pennsylvania Culture Area (or Region) and Its Core:** The Pennsylvania Culture Area was first presented by Wilbur Zelinsky (1977), and its Core was mapped by Joseph Glass (1986). Many factors went into determining the limits of each area. Notice that they straddle the division between the Piedmont and Ridge and Valley landform regions. They extend out of Pennsylvania primarily via the Cumberland Valley (Great Valley) and into Virginia's Shenandoah Valley.

Glass examined the other major element of colonial and early American culture, representing rural as opposed to town life, by examining farms. By studying farms across Zelinsky's Pennsylvania Culture Area, he identified the prototypical barn, farmhouse and farmstead. The Pennsylvania barn (see Figure 17.2) is characterized by: an overall rectangular floor plan (as originally built) with an ordinary gabled roof (one slope on each side, not two as so often depicted in barns); two levels, approached on one side at a lower ground level, and on the other side by a banked earth ramp to the second level; a forebay (the second floor of the barn overhanging the ground floor entrance) that is fully cantilevered; and all siding, where present, attached vertically. The Pennsylvania farmhouse he describes as a two-and-a-half story building whose front has four windows and/or doors on the first floor and four windows on the second floor, whose ends have two windows per floor (and at least one chimney per house), a gabled roof, and a front porch. Finally, the farmstead (the cluster of farm buildings including

the farmhouse and barn) is usually organized so that the overhanging forebay, opening onto the barnyard, faces south, the front of the farmhouse faces the road, and some part of the farmhouse had an unobstructed view of the barnyard (Glass 1986, 183-4).

In addition to the appearance of the farm, the entire operation of the colonial and early American farm owes much to the early Pennsylvanians. This will be described in greater detail in Chapter 20, but includes the selections of field crops, orchard products and animals, as well as the methods for raising them. Over time those selections and methods changed, and farmers developed new breeds which would prove to be more productive under the natural and economic conditions.

Glass then mapped the occurrence of these elements of the Pennsylvania farm and identified a Pennsylvania Culture Region Core, the smaller area outlined in Figure 17.1 (Glass 1986, 199). He noted that the Core does not include Philadelphia (which is on the eastern fringe of Zelinsky's Pennsylvania Culture Area because at the time of the Core's development, farmers were very mistrustful of Philadelphia's power and motives under British colonial rule in general and Penn's control within this colony, and felt stronger market connections via easier river travel at the time (see Chapter 24) with Baltimore (Glass 1986, 198).

Thus, the Pennsylvania Culture Area's main cultural contributions were the dialect of English (which really boils down to how a few letters and letter combinations are pronounced), the system of agriculture, especially as reflected in local food preferences, house construction appearances and techniques, and city street plan ideas.

Pennsylvania's Other Regional Cultures

Cultures change both geographically and over time. We saw the main reason for geographical variations in culture in the last chapter: varying locations and combinations of ethnic groups. We concluded that it is easy to map one representation of a culture group, but is far more difficult to map complex sets of culture elements, as Glass did. It is easier to map cultural origins by examining elements that are permanent or whose later modifications are recognizable, than it is to map their modern expression.

In order to see the Pennsylvania culture's historical evidence on our landscape, we have to peel off layers of later human developments. Some of these later developments represent cultural trends that have come and gone, the way clothing styles do today. Examples from the 1800s include the building of factories in or near existing cities and towns, or the conversion of rural land into a factory or mining operation complete with housing and commercial enterprises for the workers. The cultural identity of each settlement was a combination of the pre-existing culture (if any), the corporate culture imposed on the workers, and the cultural traits of all of the new workers who inevitably arrived, some from elsewhere in the US and the rest from overseas.

Some other historical trends and events also influence the cultural landscapes of other parts of Pennsylvania. For example, the counties of Tioga, Bradford, Susquehanna, Sullivan and Wyoming are referred to as the Northern Tier. This area of Pennsylvania is culturally closer to New England. Part of the reason, as we saw above with Philadelphia versus Baltimore, had to do with transportation difficulties. Another part is that during colonial times, northern Pennsylvania was claimed by the Connecticut colony because it represented the projection of Connecticut's boundaries westward (interrupted by New York's claim on the Hudson River valley). In an effort to assert the validity of its claim, Connecticut leaders arranged for loyal settlers to move to the Northern Tier area, and even established a Connecticut

Figure 17.2 **A Pennsylvania Barn:** From one side of the barn the ramped earth leads to the entrance into the second level. On the other side of the barn the lower level entrance is used by the dairy herd. The second level is extended (cantilevered) over the lower level entrance in order to provide some weather and sun protection. This barn is constructed of a wood frame resting on a stone foundation, with siding boards applied vertically. Other barns also considered to be Pennsylvania barns may have the end walls or all four walls constructed of stone.

Figure 17.3 **A Pennsylvania House:** The perfect vertical alignment of windows and front door(s), the smaller windows indicating a third floor, the chimneys positioned at the ends of the house, the front porch roof, and other clues, tell us that this is the house style that evolved here in Pennsylvania during the colonial period and into the middle 1800s.

county in the area (Dykstra 1989, 81). As a result of their activity, many towns in that area have street patterns and other elements characteristic of New England.

The other area of Pennsylvania, whose culture appears different from that of the Pennsylvania Culture Area, is western Pennsylvania. Pittsburgh and nearby areas of the Appalachian Plateau are most notably different in their choices and pronunciations of a number of words that southeastern and even south central Pennsylvanians have difficulty recognizing. Some examples are given in Table 17.1. By and large, though, western Pennsylvania's culture is, at its core, only slightly modified from the Pennsylvania culture.

Other instances of non-Pennsylvanian culture are present in the state's larger cities. Philadelphia, which escaped classification so far, is different because it is so metropolitan and cosmopolitan. Pittsburgh, Harrisburg and other cities have developed some similar traits. But within these complex social cityscapes are sets of neighborhoods, many of which carry very culturally cohesive identities. The easiest example, perhaps, is Philadelphia's Chinatown (see Chapter 34).

Pennsylvania Culture Today

Enclosing all of these smaller culture regions is one more layer of influence on behavior and culture: the state's boundaries and government. The boundary is an arbitrary line, but one which can function as a barrier to cultural exchange in both directions, as the example of the Northern Tier showed earlier. The government in power at any time is both a reflection of the will of the people and an influence on their lives. Party differences and election results will be discussed in Chapter 19, but in this chapter it is appropriate to point out human characteristics that are consistent within our borders and that change when we leave Pennsylvania headed for our neighboring states. These

Table 17.1 Pennsylvania Dialects

	Pittsburgh	Philadelphia
Different names for the same thing *		
rubber band	gum band	rubber band
all of you	yinz	yooz
interfering	nebby	nosey
to tidy	to redd up	to tidy
carbonated soft drink	pop	soda
Different pronunciations of the same word **		
aunt	ahnt	ant
downtown	dahntahn	downtown
route	rowt	root
steel(ers)	still(ers)	steel(ers)
tire	tahr	tire
wash	worsh	wash
water	water	wooder

Sources: * Johnstone 2004.
        ** Anonymous 2005.

cultural elements reflect the degree to which Pennsylvanians identify with being Pennsylvanians.

Some cultural elements are reflected in state laws (or the absence of state laws) on specific issues. These can be as culturally (and politically) volatile, such as stands on abortion or gambling. They can also be economic issues that reflect where our priorities lie, such as the amount of money to be spent on education or road repair, or the willingness of our elected representatives to set limits on medical malpractice lawsuits. Pennsylvania was one of the last states to repeal its laws prohibiting setting any number of restrictions on business on Sundays.

The character of Pennsylvanians is a difficult cultural element to generalize, but several have done so. For example, Lewis notes that the vast majority of our very substantial practical successes in agriculture, forestry, mining and transportation engineering are not generally attributed to any individuals, but it has to be more than coincidence that many practices were carried from here to other parts of the country. He also notes that we have produced few if any noteworthy writers, architects, philosophers, social reformers or political leaders. We instead seem to have specialized in industrial innovators (George Westinghouse, Robert Fulton and Edwin Drake), corporate/financial visionaries (Milton Hershey, Andrew Mellon and Andrew Carnegie) and merchants (F.W. Woolworth and John Wanamaker) (Lewis 1995, 9). To that we can even add actors (James Stewart, Bill Cosby and Will Smith) and popular musicians (Stephen Foster in the 1800s, and many Philadelphians in the 1960s rock-and-roll era or today's rap scene).

Pop Culture

One hazard in attempting to describe American or Pennsylvanian cultural elements today is the presence of yet another level or layer of culture often referred to as popular culture, or Pop Culture. Pop Culture is an international blending of American culture with other world cultures, remarkably uniform no matter where you go in the world. Think of the elements that look the same no matter where you travel: fast food restaurants, shopping malls, car dealerships, styles of suburban houses, clothing choices (sometimes in spite of climate differences). Pop Culture explains why music genres, automobile styles, architectural choices and even particular foods

such as McDonalds and Pepsi so easily cross international boundaries. It predates personal computers, but they (and especially the Internet) might be its defining symbol. Much of Pop Culture originates in the US, but not all. Those who study culture, like cultural geographers, anthropologists and sociologists, would say that the US was Pop Culture's hearth so, by extension, Pennsylvania must have had a significant influence.

Pop Culture uses the broadcast and print media to build demand for its products and styles. It brings ethnic foods or clothing to places where the ethnic group never lived. It means that Pennsylvanians can go anywhere and not feel completely alien, and that people migrating to Pennsylvania from other states or countries can learn to fit in more quickly. Pop Culture provides benefits, such as the ability to find replacement parts for any make of automobile, computer or vacuum cleaner anywhere in the country. It also raises outcries, as when chain retailers put family-built and -run stores out of business, or consume large amounts of once-productive farmland.

Elements of more traditional cultures are found in Pop Culture, but they are mixed incongruously, and are so commercialized and common that we easily forget their original associations. Can we recognize elements of the Pennsylvania Culture that have become absorbed into Pop Culture? Here are a few examples:

- the dialect of American English most common in the spoken media,

- the mainstream American diet of meats, potatoes and vegetables, modified in many restaurants for the sake of efficiency or convenience,

- the many Pennsylvania inventions that went on to become standardized products or businesses: the radio station, the oil well, the steamboat and bubble gum, to name just a few.

Conclusions

Pennsylvania has indeed contributed a disproportional part of what would have to characterize American culture. There may be few artifacts of that culture that are still recognizable in today's culture, but the connections to our past are obvious enough. The origins of many of those artifacts are still part of Pennsylvania's cultural landscape. Since

American culture can be argued to be the basis for Popular Culture, Pennsylvania carries a rich legacy indeed.

Bibliography

Anonymous 2005. Pittsburghese. Pittsburgh, PA. Internet site: <http://www.pittsburghese.com/>, visited 1/19/05.

Dykstra, Anne Marie 1989. "Pennsylvania's Past: Exploration and Settlement." in Cuff et al.: The Atlas of Pennsylvania. pp. 80-81.

Glass, Joseph W. 1986. The Pennsylvania Culture Region: A View from the Barn. Ann Arbor, MI, UMI Research Press.

Johnstone, Barbara 2004. Pittsburgh Speech and Society. Pittsburgh, PA, Carnegie-Mellon University. Internet site: <http://English.cmu.edu/pittsburghspeech>, visited 1/19/05.

Lewis, Peirce 1995. "American Roots in Pennsylvania Soil." Chapter 1 in Miller, E. Willard (ed.) A Geography of Pennsylvania. pp. 1-13.

Pillsbury, Richard R. 1970. "The Urban Street Pattern as a Cultural Indicator: Pennsylvania, 1682-1815." Annals of the Association of American Geographers. Vol. 60 (September), pp. 428-446.

Zelinsky, Wilbur 1977. "The Pennsylvania Town: An Overdue Geographical Account." Geographical Review. Vol. 67, no. 2 (April), pp. 127-147.

Zelinsky, Wilbur 1995. "Cultural Geography." Chapter 9 in Miller, E. Willard (ed.) A Geography of Pennsylvania. pp. 132-153.

Chapter 18

Government

No chapter, perhaps, will be more fundamentally linked to this textbook's main purpose: helping you to become a more effective citizen of Pennsylvania. The organization of the US government was likely studied extensively in your school. The study of state government was probably less complete. It is even rarer that you would have studied your county and local government's organization and operation.

One of the founding principles of the U.S. Constitution is that the federal government should not interfere in the operations of state governments unless it clearly involves the national interest. At the same time, most states have looked at the federal government as a model of how they should be organized. Pennsylvania is one whose organization is remarkably similar to the US government. The structure of local governments, however, does not look anything like our federal and state governments. Local governments in Pennsylvania include counties, cities, boroughs and townships, and one "town."

An additional question that will bear asking after reviewing this organizational structure is "How well does it work?" or, to put it differently, "Does the actual practice reflect the intent?"

US Government

Most of the structure of the US government was laid out in the US Constitution. The Constitution called for the separation of powers reflected in the duties and responsibilities of the executive, legislative and judicial branches of the federal government.

Recall that the executive branch includes the President and Vice President, the President's Cabinet and each cabinet member's (Secretary's) entire Department, and other non-Cabinet agencies such as

the Environmental Protection Agency and the Office of Management and Budget. The legislative branch includes the two houses of Congress, as well as the various support staff positions and agencies that they have created. Pennsylvania sends two senators to Washington, DC, as does every state. We also send nineteen representatives to the US House; because their districts are supposed to include roughly equal numbers of citizens, their boundaries are reviewed and adjusted (if necessary) after each census (see Figure 18.1). The judicial branch consists of the Supreme Court, Courts of Appeal and District Courts and, again, several support agencies. All of Pennsylvania is part of the Third District, centered in Philadelphia, and appeals of its decisions go first to the Court of Appeal and then, potentially, to the Supreme Court.

Even though the structure of the government was laid out early in our nation's history, the formation of many specific offices and the adoption of many *Environment* specific responsibilities reflect major events at later *laws* times. For example, at the beginning of the 1960's federal responsibility for environmental protection was scattered among many agencies, poorly coordinated and inconsistently enforced. In 1969 Congress passed, and President Nixon approved, the National Environmental Policy Act, which created the US Environmental Protection Agency when it went into effect in 1970. The same trend continues today, as the recent formation of the US Department of Homeland Security shows.

Federal government departments and agencies only have jurisdiction over issues that have been defined as federal in nature. The framers of the Constitution intended that issues that could be resolved at lower levels should be handled by state or local

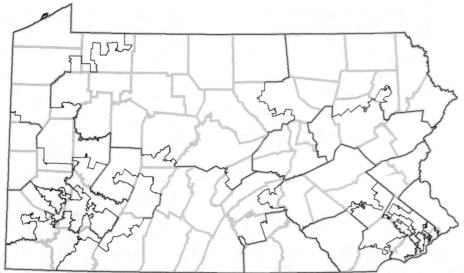

Figure 18.1 **US House of Representatives Districts:** Note that in many areas these district lines follow county boundaries. In other areas, the locations of the boundaries are the result of much analysis to balance population, but also to influence outcomes of political races.

governments. The gray area in this separation is when an issue cuts across state boundaries, but is not of nation-wide importance. An example of this, which will be examined more completely in Chapter 36, is the pollution of the Chesapeake Bay, which requires actions in all of the states within its watershed. The solution in such situations is to form a regional agency, such as the Chesapeake Bay Foundation, funded with tax dollars from both the federal government and the member states.

State Government

The vast majority of issues requiring government involvement are defined at state or lower levels, which means that the definition and interpretation of the government's role can vary significantly from one state to the next, one county to the next, etc. In Pennsylvania's case, much of the structure for resolving within-state issues looks like the federal government's structure. The differences between the Pennsylvania and US governments are usually because the counties of Pennsylvania do not have the same power relative to the state government as the states of the US have relative to the federal government. County governments are relatively weak in Pennsylvania, as we shall see later in this chapter.

Pennsylvania's government has the same three primary branches as the federal government, similarly defined. The executive branch consists of the Governor, Lieutenant Governor, the state Cabinet and cabinet departments, and various executive offices and agencies. The biggest difference between the state governor and the US president is that the former is

elected by popular vote, not by anything resembling the federal Electoral College. Even the list of cabinet departments contains many with similar responsibilities. For example, the federal Department of Homeland Security has a counterpart in the Pennsylvania Department of Homeland Security, as do departments of Education, Agriculture and Transportation. States, however, have a much longer history of involvement with natural resources and environmental issues. While the creation of the Environmental Protection Agency in 1970 marked the beginning of environmental reform in the federal government, it is not a cabinet department. Its state counterpart, the Department of Environmental Protection (and its predecessor the Department of Environmental Resources), has been part of the governor's cabinet since it was formed in 1972.

The legislative branch of the state government consists, as in its federal counterpart, of a House of Representatives and a Senate, the former with many more members than the latter. While the federal House has 435 members, Pennsylvania's has 203, and both the federal and state Senates have 50 members. In fact, though, the state Senate and House are very similar in that both have territories whose boundaries are redefined whenever the population distribution among their districts becomes unbalanced (see Figures 18.2 and 18.3).

In the judicial branch occur the most significant differences between federal and state governmental structure. Both are headed by Supreme Courts, the federal version featuring nine justices while the state

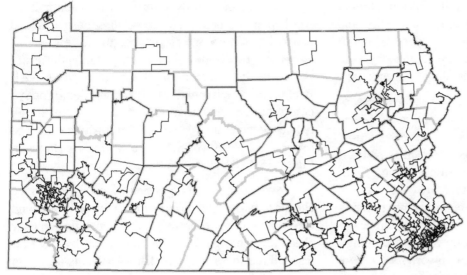

Figure 18.2 **Pennsylvania House of Representatives Districts:** These district lines are much less likely to follow county boundaries because of the number of representatives, and due to population distribution within Pennsylvania. Once again, the locations of the boundaries are the result of analyses to balance district populations, with some political influence also playing a role.

version features seven. Both bodies hear appeals of cases that define laws appropriate to their level of government. Beneath the state Supreme Court is the Superior Court, which is also primarily an appeals court, and the Commonwealth Court, which hears charges and appeals brought against or by the state. The courts whose cases are appealed to the latter two courts within Pennsylvania are the Courts of Common Pleas. The Common Pleas courts' territories are organized into 60 judicial districts, which are either entire counties or pairs of counties. In each district, the judges are elected to their positions (PA AOPC 2002). These are the courts for which you may be called to be a juror, because most cases are jury trials. At the lowest level of jurisdiction is the group of courts known as Special Courts. Most of these are the local courts of the Magisterial District Judges, which were

known until January 2005 as District Justices (PA AOPC 2005). There are 550 such judges across the state. The Magisterial District Judges hear most cases first in order to decide which should be forwarded to the Courts of Common Pleas, and pass judgment in cases brought for traffic violations and small claims. The Special Courts also include Philadelphia's Municipal Court and Traffic Court and Pittsburgh's Magistrates Court (PA AOPC 2002).

Local Government: Counties and Municipalities

Counties are a level of territorial division that most Pennsylvanians identify strongly with. At the beginning of Pennsylvania's history, the counties were created by William Penn and his colonial co-leaders. For a territorial area to achieve county status meant a

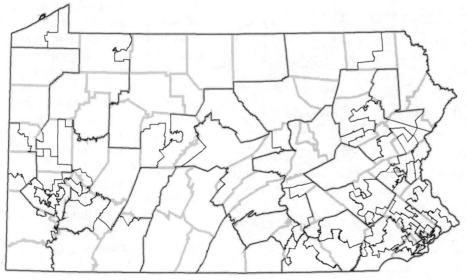

Figure 18.3 **Pennsylvania Senate Districts:** Because there are fewer senators, these district lines are more likely to follow county boundaries. They do get more intricate in and near the major cities.

great deal of effort to attract sufficient population, organize communication and economic activity, and make a strong case to the colony's leaders. Many of the earliest counties were much larger than today, with later counties being created by dividing a larger county.

A county's main roles are served as local arms of the state government in such functions as the police and judicial system, social welfare, voter registration, property assessment and emergency management. At election time, such positions as District Attorney, Clerk of Courts and Public Defender are likely to be prominently contested. The county government's administrative power is usually (see the section below on Home Rule) held by a group of three commissioners. Counties are also empowered to collect income taxes, and may spend that money on solutions to their regional problems, such as roads, public transportation and local parks and library systems.

The county has limited authority beyond those functions. One illustration of their lack of power or recognition is that counties do not figure in the apportionment of representatives to the state legislature. Another illustration is that, even though county government is in an ideal position to make decisions regarding land uses, such as by limiting where new housing developments should go or controlling the uses to which a property can be put, and by resolving such issues when there are differences between neighboring municipalities, they do not have that power. They can advise and coordinate, but only to the extent and effectiveness of the municipalities' willingness to listen to them and accept their advice.

Much of the locally significant decision making was kept in the hands of the townships, boroughs (and town) and cities, which are the areas into which counties are divided. In Pennsylvania these are collectively referred to as municipalities. All Pennsylvanians live within the jurisdiction of both a county and a municipality. In addition to the 67 counties, there are 56 cities, 961 boroughs and one town, and 1548 townships, all of which are shown in Figure 18.4 (PA DGS 2003, 6-3).

Townships are either First Class (if they have chosen to be designated as primarily urban) or Second Class (if they are primarily rural). Depending on their class, township government can be led by from (at least) three to as many as fifteen (if it is an urban township divided into wards) commissioners. Many townships adjacent to cities have gone from very rural to densely suburban or even urban over the last 50 years or less. Some have not adapted their government, within the range of options allowed by state law, to reflect the greater amount of work and decision making required with busier conditions (PA DGS 2003, 6-6).

Boroughs are small urban concentrations. Most of our cities were boroughs at an earlier stage, and most boroughs started out as towns or villages. To become a borough, the small town (which would have been part of a township) "incorporates." Boroughs are headed by seven council members elected at-large, or by councils whose members represent their wards. An elected mayor also holds office, but has limited power compared to the council. Most boroughs have an appointed borough manager to run such day-to-day

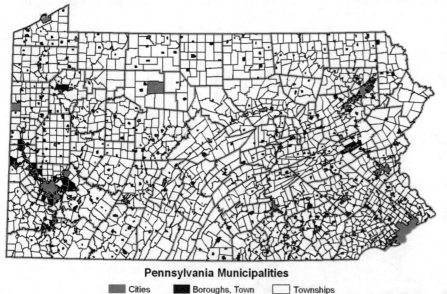

Pennsylvania Municipalities

Cities ▪ Boroughs, Town □ Townships

Figure 18.4 **Pennsylvania's Counties and Municipalities:** Most townships are relatively large, and most boroughs are smaller and more compact. The large borough in Elk County is St. Marys, which outgrew its original incorporated boundaries and merged with two neighboring townships. It may have gone against the typical Pennsylvania size-to-population ratio for a borough, but it is following the recommendations of scholars who argue that such larger units will be more effective at planning their future.

110

operations as garbage, water and sewer services, and all will either have their own police chief or be part of a regional police force (PA DGS 2003, 6-5). In addition, but equivalent, to the boroughs, there is one "incorporated town" in Pennsylvania: Bloomsburg.

Cities are classified and governed differently, depending on their size. Philadelphia is the state's only First Class city, while Pittsburgh and Scranton are the only Second Class cities. Both classes have governments organized around an executive branch headed by a mayor with broad administrative powers, and a legislature-like council whose members represent city wards (Philadelphia also has at-large city council members) (PA DGS 2003, 6-5). In 1854 the City of Philadelphia was expanded to encompass Philadelphia County. Later in the 1800s statewide legislation prevented any city from annexing surrounding communities or land without that area's consent (as recorded in a referendum).

The rest of Pennsylvania's cities are Third Class cities. State law provides for three choices of borough government structures. In the first, most common, system, the mayor is elected separately but sits as one of five "commission" members, which makes decisions by majority vote. A second system works like the higher class cities, with the mayor as the city's executive separate from council (legislature). The third form of city government operates under an elected council only, with no mayor; instead the appointed city manager controls the day-to-day functions adopted as ordinances by city council.

Other government-like bodies also operate, making decisions and collecting taxes or fees. These include the 501 school districts and the many (200 or more) specialized "authorities," responsible for such functions as public transportation, water or sewer systems within their regions.

The systems of government at the county and municipal level described above are prescribed by state legislation. The state's prescription includes both their structure and their responsibilities. In 1968, however (1952 for Philadelphia), the state gave local governments another choice, which it called Home Rule. Under this option, as long as the county or municipality was performing all the functions required by the state and not violating any laws, it could create its own unique government structure. Adopting a Home Rule Charter required local effort, and also required approval from the Pennsylvania legislature. Figure 18.5 shows the counties and municipalities that have done so. Note that many of them cluster around the three largest cities, Philadelphia, Pittsburgh and Scranton (PA DCED 2003). *Home Rule*

Does It Work?

A recent study conducted under the auspices of the widely-respected Brookings Institutions declared Pennsylvania's governmental structure to be more of a hindrance than a facilitator in solving problems today. Our system of government was built under very different economic, political and even population-distribution circumstances, of course. It reflects the values of our colonial past as much as our American past. Rusk, the author of the Brookings study, terms Pennsylvania a "Little Box" state because of the number and small populations of our municipalities. The effect of these little boxes is made worse by the

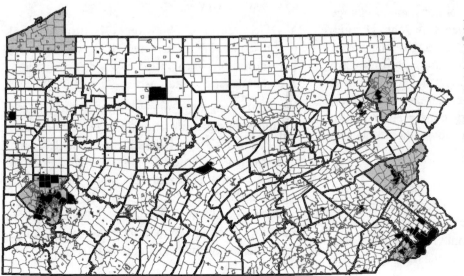

Figure 18.5 **Pennsylvania's Home Rule Governments:** Seven counties (including Philadelphia) and many more cities, boroughs and townships have devised their own system of local government. Clustered around the larger cities, they seem to reflect unique situations in the urbanization process.

111

cumbersome process it would take to redraw any of their boundaries, especially since that progress requires the unlikely prospect of nearly unanimous consent (Rusk 2003).

Many of our government-related problems, Rusk declares, revolve around the difference between the "real" size of cities and the size of the area over which they have jurisdiction. For example, Philadelphia plus its suburbs are effectively one place, but cannot act like one because the city's inability to annex the urban or suburban parts of its neighboring counties effectively forces them into adversarial rather than cooperative roles. In many other states cities are still able to annex land (Rusk 2003).

Another part of the problem, as noted earlier, is that counties have little say in land use decisions. As a result, many decisions made by townships or boroughs create problems outside their borders. Once again, in many other states counties have many more powers over the land development process than they are granted in Pennsylvania (Rusk 2003).

Such problems as strong disparities between the economies of cities and their suburbs, and ethnic and racial imbalances between the same areas, only serve to encourage sprawl and to isolate minorities from greater participation in the economy. Central cities' mean household incomes are declining, and their bond ratings (a measure of stability as perceived by financial institutions) are weakening. Even boroughs, as smaller versions of cities facing the same restrictions in Pennsylvania, have lower mean incomes than if they included suburban areas of neighboring townships (Rusk 2003, 9-12).

Rusk (2003) presents two options: either change the system of local governments in Pennsylvania, forcing small under-populated boroughs (the littlest boxes) to merge with one or more neighboring townships, for example, or set new rules defining what each type of government (county, municipality, etc.) is responsible for. The strongly conservative culture of Pennsylvania is unlikely to allow either to occur.

Conclusions

Pennsylvania government holds some features left over from its colonial formation under William Penn who, for all his efforts toward religious and social freedom, favored strong governments, both at the state level and at the very local level. The system is complex, especially when viewed in addition to the structure of the federal government.

In addition to its complexity, Pennsylvania's government is both disjointed, meaning that different levels of government and different governments at the same level tend to go their own ways, and resistant to structural change. Proposals to impose some sort of change are increasingly heard. However, such radical ideas are not likely to pass in conservative areas of Pennsylvania. In the next chapter, such political realities will be examined.

Bibliography

PA AOPC 2002. "The Structure of Pennsylvania's Unified Judicial System." in Report of the Administrative Office of Pennsylvania Courts 2001 [annual report]. Harrisburg, PA, Administrative Office of Pennsylvania's Courts. Internet site: <http://www.courts.state.pa.us/Index/Aopc/AnnualReport/annual01/07struct.pdf>, visited 10/21/04.

PA AOPC 2005. Pennsylvania's Unified Judicial System. Harrisburg, PA, Administrative Office of Pennsylvania's Courts. Internet site: <http://www.courts.state.pa.us/index.asp>, visited 1/21/05.

PA DCED 2003. Home Rule in Pennsylvania (7th ed.). Harrisburg, PA, Pennsylvania Department of Community and Economic Development, Governor's Center for Local Government Services. Internet site: <http://www.inventpa.com/docs/Document/application/pdf/a882c48f-b26e-4cc0-9b3d-0681c5662df4/home-rule.pdf>, visited 10/18/04.

PA DGS 2003. The Pennsylvania Manual. Harrisburg, PA, Department of General Services. Internet version: <http://www.dgs.state.pa.us/pamanual/site/default.asp>, visited 1/21/05.

Rusk 2003. "'Little Boxes' – Limited Horizons: A Study of Fragmented Local Governance in Pennsylvania: Its Scope, Consequences, and Reforms." Background paper for the project: Back to Prosperity: A Competitive Agenda for Renewing Pennsylvania. Washington, DC, Brookings Institution, Center on Urban and Metropolitan Policy. Internet version: <http://www.brookings.edu/pennsylvania>, visited 10/18/04.

Chapter 19

Politics

In the previous chapter we considered how our system of leadership is organized. In this chapter we will consider who our leaders are and have been, and how they reflect our preferences, and therefore our culture.

It has been facetiously rued that Pennsylvania has not produced political leaders on par with our economic importance. The highest-ranking politician from Pennsylvania in federal politics was James Buchanan, who will never be mentioned as one of our greatest presidents. Other Pennsylvanians in the federal government have achieved positions of strength, but not positions of prominence. Does this political reality on the national stage somehow reflect other aspects of Pennsylvania's culture? Does politics within the state show similar cultural characteristics? Since the foundation of American government is the American people, how do our political choices reflect our collective political attitude? Finally, as we always ask, is there any geographical pattern to Pennsylvania politics?

These questions will be addressed in two ways. First we will consider which elements of our culture and economy are politically significant. That approach will allow us to generalize about which issues are likely to be keys to understanding Pennsylvania voter preferences. Then we will examine maps of election results from some recent state and federal elections. We will see that patterns of voting give a strong idea of the geographical dimensions of political behavior in Pennsylvania, and of what kinds of issues are likely to determine election outcomes.

Culture and Politics

When US presidential elections are fought, all of the attention is on who is going to take each state, such as Pennsylvania, since the Electoral College system decrees that all of the state's 21 (the number based on 2000 census results) electoral votes go to the winner of that state's popular vote. Who is elected is due to a variety of factors related to the candidate: his or her party, personality, campaign finances, home territory and the hot issues of the moment.

The parties are, numerically, nearly equal in Pennsylvania. The Board of Elections in the Pennsylvania Department of State reported that in the fall of 2003 there were 3,677,488 registered Democrats (47.7 percent of the total), 3,204,440 registered Republicans (41.6 percent), and 828,673 voters allied with other parties (10.7 percent). There is always a minority of party-declared voters who are willing to cross party lines, and this factor has been significant in past Pennsylvania elections (PA BCEL 2003).

Pennsylvania has not always spoken with such a divided political voice, though; this has only developed since the 1930s (Williams 1995, 155). Until that time, Pennsylvania was a very conservative state everywhere. With the development of Depression-era welfare programs and the beginning of the decline of several major industries, the biggest cities started voting Democrat in ever larger numbers. The big shift toward more Democrat-leaning cities also coincided with the growth of suburbs, many of which drew wealthier Republican-leaning urbanites out of cities.

Table 19.1 Pennsylvania's Historical Presidential Election Results

Year	Democrats Candidate	%	Republicans Candidate	%	Other %	Winner's Party	Electoral Votes
2004	Kerry	51.0	Bush GW	48.5	0.6	D*	21/538
2000	Gore	50.6	Bush GW	46.4	3.0	D*	23
1996	Clinton	49.3	Dole	40.1	10.8	D	23
1992	Clinton	45.2	Bush GH	36.1	18.7	D	25
1988	Dukakis	48.4	Bush GH	50.7	0.9	R	25
1984	Mondale	46.0	Reagan	53.3	0.7	R	27
1980	Carter	42.5	Reagan	49.6	7.9	R	27
1976	Carter	50.4	Ford	47.7	1.9	D	27
1972	McGovern	39.1	Nixon	59.1	1.8	R	29
1968	Humphrey	47.6	Nixon	44.0	8.4	D*	29
1964	Johnson	64.9	Goldwater	34.7	0.4	D	29/538
1960	Kennedy	51.1	Nixon	48.7	0.2	D	32/537
1956	Stevenson	43.3	Eisenhower	56.5	0.2	R	32/531
1952	Stevenson	46.9	Eisenhower	52.7	0.4	R	32
1948	Truman	46.9	Dewey	50.9	2.2	R*	35
1944	Roosevelt F	51.1	Dewey	48.4	0.5	D	35
1940	Roosevelt F	53.2	Willkie	46.3	0.4	D	36
1936	Roosevelt F	56.9	Landon	40.8	2.3	D	36
1932	Roosevelt F	45.3	Hoover	50.8	3.8	R*	36
1928	Smith	33.9	Hoover	65.2	0.9	R	38
1924	Davis	19.1	Coolidge	65.4	15.6	R	38
1920	Cox	27.2	Harding	65.8	7.1	R	38
1916	Wilson	40.2	Hughes	54.3	5.5	R*	38
1912	Wilson	32.5	Taft	22.5	45.1	Other*	38/531
1908	Bryan	35.4	Taft	58.8	5.8	R	34/483
1904	Parker	27.1	Roosevelt T	68.0	4.9	R	34/476
1900	Bryan	36.2	McKinley	60.7	3.1	R	32/447

* In these elections Pennsylvania did not support the ultimate winner.
Sources (election results): Cox 2004, PA DOS 2005.
Source (electoral votes): US NARA 2004.

Table 19.1 shows the results of the last 27 presidential elections as percentages, along with their Electoral College vote totals. Republicans won easily until the first election of the Depression in 1932. Since 1948 it seems to have been an evenly balanced see-saw, with each Republican stretch matched by a one for the Democrats, and half of the elections decided by a difference less than five percent. Different factors weighed most heavily in different election.

The effects of such political realities in Pennsylvania emerge in much more than just presidential election voting results. Every popular election at every scale and every vote for new policies or other issues in the federal and state legislatures can be analyzed for such patterns and factors.

Economic and social aspects of issues tend to play a very large role in influencing voters' preferences. Attempts to deal with urban social issues related to housing, education, welfare and jobs by shifting their costs to the state are associated with Democratic voting. The dominant issues supported by Republican usually involve lowering taxes, guaranteeing individual freedoms, and shifting responsibility for solving urban social problems onto the individual cities.

In a very key way, then, support for and opposition to any issue comes from certain easily-identified groups of voters, especially the urban vs. rural vs. suburban populations of Pennsylvania. The urban residents tend to vote Democrat, while the rural and suburban population leans toward the Republicans.

However, that is not to say that suburban and rural dwellers vote the same way for the same reasons, or on the very same issues. Income-related issues may see agreement between urban poor and rural poor voting against wealthy suburban and urban dwellers, for example. Many suburban neighborhoods are dominated by the lower middle class, especially near the larger cities, so even the suburbs can be split. Because many election-related issues change from one year to the next, such factions are usually temporary. For issues to become successful legislation, coalition building becomes very important. The safest strategies spread the burden proportionately, or even onto tourists or other cross-boundary visitors.

The Geography of Pennsylvania Politics

As in most states of the US, the role of the Electoral College in presidential elections hides the most interesting political factors, which are the geographical ones. Ever since the transition occurred to voter registration totals more evenly balanced between the two parties, the political geography of Pennsylvania has become very dynamic. All of the issues which divide urban, rural and suburban voters are, at their heart, geographical issues. Each political issue that plays out differently depending on its scale (local, regional, statewide or national) is also geographical. This means that additional explanations

for voting outcomes can come from looking at those votes on maps.

The voter registration map in Figure 19.1 shows a striking dichotomy. In 46 counties out of 67, the Republicans have the greater percentage; in 25 of those, the difference is at least 10%. Notice that the counties with the highest presence of Democrats also contain the largest cities. This pattern emphasizes the numbers of voters and the intense concentrations of political power in our cities. On the other side, the counties south and west of Pittsburgh expect to vote for the Democrats. What issues could unite that predominantly rural area more closely with the cities than with other rural areas? The explanation goes back to the period of economic dominance for the iron and steel industry, which affected Pittsburgh very strongly, as well as the smaller towns that had become linked to and dependent on Pittsburgh's economy. They have also remained politically cohesive during the decline of that industry.

At least one source describes the pattern depicted in Figure 19.1 as Pennsylvania's political "T," because of the shape of the Republican dominated area (PEL-East 2004). Williams describes the challenges of such a geographically fragmented political landscape using a comment by Senator Arlen Specter who likened campaigning in Pennsylvania to trying to win six different states (1995, 154).

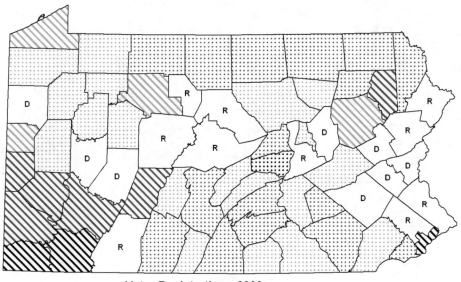

Figure 19.1 **The Parties of Registered Voters around Pennsylvania:** The state's largest urban centers are strongholds for Democrats, while the rural areas are strongly Republican. Counties with moderately sized cities surrounded by extensive suburbs, and Philadelphia's suburban counties are less clearly categorized. Some suburbs are working class areas that tend to favor Democrats, and others are wealthier and tend to vote for Republicans. (Data source: PA BCEL 2003)

Voter Registration: 2003

Democrats	Republicans
D 41.1 - 50.0	R 41.1 - 50.0
50.1 - 57.0	50.1 - 57.0
57.1 - 67.0	57.1 - 67.0
67.1 - 75.5	67.1 - 75.5

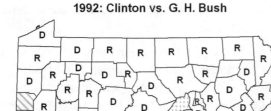

1992: Clinton vs. G. H. Bush

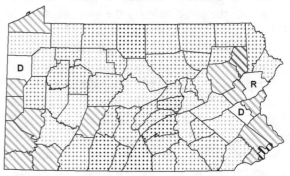

2000: G. W. Bush vs. Gore

Percentage of Voters

Democrats		Republicans	
D	36 - 50	R	36 - 50
	51 - 57		51 - 57
	58 - 67		58 - 67
	68 - 80		68 - 80

Figure 19.2 Election Results for Recent Presidential Elections in Pennsylvania Counties: When Bill Clinton beat George H. Bush in 1992, he narrowly won many normally-Republican counties. On the other hand, even though George W. Bush lost in Pennsylvania in 2000, he won in more normally-Democratic counties than he lost in normally-Republican counties. (Data source: Cox 2004)

Examinations of voting patterns in recent elections also bear this out (see Figures 19.2 and 19.3). In 1992 Bill Clinton beat incumbent George H. Bush in an election that featured a relatively strong third candidate, H. Ross Perot. The Democrats prevailed in many counties where the Republicans usually fared best, by winning with less than 50 percent of the vote. The map of voting results from the 2000 election, one which did not feature an incumbent running, looks much more like the voter registrations map (Figure

1994: Ridge vs. Singel

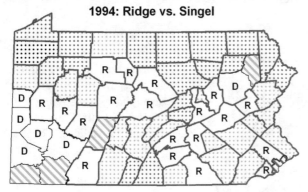

2002: Rendell vs. Fisher

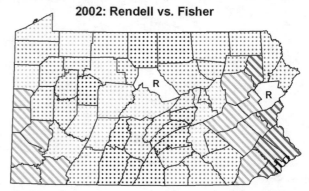

Percentage of Voters

Democrats		Republicans	
D	37 - 50	R	37 - 50
	51 - 57		51 - 57
	58 - 67		58 - 67
	68 - 80		68 - 80

Figure 19.3 Election Results for Recent Pennsylvania Governor Elections across Pennsylvania Counties: Election returns for governor races are just as hard-fought in Pennsylvania as the races for President. Regional biases as well as issues can be factors in election results. These two elections illustrate that pattern since the winners were strongly identified with their regions: Ridge from western Pennsylvania and Rendell from Philadelphia. (Data source: Cox 2004)

19.1). Only eight counties voted contrary to their registration-based preference; in six of those counties the preference is a majority based on less than 50 percent of the county voters.

Elections for the Pennsylvania governorship demonstrate that regional biases can outweigh party preferences (see Figure 19.3). In 1994 Tom Ridge, a Republican, won after Democrat Robert Casey's term, and in 1998 Ridge was re-elected. Tom Ridge, from Erie, failed to take only the staunchest of the Democrats' counties. In 2002, on the other hand, Democrat Ed Rendell, the former Philadelphia mayor, carried many normally-Republican counties in the eastern part of the state, where his name and reputation were better known, but lost several normally-Democratic counties near Pittsburgh, Republican Mike Fisher's home territory.

Harrisburg: Our Political Center

Harrisburg was chosen as Pennsylvania's state capitol for almost the same reason that Washington, DC was chosen to be our nation's capitol. By the early 1800s it was obvious that, as important as Philadelphia was, it was not well situated to connect all of the state's constituencies, especially those oriented toward Lancaster, Reading, York and the new trading center of Pittsburgh. In 1812, the state legislature began its first sessions in Harrisburg.

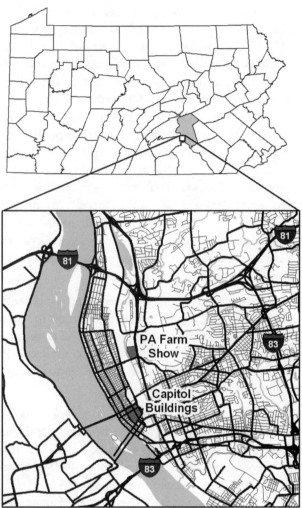

Figure 19.4 **Harrisburg: Our Capitol City:** Harrisburg's location is probably close to the balancing point for Pennsylvania's population distribution. In addition to the state's legislative activities, it also hosts the State Museum and the annual Pennsylvania Farm Show.

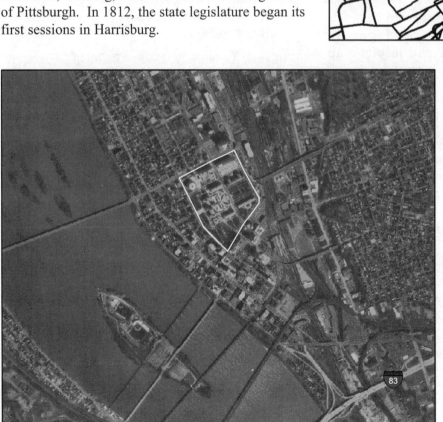

Figure 19.5 **Aerial View of Harrisburg:** The monumental sizes of Pennsylvania's main government buildings is visible in this photograph. For a sense of scale, notice the two baseball fields (almost 300 feet from home plate to the outfield) on City Island, in the Susquehanna River (USGS 1999a, USGS 1999b).

117

Harrisburg's "strategic location" could describe its entire history. It started life as a trading post in the early 1700s, founded by John Harris, Sr. John Harris, Jr. continued the role by establishing a ferry service across the Susquehanna and renaming the village Harris's Ferry. He had the more ambitious urban street plan laid out in 1785, and by 1791 it was incorporated as a borough. By 1860, when it was incorporated as a city, it was a well-established transportation center. Harrisburg was a key stop on The Pennsylvania Canal in the late 1830s, and the Pennsylvania Railroads had also established a major facility there. A significant Civil War battle at nearby Camp Hill contributed to the city's selection as a site for the National Civil War Museum. Nearby is the US Army War College, a US Army Depot and a US Navy Depot.

The state government has had a monumental presence in Harrisburg since 1906, when the state capitol complex of buildings housing the governor's offices and the state legislature were completed. Its economy has also drawn on a major heavy manufacturing sector, as well.

Like many cities, Harrisburg's economy took a downturn following World War II. Many residents moved out into surrounding suburbs, including across to the west bank of the Susquehanna. By 1981 it had been labeled "the second most distressed city in the nation." Beginning in the 1970s a series of aggressive comprehensive plans has helped to clean up, modernize and revitalize the city's center.

Conclusions

The 2002 gubernatorial election shows that residents of Philadelphia and Pittsburgh do not always agree on solutions to urban issues. Because the ethnic make-up of their citizenry and the particular industries that were strong in the past differ, some solutions favor one of the two cities over the other. Successful coalition building usually requires that economic solutions give equitably to both places.

Bibliography

Cox, Harold E. 2004. Pennsylvania Presidential Election Returns. Wilkes-Barre, PA, Wilkes University: The Wilkes University Election Statistics Project: Dr. Harold E. Cox, Director. Internet site: <http://wilkes-fs1.wilkes.edu/~hcox>, visited 10/25/04.

PA BCEL 2003. Official Voter Registration Data, November 2003. Harrisburg, PA, Pennsylvania Department of State, Bureau of Commissions, Elections and Legislation. Internet site: <http://www.dos.state.pa.us/bcel/cwp/view.asp?a=1099&Q=441857&PM=1>, visited 8/5/04.

PEL-East 2004. Pennsylvania's Geo-Politics. Philadelphia, PA, Pennsylvania Economy League, Inc., Eastern Division. Internet site: <http://www.peleast.org/geo.htm>, visited 7/21/04.

USGS 1999a. Digital Ortho Quarter Quad (aerial photograph): harrisburg_east_pa_sw. Downloaded in TIFF format from http://www.pasda.psu.edu/access/doq99list.cgi, 8/30/04. US Geological Survey, Washington, DC.

USGS 1999b. Digital Ortho Quarter Quad (aerial photograph): harrisburg_west_pa_se. Downloaded in TIFF format from http://www.pasda.psu.edu/access/doq99list.cgi, 8/30/04. US Geological Survey, Washington, DC.

US NARA 2004. Historical Election Results: Electoral Votes by State. Washington, DC, US National Archives and Records Administration, Office of the Federal Register. Internet site: <http://www.archives.gov/federal_register/electoral_college/votes/votes_by_state.html>, visited 10/25/04.

Williams, Anthony V. 1995. "Political Geography." Chapter 10 in Miller, E. Willard (ed.) A Geography of Pennsylvania. pp. 154-164.

Section C: Chapters 20 to 31

The Structure of Economic Activity

The economy is all about money, jobs, products and services, of course, but it is necessary to organize the discussion of all of this. Economists have developed a few terms which will be useful to learn. These simple terms serve to give structure to the economy, and also remind us of how the economy developed historically. A geographer, as you have seen, adds questions about *where* each element of the economic system is located and why.

During the pre-economic stage of development, humans were hunters and gatherers. They *found* building materials, means for clothing themselves and for preparing food, in addition to finding the food itself. In the first stage of developing an "economy" all civilizations began by perfecting some forms of agriculture. Domesticating plants and animals led to other forms of planned or organized "extraction" of environmental materials, namely forestry, mining and fishing; it also led to sedentary living in communities that included some who specialized in agriculture or one of those other activities, and obtained the goods they did not produce by some form of exchange. This extraction-based activity is called the "Primary" level of economic activity (see Table C.1).

Table C.1: Levels of Economic Activity

Economic Level	Function	Examples
Primary	Extraction	Agriculture (including crops and livestock), Mining, Forestry, Fishing, Hunting
Secondary	Processing and Manufacturing	Energy production, Transportation, Food processing, Manufacture of durable goods and clothing, Construction
Tertiary	Retailing and Services	Wholesaling, Warehousing, Retail sales, Real estate, Financial services, Environmental services, Health care, Administration, Education, Entertainment and recreation
Quaternary	High-tech Goods and Services	Insurance, Information processing, High-tech manufacturing, Communication

In America, the pre-European natives had domesticated corn and various members of the squash and melon families. The vast majority of the first Europeans, over 90% of them, were farmers feeding their families on the food they could grow or raise. Each had to have sufficient land, even many of those living in Philadelphia. Thus, economic development during colonial times was focused on this Primary level of activity.

When the Industrial Revolution began, delayed in America many decades after its European origins, workers were often housed, clothed and fed by the factory owner. Gradually, however, the wage-based system developed, which in turn inspired farmers to produce even more excess goods for sale. Although this cash economy was present in Philadelphia right from the beginning, in order to feed the traders and merchants along the waterfront, for example, it developed into a more complex system of specialization as time progressed.

Industrialization, the "Secondary" level of economic activity, usually refers to factory production of material goods, such as the heavy industry that characterized Pennsylvania and the US toward the end of the 1800s. However, the term "industry" can also be used to describe the changes in Primary-level activities in the early 1900s; farms, forests and mines are often exploited using factory-inspired systems of production. The same principles applied to the development of transportation and energy production in the late 1800s and early 1900s.

With the development of industrialization, another kind of specialization was also needed. Wholesalers and retailers brought goods from producer to consumer, banks became more common, and then governments got more serious about overseeing several aspects of the system in order to protect the public, provide infrastructure for the common good, and attract new settlers. All of these activities represented new employment opportunities, at what is called the "Tertiary" (third) level of activity. Much of this development was geared toward a growing middle class.

By the end of the 1800s, economic terms such as investing, monopolies and diversification (as the opposite of specialization) became more widely used. The early 1900s were dominated by the development of new products, the search for greater production efficiency and resources, the new work laws that ended up creating more leisure time and more disposable income for greater numbers of people, and interconnecting more people worldwide.

The next big economic change came about in the 1960s, even though its precursors date back to before World War II. The earliest electronics used vacuum tube-based technologies to make radios and televisions and the related broadcasting equipment. By the 1950s solid state circuitry and circuit elements (such as transistors) replaced tube-based circuits. By the 1970s, integrated circuits miniaturized the solid state circuits, ushering in the "high-tech" revolution. These advanced technologies, considered by some economists to be an additional, "Quaternary" (fourth), level of economic activity coincided with and helped to create the "Information Age." Quaternary activity refers to both the manufacturing breakthroughs that emerged and the expanded space exploration, communication and information processing activities related to them.

The US Bureau of the Census, in addition to its population census responsibilities, also counts and surveys the companies of the US in its Economic Census every five years. Companies are categorized according to their primary function. Those function categories are enumerated in a complex numerical system known as the North American Industry Classification System (NAICS). Broad categories are designated by two-digit numbers, sub-categories are formed by adding a third digit, and the system continues until a six-digit code is possible. The two-digit codes are laid out below in Table C.2.

Table C.2: NAICS Two-digit Codes (US Census Bureau 2004)

11	Agriculture, Forestry, Fishing and Hunting
21	Mining
22	Utilities
23	Construction
31-33	Manufacturing
42	Wholesale Trade
44-45	Retail Trade
48-49	Transportation and Warehousing
51	Information
52	Finance and Insurance
53	Real Estate and Rental and Leasing
54	Professional, Scientific and Technical Services
55	Management of Companies and Enterprises
56	Administrative and Support and Waste Management and Remediation Services
61	Educational Services
62	Health Care and Social Assistance
71	Arts, Entertainment and Recreation
72	Accommodation and Food Services
81	Other Services (except Public Administration)
92	Public Administration

Bibliography

US Census Bureau 2004. North American Industry Classification System (NAICS). Washington, DC, US Department of Commerce, Bureau of the Census. Internet site: <http://www.census.gov/epcd/www/naics.html>, visited multiple times.

Chapter 20

Traditional Agriculture

The key difference between traditional and modern agriculture in America is that traditional agriculture evolved from a system organized around sustaining a family, while modern agriculture is much more focused on productivity and profit. The traditional farm combined many aspects of agriculture: grain crops, orchards, fruit and vegetable gardening, and animal husbandry for every farm animal from steer to hen.

The original European settlers in Pennsylvania were involved in a constant struggle to re-learn how to grow foods they were familiar with, to learn how to grow new American foods such as melons, tomatoes and potatoes in a climate and soils unlike those they were used to, to clear new land of its natural forest cover, and also to deal with natives or tax collectors or neighbors. News and information traveled slowly and different culture groups contributed different approaches to solving problems. The development of a Pennsylvanian agriculture went hand-in-hand with the development of the Pennsylvania culture.

Different areas of the colony tended to be populated by different culture groups, but few areas were purely English, German or Scots-Irish. Lemon observed that English farming methods were generally preferred, and that Germans did not have a reputation for being innovative (1972, 183).

The Traditional Farm

Traditional farming was labor intensive, some of it animal labor (mostly from horses) but most of it human labor. The successful farm housed a large family and covered, on average, 125 acres (see Table 20.1). Of that space, a little more than half was planted or pasture land in any given year, with additional acreage in fallow.

That left nearly half of the farm in woodland, which possibly even covered part of the pasture land. The land also held the farmer's home, barn and other storage buildings (Lemon 1972, 167).

Cropland was planted in several different grains; the most important was wheat, the original seeds of which were imported from Europe. Most grain planting was done in the spring for a fall harvest, but

Table 20.1 Typical Colonial Farm Acreages
(Lemon 1972, 152-3)

Grains (46 acres)	
Wheat	8 acres
Indian corn	8
Oats	4
Rye	2
Barley	2
Buckwheat	2
Hay	20

Other Crops (7 acres)	
Potatoes, turnips	3 acres
Flax, hemp	2
Fruit trees	2
Other vegetables and tobacco	variable

Livestock Pastures (19 acres)	
Cattle	15 acres
Horses	3
Swine	nominal
Sheep	1
Poultry	nominal
Bees	nominal

Total: 72 acres out of an average farm size of 125 acres.

some, especially the rye, was planted in the fall to be harvested in early summer. A significant part of the rye's value was its ability to be distilled into whiskey, which was preferred over beer in southeastern Pennsylvania. Grains that were ground into flour (for consumption and for sale) grew alongside grains that could be stored as winter fodder for the animals. Some areas planted with a variety of grasses were left as pasture, and others grew into tall-grass meadows to be cut as hay (Lemon 1972, 154-160).

Among animals, the largest in number were cattle (both beef and dairy) and hogs. Fewer than ten each of cattle and hogs were kept on the average farm. A few horses helped with labor needs. About two thirds of the early colonists' meat consumption came from swine, with the rest from beef and a small amount of poultry. The latter, mostly chickens but possibly turkeys, geese and ducks also provided eggs. Sheep sometimes, but not typically, were raised for both meat and wool (see Figure 20.1). Bees were kept primarily as pollinators (Lemon 1972, 160-167).

Vegetables and fruit include both ground and orchard produce. The most important vegetables were cabbage, potatoes and turnips (Lemon 1972, 155). Corn was not usually eaten, but was fed to hogs (Lemon p. 157), as were many of the potatoes and turnips. Fruits included apples most commonly, but also peaches and cherries (Lemon p. 158). The only other vegetables and fruits named by Lemon are pumpkins (p. 159), peas and beans (171) and sweet potatoes (172).

Tobacco was another type of crop grown both for the farmer's use and for sale. Flax, also a typical crop, produced cooking oil as well as fibers for rope making (Lemon 1972, 157).

This variety of production usually helped the average farmer to produce excess. Some of that excess was destined for Britain, as "flour, wheat, corn, flaxseed, meat, lumber, iron, and small amount of other commodities." The peak years for exports were the early 1770s, with wheat and flour the dominant products (Lemon 1972, 181).

Amish Farming

Many cultural ideas, such as food preferences and valued economic occupations, are associated with agriculture or its products; indeed the meaning of the word "culture" has its roots in agriculture's concept of cultivation. One group whose very existence strongly links culture and agriculture is the Amish (actually "Old Order Amish"). There are three active Amish settlements in Pennsylvania today: in Mifflin and Somerset Counties, as well as the oldest New World settlement centered in Lancaster County.

The Amish are technically a subgroup of the Mennonites, who are much more centrally organized and modern (expect for some Old Order Mennonite congregations). The Amish and Mennonites are both

Figure 20.1 **Traditional Livestock Raising:** This photo was taken at Landis Valley Farm Museum, just northeast of Lancaster. This museum preserves a nineteenth century "Pennsylvania Dutch" farm, with historically accurate buildings, animal breeds and a large collection of plant seeds.

Anabaptists, Protestants believing in voluntary adult baptism rather than the baptism of infants. The difference between the two groups is that the Amish believe more strongly that sinners who are shunned and do not repent must be cut off completely. The division between the two sects occurred in Europe long before both groups migrated to Pennsylvania in hopes of finding freedom from religious persecution (Glass 1979).

Amish settlers began arriving in Pennsylvania in the 1720s or early 1730s (Glass 1979, Meyers 1990). Most of the early Amish headed for Lancaster and Chester Counties. When no more land was available in an area, members of the group moved on. Part of the challenge comes from the Amish practice of having over seven children per family, on average, and yet being unwilling to divide a farm property among male heirs. Younger sons look to buy property locally, often on less fertile soils these days, move into non-agricultural jobs (woodworking or harness-making, for example), or move cross-country. From the Lancaster County hearth, the Amish have migrated to several other states (the largest numbers now live in Ohio and Indiana) as well as Canada. Only the Somerset County Amish settlement includes many who bypassed Lancaster County completely.

The Amish lifestyle is driven by the ideals of producing from the land, and keeping their personal and community lives separate from non-Amish. The Lancaster County Amish have been notoriously unsuccessful in the latter because non-Amish have built up a thriving tourist industry based on sight-seeing in Amish areas. The separateness includes separate schooling, refusal to allow electricity and telephone lines into their homes, and refusal to operate motorized machinery for transportation or for fieldwork. On the other hand, because much of their dairying and other businesses requires them to interact with non-Amish, they have motorized (not electric) refrigeration units to hold milk sold to dairy companies, and telephone booths (often shared between neighbors) at the road end of their driveways (see Figure 20.2).

The Amish separateness has become what appears to outsiders a stagnant lifestyle. It is not, however, an attempt to preserve the lifestyle of the original Amish. For one thing, many current Amish practices were typical of rural people from all backgrounds in the middle 1800s, not the 1700s when they first arrived.

Figure 20.2 **Amish Telephone:** Since the Amish will not let most modern technologies into their homes they are sometimes forced to compromise. This box is near the end of a long driveway back to the farmhouse, and probably sees only occasional use.

The large families, the farming methods, the mistrust of banks and government and, of course, the travel by horse and buggy would have been widespread back then. Today, they constantly have to make decisions about which non-Amish practices and goods are acceptable. For example, their children are seen on scooters (and sometimes roller skates) some of which are very modern in design. Some adaptations are imposed on them, such as safety signals on their horse-pulled buggies, and others are made by choice. In Lancaster County, more buggies are incorporating fiberglass body materials and LED lighting and signaling systems (Strausbaugh 2004, A4).

The system of agriculture practiced by the Amish is similarly fraught with contradictions. Their farms are built around the mixed farming model described in the first part of this chapter. Crop farming is mechanized compared to standard practices of the early 1800s, but of course not by today's conventional technologies since equipment is pulled by teams of horses or mules. Even though the purpose of the mixed farming model is to produce all of their food needs, they also attempt to produce sufficient surplus for sale, either at road-side stands and local farmers' markets (see Figure 20.3), or under agricultural contracts, and at agricultural auctions, with commercial buyers, tobacco companies for example.

The main feature that makes the Amish successful is that they have realized the necessity of creating clustered communities. Even if a few non-Amish live within their rural areas, there must be a sufficient number of Amish within an easy travel distance to keep the Amish farm-related businesses profitable, to assure that all the occasionally-needed skills and labor (barn construction, for example) are there, to keep sufficient numbers of children in each one-room school, and to have sufficient families that young men and women will marry within the faith and carry on the traditions. Glass cites a 1960 source which had counted two thirds of Lancaster County Amish as having only five different surnames: Stoltzfus, King, Beiler, Fisher and Lapp (Glass 1979).

Conclusions

Traditional farms and farming methods are still visible on the landscape today. The conservative nature of Pennsylvania's rural areas (see Chapter 19) reflects reluctance to change as well as a distrust of cities and government. The Amish may take that reluctance to an extreme, but have so far adapted effectively.

Even more profound is the impact that early Pennsylvania agriculture had on the agricultural preferences and practices for many succeeding generations of farmers from Pennsylvania through the Midwestern US. The methods have become modernized, as the next chapter will show, but the farm products and the food products extracted from them have not changed significantly.

Bibliography

Glass, Joseph W. 1979. "Be Ye Separate, Saith the Lord: Old Order Amish in Lancaster County." In Cybriwsky, Roman A. (ed.): The Philadelphia Region: Selected Essays and Field Trip Itineraries. Washington, DC, Association of American Geographers.

Lemon, James T. 1972. The Best Poor Man's Country: A Geographical Study of Early Southeastern Pennsylvania. Baltimore, Johns Hopkins University Press.

Meyers, Thomas J. 1990. Amish. Canada, Mennonite Historical Society of Canada. Internet site, created 1998, copyrighted 2004: < http://www.mhsc.ca/index.asp?content=http://www.mhsc.ca/encyclopedia/contents/A4574ME.html>, visited 7/25/04.

Strausbaugh, Judy A. 2004. "Using Horse Sense in Buggy Safety." Sunday News, August 29, 2004. Lancaster, PA, pp. A1, A4.

Figure 20.3 **Roadside Stand:** Stands like these dot the rural countryside in areas where Amish and other family-owned farms are common. Customers get better deals than the grocery store gives, and the farmer gets more than a commercial buyer would pay.

Chapter 21

Modern Agriculture

Modern agriculture has the same inputs and outputs as traditional agriculture, and may appear similar to traditional agriculture to the casual observer (see Figure 21.1), but all of the additional processing bears little resemblance. Pennsylvania is still among the most productive states, in part because the state is well situated for productive soils and climate, in part because most of its farmers have readily adapted new methods, and in part because the state is part of the huge east coast market for its produce.

Early in the 1900s many biologists focused on improving their understanding of food plants and animals, and began scientific experimentation which greatly improved yields and productivity. This has become known as the Green Revolution. Much of the improvement in plant yields came from the development of synthetic fertilizers and pesticides, and from interbreeding of different strains of plants. More recently, the biologists have been genetically re-engineering the plants, in order to virtually customize each field to its exact growing conditions. Much more sophisticated farm equipment was also developed. The financial result has been to make the cost of a year's crops much higher, but to increase the potential reward at harvest time as well.

The modern family farmer must be knowledgeable in everything from crop and animal biology, technology and environmental regulations, to business

Figure 21.1 **Modern Piedmont Farm:** Many farms have changed little since the 1800s in the outward appearance of their buildings and fields. However, every aspect of their operation has been scrutinized for possible improvements. The result is larger livestock, greater milk production per cow, and higher crop densities and yields. The buildings hold much more sophisticated and expensive equipment than a hundred years ago.

management. The bottom line challenge is for the farmer to make enough profit to support the family. Most have met this challenge by specializing, by diversifying, or by entering into agreements with agricultural cooperatives or with corporations who will buy their produce.

On the other hand, many farms in the US today are owned by companies rather than families. These corporate farms are managed by professional teams of specialists. The larger corporations that they are usually part of are often active in other agriculture- or food-related businesses. Having such diverse sources of income means that the corporation is not as susceptible if one year's weather is poor or if a new disease or pest ruins the crop. Interestingly, this business approach to farming has not caught on very widely in Pennsylvania. In 2002 over 91% of the state's farms were owned by individuals or families, and only 6% were owned by corporations or partnerships (US NASS 2004b, Table 58).

Pennsylvania Farm Production

All agriculture in Pennsylvania is monitored by the Pennsylvania Department of Agriculture and the US Department of Agriculture, especially through the state's main agricultural research facilities operated in conjunction with the Pennsylvania State University College of Agricultural Science. These organizations primarily assess performance by measures of production (as measured in bushels, tons, gallons), yield (production per acre or per cow) and sales or profits. Farmers report their information annually, and nation-wide detailed censuses are conducted every five years.

Because agriculture is very sensitive to natural conditions, production and yield totals will vary from year to year. More limited production in one year will lead to higher prices, but in another year higher production throughout a region will flood the market and decrease the prices farmers receive. Making comparisons based on sales totals eliminates some of that variability. We can learn a lot by examining county sales totals for different agricultural products.

Crop growing generates the lowest sales totals, but is vital to feeding Americans. Some are eaten as grains, or similar foods, while others provide food for animals. "Crops" include grains (such as corn and wheat), oilseeds (such as soybeans and flaxseed) and other crops, plus hay for animal fodder. Good soils

and higher populations nearby both help attract this type of farming. Since grains, dairying and meat production all used to be part of the typical general farm, the crop farms still tend to be located near the dairy and livestock operations. Significant portions of the corn grown, for example, ends up as cattle feed. Crops, such as that same corn, produced for direct home use are a very small part of the total; much more of the grains and other crops for human consumption will be processed first. In this business Pennsylvania falls well behind the giant farms of the midwestern US.

Figure 21.2 shows that crops grow best on the alfisol soils in the Piedmont and in the Ridge and Valley valleys. A secondary area of crop production is on the Central Lowlands landform region along Lake Erie and on adjacent areas of the Appalachian Plateau. Given its mixed farming origins, nearness to cities both in Pennsylvania and nearby explains most of the more productive crop farming counties. Trenton and Camden, New Jersey, Wilmington, Delaware and Baltimore, Maryland help to stimulate farming in the southeast. Cleveland and Youngstown, Ohio and to a lesser extent Fredonia and Jamestown, New York play similar roles in the west.

While most crops are consumed as staple grains and grasses, by humans and livestock, several specialized crops are also included in the totals. Beans, potatoes, sunflower seeds and tobacco (see Figure 21.3) add small but significant value to the crop sales totals.

Fruits and vegetables are grown as field crops (but not included above) or in orchards and vineyards. They vary in shelf life but most must be eaten or processed (frozen, canned or other processing) soon after picking. The areas of highest sales in Figure 21.4 illustrate three different scenarios. In Erie County, a large portion of the output in this category comes from vineyards growing Concord, Niagara and other grapes. This supports a significant wine industry and also is part of a significant region for grape juice, grape jelly and related products in the US. Much of Adams County is located in the Gettysburg-Newark Lowland, the portion of the Piedmont landform region adjacent to the New England and Ridge and Valley regions. Gettysburg, the county seat of Adams County, is the center of a very large orchard industry focused on growing apples. Lancaster County farmers devote a significant acreage to fruits and vegetables, most of

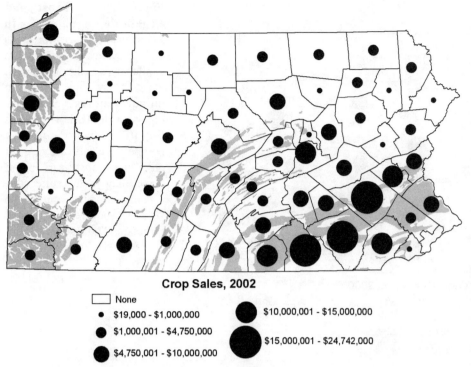

Figure 21.2 **Crop Sales in Pennsylvania Counties, with Alfisol Soils:** "Crops" include the grains wheat, corn, oats, rye, barley, sorghum, soybeans, tobacco and potatoes. Note the correspondence between the higher sales and the alfisol soils. Berks, Lancaster, York and Adams Counties form a belt close enough to Allentown, Philadelphia and Baltimore to have a large market, yet far enough away to be less developed. Montgomery, Bucks and Chester Counties are even better located and have plentiful alfisol soils. However, these three counties have lost a great deal of farmland to urban/ suburban development. (Data source: US NASS 2004a, Table 2)

Crop Sales, 2002

☐ None
• $19,000 - $1,000,000
• $1,000,001 - $4,750,000
● $4,750,001 - $10,000,000
● $10,000,001 - $15,000,000
● $15,000,001 - $24,742,000

which are ground crops. Most of this produce, like that of counties to its northeast and east, is destined for the Philadelphia metropolitan area and other nearby major urban/suburban areas outside of Pennsylvania. The produce is picked, sold and eaten fresh. The variety is enormous, ranging from strawberries and blueberries, to pumpkins, squash and melons, to beans, and to soft orchard fruits including cherries and

peaches. The same generalizations can be made throughout the state: production is limited to particular soils, and products grown near major cities are more likely to include foods for fresh consumption while those grown farther away will more likely require processing and packaging.

Dairy production in Pennsylvania is focused mostly on milk for bottling, but also includes

Figure 21.3 **Tobacco Drying in Lancaster County:** Tobacco is still a viable cash crop (though usually not a farmer's main crop) for many farmers. As smoking habits change, though, the total production in Pennsylvania is declining.

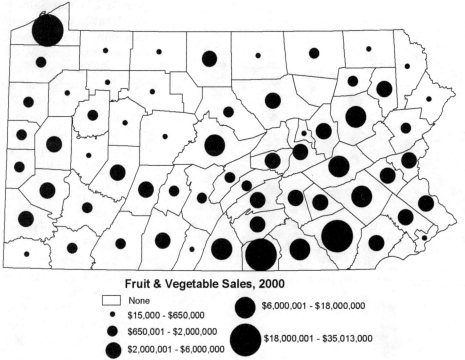

Figure 21.4 Fruit and Vegetable Sales in Pennsylvania Counties: Fruits and vegetables are attracted to similar areas as the other crops, but some are able to thrive in unique settings. Specialty regions such as Erie and Adams Counties can enjoy a competitive advantage over other areas most years, but if that leading product suffers due to natural conditions during one year, the whole region feels the effects. (Data source: US NASS 2004b, Table 2)

Fruit & Vegetable Sales, 2000

☐ None
· $15,000 - $650,000
● $650,001 - $2,000,000
● $2,000,001 - $6,000,000
● $6,000,001 - $18,000,000
● $18,000,001 - $35,013,000

processing the milk into cheese, butter, ice cream, yogurts and other products. Because milk, like some of the fruits and vegetables described above, has a short shelf life, its production location is more market-oriented. Counties with significant production but located farther from the population centers, are more likely producing processed dairy products. The map

in Figure 21.5 illustrates this, as well as the importance of local feed.

Meat consumption and production have gone through some interesting changes over the centuries. Since the early 1900s beef has been the meat consumed more than all others. One focus of the Green Revolution was to find ways to optimize the diets of

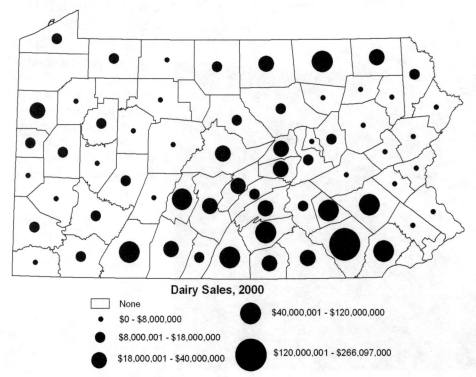

Figure 21.5 Sales of Milk and Other Dairy Products from Cows in Pennsylvania Counties: Only Chester County is a heavily suburban/urban county with significant dairy production. However, the counties of central and south-central Pennsylvania, led by Lancaster County, are within delivery distances of major cities such as Philadelphia and Baltimore. The Northern Tier counties of Tioga, Bradford and Susquehanna are probably serving the New York City region or Binghamton and Elmira, NY. (Data source: US NASS 2004b, Table 2)

Dairy Sales, 2000

☐ None
· $0 - $8,000,000
● $8,000,001 - $18,000,000
● $18,000,001 - $40,000,000
● $40,000,001 - $120,000,000
● $120,000,001 - $266,097,000

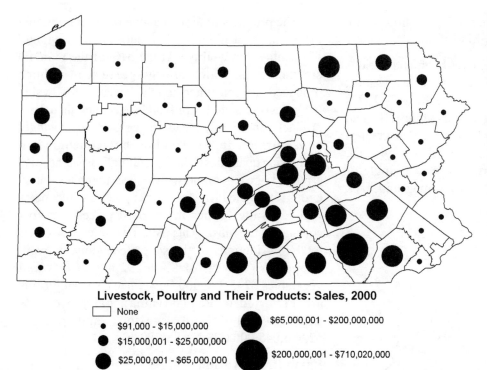

Figure 21.6 **Sales of Livestock, Meat and Related Products in Pennsylvania Counties:** The patterns here look like those in Figure 21.5, but the total values are much higher. Like dairy products, meat products require careful handling and relatively quick delivery, though slightly longer than the dairy goods. (Data source: US NASS 2004b, Table 2)

Livestock, Poultry and Their Products: Sales, 2000

	None
•	$91,000 - $15,000,000
●	$15,000,001 - $25,000,000
⬤	$25,000,001 - $65,000,000
⬤	$65,000,001 - $200,000,000
⬤	$200,000,001 - $710,020,000

all livestock—pigs, chickens, turkeys and especially beef cows—in order to maximize the meat output (and egg production from laying chickens, or layers) in as short a time as possible (or for as long as possible for the layers). Beef cows now eat more and exercise less than their ancestors. The same is true for broiler (meat) chickens, raised in modern chicken houses, in cages stacked several rows high, crowded with

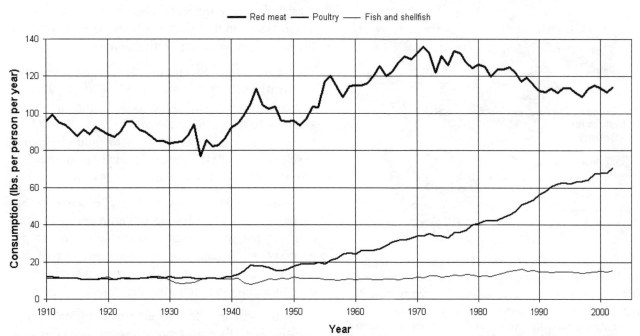

Figure 21.7 **Meat Consumption History:** Increasing affluence and productivity kept beef and swine consumption (red meat) per person rising from the beginning of the Green Revolution through to the 1970s. Its decline since then has been to the benefit of the chicken (broiler) farmers. Seafood consumption has changed little, even surviving decrees of the church in the 1960s that Catholics no longer had to eat seafood on Fridays. (Data source: USDA 2002)

129

chickens. Layers are raised in similar conditions with their diet optimized for egg production.

Costs now favor chicken over beef, and various health reports have claimed benefits for white meat (poultry) over red meat (beef). Since the early 1970s poultry consumption has doubled while red meat consumption has declined by 15 percent. The greatest livestock production is in south-central Pennsylvania, shown on the map in Figure 21.6 as a triangle of counties from Somerset up to Lycoming and back down to Chester.

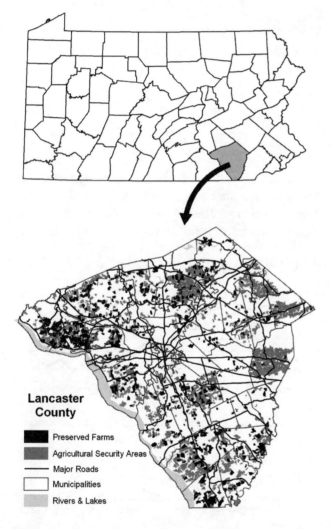

Figure 21.8 **Lancaster County, PA:** Lancaster is the most productive non-irrigated county in the US. Lancaster City is where all roads meet near the center of the county. The lighter gray shading shows clusters of farms designated as Agricultural Security Areas. Only farms in Agricultural Security Areas are elegible for preservation. The preservation of a farm is accomplished by the purchase of its development rights by an organization committed to protecting farmland. (Data source: LC APB 2005)

The products described above represent not only the full range of agricultural output, but also a pyramid of quantity and quality. At the base of the pyramid are the crops, which are the greatest in volume produced and consumed and the lowest in food value (Calories) per pound. Fruits and vegetables, then dairy products, and finally meats and eggs are increasingly concentrated in food value, but consumed in decreasing quantities. The agricultural outputs are also arranged like a pyramid with respect to money value. Crops return the lowest value per pound, while meats will fetch the farmer the highest prices and cost consumers the most. This pattern is reflected in their total contributions to the state's economy, as well. Crops earn a total of $306,292,000, fruits and vegetables $235,306,000 (lower because many fruits are imported to Pennsylvania), dairy products $1,393,992,000, and livestock and poultry $2,936,045,000 (US NASS 2004b).

Farm Preservation

The best farmland in Lancaster County and other agriculturally productive areas of the state is frequently under the greatest pressure from developers. These developers give the family farmers another way to avoid financial risk: selling the farm. A related issue is whether that farmer has children who work on the farm and want to own and operate the farm after the present farmer retires; this is becoming less common. The pressure in that situation is on the farmer to sell the farm in order to have retirement income.

Farms close to already-developed areas are frequently selling for around $10,000 per acre in Lancaster County, which means that selling a farm as small as 100 acres in size would bring $1,000,000 to the farmer. A developer intending to build 150 homes to sell for $200,000 apiece on that land can more easily afford to offer that price than can other farmers.

Since 1980 there has been an alternative to this scenario. Local and state governments have the authority to decree that owning the right to develop a property can be separated from the ownership of that property. Only the property's owner can separate the two, by selling the property's *development rights*.

Who will buy the development rights? That question is really a matter of whose interests (beyond the farmers') are served by the purchase. The public benefits from more efficient food production, a more scenic landscape (in Lancaster County, the agricultural

130

landscape contributes to a sizable tourist economy) and such environmental positives as greater groundwater absorption and less housing- or business-related pollution. Because of those public benefits, state and county governments have taken the lead in passing legislation and providing funding. The money comes from state and county taxes, and some federal funding to the state. The state disburses its money through county agencies; in Lancaster County's case it is the county Agricultural Preserve Board. Non-profit public interest organizations also get involved; for example, the Lancaster Farmland Trust accepts donations and membership fees toward their preservation efforts and the Brandywine Conservancy has preserved farms in both Lancaster and Chester Counties (see Figure 21.8).

The state sets guidelines that the farmers must meet to apply for the funds. First, the farm must be on land zoned as Agricultural by its municipality. Second, it must be part of what is termed an Agricultural Security Area, a designation granted by the state to groups of farms that will be guaranteed tax benefits for their agriculturally zoned area. The farmer must then apply to the preferred agency. The county agencies and non-profits maintain waiting lists of farms arranged in order starting with the best land at the greatest risk. The best land has higher quality and more productive soils than the other farms on the list. Once accepted, the farmer and the agency negotiate the value of the development rights.

The farmer who preserves a farm from development in this way will be paid perhaps $1,500 to $3,000 per acre (again, depending on location), and also gains the assurance that developers cannot touch his land. The income relieves financial pressures, even though it is a one-time-only deal. The Lancaster Farmland Trust can work with Old Order Amish and Mennonite farmers whose beliefs prevent them from working with the government. In some cases the farmers even donate their development rights to such organizations. Ownership of the development rights rests with the county or non-profit agency. Developers, on the other hand, do not enter into such purchases, because owning the development rights without owning the land is just as bad as owning the land without the development rights: it gives them too little.

Lancaster County has had the greatest success in preserving farmland from development. Since it started purchasing farmland development rights in

1983 the county Agricultural Preserve Board has preserved over 529 farms, including over 45,000 acres (LC APB 2004). The Lancaster Farmland Trust has preserved 185 farms covering 11,535 acres (Lancaster Farmland Trust 2004). The map in Figure shows their locations. Debates over the ideal strategy vacillate between those favoring trying to create large contiguous areas of preserved land so that developments don't break up the area, versus those favoring "making a start" in as many different areas as possible. Confounding the decision process is that a preserved farm attracts development interest in neighboring properties because the view toward the farm is virtually guaranteed to remain unspoiled.

Being a farming-intensive county raises other concerns, particularly environmental ones. Even though irrigation is generally unnecessary, fertilization for such intense use is very important. It propels Lancaster to be the highest-value county on all four of the product maps earlier in this chapter. Heavy fertilization has meant over-fertilization, which means that the excess fertilizer contaminates the water-table groundwater and the run-off that ends up in the Chesapeake Bay (see Chapter 36).

Conclusions

Modern agriculture is high-tech (potentially, at least), but farmers tend to be a conservative lot. Farming is, literally, the lifeblood of any society, and yet it is not high on the list of desirable occupations because the hours are long and the pay is not great. The best farmland, for the last hundred years at least, has been located where there is the greatest demand for new housing, roads and other development: as close to the city as possible. These contradictions seem to give agriculture an uncertain future, and yet governments at all levels–municipal, county and state– are working very hard to guarantee that farmland will be preserved forever.

Bibliography

Lancaster Farmland Trust 2004. What's New. Lancaster, PA, Lancaster Farmland Trust. Internet site: <http://www.savelancasterfarms.org/savelanc/cwp/view.asp?a=3&q=570958>, visited 10/28/04.

LC APB 2004. Agricultural Preserve Board. Lancaster, PA, Lancaster County Agricultural Preserve Board. Internet site: <http://

www.co.lancaster.pa.us/lanco/cwp/view.asp?
a=371&Q=384772&tx=1>, visited 10/28/04.

LC APB 2005. Personal communication. Lancaster,
PA, Lancaster County Agricultural Preserve
Board. 1/27/05.

USDA 2002. Food Consumption (per capita) Data
System. Washington, DC, US Department of
Agriculture, Economic Research Service. Internet
site: <http://www.ers.usda.gov/Data/foodcon-
sumption/>, visited 6/24/04.

US NASS 2004a. 2002 Census of Agriculture –
County Data: Pennsylvania. Washington, DC, US
Department of Agriculture: National Agricultural
Statistics Service. Internet site: <http://
www.nass.usda.gov/census/census02/volume1/pa/
index2.htm>, visited 7/22/04.

US NASS 2004b. 2002 Census of Agriculture – State
Data: Pennsylvania. Washington, DC, US
Department of Agriculture: National Agricultural
Statistics Service. Internet site: <http://
www.nass.usda.gov/census/census02/volume1/pa/
index1.htm>, visited 10/28/04.

Chapter 22

Forestry

Forestry is the act of managing forests, including harvesting the trees for their various products. In the 1800s, logging was practiced with no regard for forest re-growth. The result, as we saw in Chapter 10, was the nearly complete clearcutting of Pennsylvania's virgin forests. Since the early 1900s, though, forestry has become the integration of tree harvesting with tree growth science on par with agriculture and environmental protection activities recognizing forests' other functions.

Once again, Pennsylvania has played an important historical role, both in the development of the wood products industry and in the understanding of forest growth dynamics. Many forestry practices were refined here in Pennsylvania, which at one time was the prime lumber source for the new world and, later, the industrial revolution.

Just look at the range of uses for harvested trees:

- wood products, such as structural lumber, fine hardwoods for cabinetry, furniture and veneers, and even firewood

- charcoal and other processed wood products such as plywood and other particle boards

- pulp for making papers of various qualities, and other cellulose fiber products such as cellophane

- tannin and other industrial and medicinal chemicals

- food products such as orchard fruits, nuts, maple syrup and flavors derived from roots

- Christmas trees and other trees for planting.

1800s Wood Production

House building and most forms of manufacturing in America in the early 1800s were heavily dependent on wood. The most valuable trees in Pennsylvania were the white pine and the oak. Stranahan wrote about the white pine: "The grain of white pine is straight and true; it does not warp or rot. It is light and easy to work, yet remarkably strong. Its durability is obvious even today: white pine stumps still dot the steep hillsides of the West Branch [of the Susquehanna] a century or more after the trees were felled. Such trees could be cut in arrow-straight lengths of 90 or 100 feet that carried neither a blemish nor a knot. . . . A legendary white pine cut in Clearfield County, near the headwaters of the Susquehanna, measured 168 feet and produced a 150-foot-long plank" (1993, 79). Such trees were virgin timber, as discussed in Chapter 10.

Harvesting had to consider demand for the product and the feasibility of getting felled trees to mills and market. Individuals involved in this business often had other responsibilities during the summer (such as the family farm), so they cut the trees during winter, moved the logs toward the nearest stream with help from the snow cover, floated them into the stream when water levels rose in spring, lashed as many together as they could manage as a raft, and rode it downstream to where they could sell them to a lumber mill or other buyer. Companies, large and small, whose sole occupation was lumbering, found additional means of transporting logs after 1864. It was then that trains were first convinced to extend tracks into the logging regions. The transportation of logs by railroad came into its most widespread use after 1900, when all of the land near streams had been logged (Schein and Miller 1995, 78-9).

The most common route (that is, the largest watershed) was along the Susquehanna River, so Williamsport (see its description below), Marietta, Columbia and Havre de Grace, Maryland became important towns in the trade (Stranahan 1993, 80).

Modern Forestry

Forests in Pennsylvania today are owned by a mixture of private landowners and federal, state or local governments. All of them were able to purchase large tracts of land at bargain prices in the early 1900s when they had been abandoned by their earlier owners. A combination of natural succession and replanting has brought the land back to rich full forests, though with somewhat different combinations of species. In the north-central Appalachian Plateau, especially, over 80 percent of the land in each of eight counties is still forested (see Figure 10.3 in Chapter 10).

Modern forestry began in the 1870s when the nation's first laws were passed making the setting of forest fires illegal. Conservation-minded individuals took up the cause of protesting the destructive logging practices and abandonment of clear-cut lands. Dr. Joseph Rothrock, in the 1880s and Gifford Pinchot in the 1890s were the leading voices. Pennsylvania created its Forestry Commission within the state Department of Agriculture in 1893, and the federal government did likewise in 1898. Pinchot was the first leader of the federal reforestation and conservation efforts, serving during parts of three presidential terms (Stranahan 1993, 108-109). In 1923 and 1931 he was elected Governor of Pennsylvania.

The forestry programs Pinchot created emphasized the multiple functions that forests played, and made replanting the forests a priority. In Pennsylvania, the state invested heavily in the purchase of abandoned forest lands, owning half of its current two million acres by 1913 (Stranahan 1993, 110). Between the Allegheny National Forest and the many State Forests and Game Lands, governments manage extensive areas. Federal laws require the US Forest Service and other federal agencies that own forest land to following the conservation principles Multiple Use and Sustained Yield. Multiple Use requires consideration of all of the contributions that forests make to our environment. Those uses include wood products and even minerals beneath the surface, but also include the forests' value for recreation and wilderness. Sustained Yield principles require that the agencies limit wood harvests to avoid extensive clear-cutting and to ensure that at least the same level of wood production can be maintained indefinitely (Holechek et al. 2003, 59). Within Pennsylvania a new round of State Forest management planning requires that all

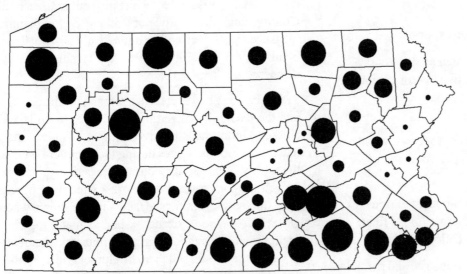

Figure 22.1 **The Economic Value of Forestry Around Pennsylvania:** Forest management and harvesting contribute nearly $670 million to the Pennsylvania economy. Forestry is still a major part of the state's economy, but even more importantly it is a significant part of the economy of many of our more rural counties. (Data source: Jacobson and Seyler 2004)

Annual Economic Contribution of Forestry Sector

- $500,000 - $3,000,000
- $3,100,000 - $7,500,000
- $7,600,000 - $15,000,000
- $15,100,000 - $24,000,000
- $24,100,000 - $36,900,000

134

decisions consider the functioning of the forest ecosystem, defined as: "the best understanding of ecological interactions and processes necessary to sustain the ecosystem's composition, structure, and function over the long term" (PA DCNR 2003). The definition does not preclude human economic uses of the forest's resources, but sets those uses as secondary to the long-term viability of the forests.

Forest lands are productive sources of income for Pennsylvanians in the forestry and wood products industry and for the state government. Figure 22.1 shows the distribution of money earned in the forestry sector of the economy directly (excluding products made from the wood, wood chips, and other materials that logging companies extract and sell). The map shows that some of the more productive areas are located in heavily forested parts of the state, such as McKean and Jefferson Counties, while others are closer to areas of major demand, either by virtue of large populations as in southeastern Pennsylvania, or because of large wood product manufacturing facilities (paper mills, furniture manufacturers, lumber distributors and others). For example, Delaware and Erie Counties have major paper product manufacturers.

Pennsylvania's forests have rebounded today to be a major resource. We are the leading source of hardwood lumber in the US, and also lead the nation in the export of hardwood lumber and wood products. Industries related to forestry activity are some of the state's leading manufacturing employers, especially since the decline of iron and steel. We are a leading producer of cabinetry, furniture, veneer, flooring and molding made from hardwoods, whose total production exceeds $5 billion per year (PA DA 2004). Pennsylvania is even a significant producer of Christmas trees in the US.

Williamsport

The forestry industry encompasses the logging companies that cut the trees, sawmills that cut them to the dimensions requested by customers, and even the manufacturers who make some of the wood products listed above. Greater economic power is wielded by individuals or companies who combine multiple operations as well as owning forest land. A good location is always an advantage, too. This was the case for Williamsport in the late 1800s (see Figure 22.2).

Williamsport was founded and named the county seat of Lycoming County in 1795 in order to attract larger numbers of settlers to the region (Stranahan 1993, 81). It was still small and insignificant in 1845 when James Perkins and John Leighton arrived with an idea to change Williamsport's role in the region's forest products industry. They created the first "log boom" in Pennsylvania by stringing a strong chain most of the way across the Susquehanna. The purpose of the log boom was to catch and hold logs at Williamsport for a longer portion of the year to help the profitability of their saw mills. Williamsport's location was ideal because the river widened and slowed down at that point; its capacity was huge (Stranahan 1993, 86-87).

The project ended up being a great financial success, since the entire West Branch of the river passes that point. Without the boom, the ability to catch and store logs was at the mercy of nature. With the boom, logs were secure, even during minor floods.

After eight years Perkins and his associates sold the boom operation to Peter Herdic. Herdic owned several sawmills and over 57,000 acres of forest. He was a tycoon who invested his earnings back into his community, and helped create what became known as Williamsport's Millionaires' Row on West Fourth Street (see Figure 22.2). When Williamsport had eighteen millionaires in the late 1800s, all in the wood industry, it boasted a higher percentage of millionaires in its population than any other city in the country. In addition, Herdic built churches, houses and a hotel, and owned the newspaper, trolley company, gas and water works and a bank. Williamsport had endured a lumber mill strike in 1872 and the nation had suffered a stock market crash in 1873. Late in the 1870s Herdic declared bankruptcy. By the 1880's the pace of lumbering and Williamsport's control over the market began to decline.

While the decline of lumbering in central Pennsylvania may have meant the end of an era in Williamsport, it did not mean the end of the town. Williamsport was already the county seat of Lycoming County, hosting many county government activities. Beginning in 1939 it also became the host for a different activity when Little League Baseball was formed there. Over the first eight years the organization expanded across Pennsylvania, in 1947 it started spreading nation-wide, and a few years later

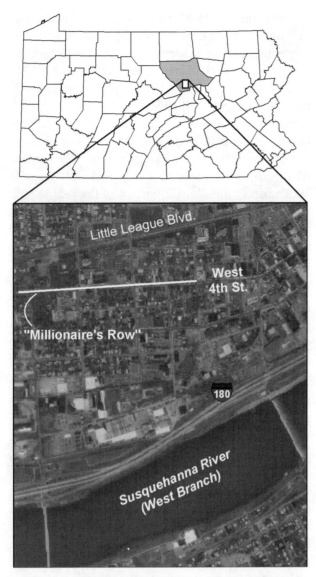

Figure 22.2 **Part of Williamsport:** The West Branch of the Susquehanna River once brought tremendous quantities of logs to Williamsport where sawmills and other wood products were the basis of a hugely productive industrial economy. (Image source: USGS 1999)

became an international phenomenon. Little League's World Series is still played every summer in Williamsport, and the organization's logo still includes a keystone as part of its design (Little League 2002).

Conclusions

Wood is not high-tech and is not infinitely customizable, as modern products based on plastic resins, ceramics and metals seem to be. Those technologies have displaced the use of wood as a production element. Pennsylvania's status in the wood industry, however, is still as one of the most productive states in the country.

Bibliography

Holechek, Jerry L., Richard A. Cole, James T. Fisher, and Raul Valdez 2003. Natural Resources: Ecology, Economics, and Policy. Upper Saddle River, NJ, Prentice-Hall.

Jacobson, Mike and Cathy Seyler 2004. Economic Contribution of Forestry to Pennsylvania. University Park, PA, The Pennsylvania State University, College of Agricultural Sciences, School of Forest Resources. Internet site: <http://rnrext.cas.psu.edu/counties/extmap.htm>, visited: 7/29/04.

Little League 2002. Little League Baseball Historical Timeline. Williamsport, PA, Little League Baseball, Inc. Internet site: <http://www.littleleague.org/history/index.htm>, visited 1/25/05.

Lycoming County 2005. Internet site <http://www.williamsport.org/visitors/mill_row.htm>, visited 1/28/05.

PA DA 2004. PA's Forest Product Industry. Harrisburg, PA, Pennsylvania Department of Agriculture: Hardwoods Development Council. Internet site: <http://www.agriculture.state.pa.us/agriculture/cwp/view.asp?a=3&q=128894>, visited 7/19/04.

PA DCNR 2003. State Forest Resource Management Plan: Executive Summary. Harrisburg, PA, Pennsylvania Department of Conservation and Natural Resources, Bureau of Forestry. Internet site: <http://www.dcnr.state.pa.us/forestry/sfrmp/execsummary.htm>, visited 1/27/05.

Schein, Richard D. and E. Willard Miller 1995. "Forest Resources." Chapter 6 in Miller, E. Willard (ed.) A Geography of Pennsylvania. pp. 74-83.

Stranahan, Susan Q. 1993. Susquehanna, River of Dreams. Baltimore, MD, The Johns Hopkins University Press.

USGS 1999. Digital Ortho Quarter Quad (aerial photograph): williamsport_pa_ne. Downloaded in TIFF format from http://www.pasda.psu.edu/access/doq99list.cgi, 9/18/04. US Geological Survey, Washington, DC.

Chapter 23

Mining

The locations of mineral resources in Pennsylvania seem to have been perfectly suited to the timing and extent of America's industrialization and economic success in the 1800s. The same minerals are still an important part of our state economy, but sources elsewhere in the US and the world have eclipsed Pennsylvania's in quantity and quality. Many of the exploration and mining methods developed in Pennsylvania, though, influenced practices elsewhere.

By the 1820s Philadelphia already had a textiles industry, settlement had already moved into every part of the state, and forests were both obstacle and resource. By the 1860s, the potential for both anthracite and bituminous coal was well established. What remained were two needs: for markets with the technology best suited to these resources, and for the means to move the resources to those markets. Along the way, mining technologies and practices were also refined, including the different approaches of underground and surface mining.

Later in the 1800s oil extraction faced the same challenges. It was a mineral whose production could succeed only when technologies best suited to its unique qualities were developed, and when appropriate means of transportation came into operation. As with coal, the early oil extraction industry in Pennsylvania learned many lessons that were applied later in other parts of the country.

The Bigger Picture

Even though oil and coal will be the focus of this chapter, many other minerals have been and are part of Pennsylvania's mining heritage (see Chapter 4). The first important one was iron. Though no longer significant in Pennsylvania, the widespread and diverse iron deposits led to the specialization and

dominance in iron and steel production in the late 1800s and early 1900's. Iron extraction required nearby sources of energy in order to smelt (remove impurities by heating) the ore. Wood and coal, also widespread in Pennsylvania, have both served that purpose. Iron was mined and processed here in significant quantities as recently as the early 1970s in Lebanon County and middle 1980s in Berks County.

Limestone and dolomite, principally, along with other minerals are quarried (surface mined) and crushed for use in road paving and cement making. This activity is very market oriented, mostly occurring in southeastern counties, though the glaciated areas of the state have great natural sources. Other forms of stone construction are based on Pennsylvania slate and dimension stones such as granite, marble and sandstone, again retrieved in quarrying operations.

Specialty minerals still in demand include sand, clay and lime. The sand is mined as industrial sand and used in glass manufacturing, once an important industry in western Pennsylvania. The clay is used in making bricks, important in steel making to line blast furnaces, as well as in building houses, although both applications are greatly reduced from the levels demanded even fifty years ago. Lime production has held its own, despite a shift in its market from steel making to water purification, pulp and paper, and agricultural fertilizer.

Pennsylvania Coal Productivity

In addition to their differences in location and mineral qualities, bituminous and anthracite coals have very significant differences in history and patterns of use (see Figures 23.1 and 23.2). Both were used as early as the second half of the 1700s, mostly for domestic or small commercial uses, such as blacksmiths. Both coals came into extensive production as a top energy source well into the 1800s. The needs for coal were as a boiler fuel in locomotives and as intense heat for iron smelting furnaces and for iron working in forges and mills.

Anthracite mining took off first, because it was closer to the industrial activity in Philadelphia, which needed to be less tied to the Delaware and Schuylkill Rivers. Boilers and furnaces were soon modified to withstand the higher temperatures than those produced by burning wood in factories and homes. Anthracite production centered in the southern anthracite field, especially in Schuylkill County, increased significantly by 1840 when a larger market for iron smelting developed. This established a demand for the fuel.

Coal extraction had to be made efficient. Europeans had been mining coal, in England and Ireland, Germany and Poland. Both the mining companies and immigrants brought that knowledge to Pennsylvania. Early mining in eastern Pennsylvania of anthracite coal was mostly by underground methods. Because the layers of bedrock are more contorted in the Ridge and Valley landform region (see Chapter 4), and all of the anthracite is in that region, there are difficult mining conditions. Today, more than ninety percent of anthracite coal is surface mined (PA DEP 2001).

Bituminous demand grew in the 1850s, when railroads reached the Appalachian Plateau, but really took off in the late 1870s with rapid increases in coke production (see Figure 23.2) as the center for iron and steel production began its westward shift. In some areas, and for a while, some steel was made using anthracite in the smelters and coke in the mills. By the 1920s coke production shifted from a mine-oriented location to a mill process, so that byproduct gases could be used elsewhere in the mill. Bituminous coal mine production reached its peak in 1918. Its use then slowed due to such factors as increased steel mill efficiency, increased competition from oil and natural gas, the development of new coal mines elsewhere in Appalachia and in the Midwestern US, and unionization of Pennsylvania miners earlier than miners elsewhere in the US.

Both anthracite and bituminous coals have also seen a great deal of use in Pennsylvania as a fuel for generating electricity. This use began in the 1870s (see Chapter 26). Some of our largest electric power stations in the state today burn coal. The coal consumption is large enough that the power companies tend to locate the generating stations near the mines.

The Mining Process

The mining process has changed almost as much as the demand for Pennsylvania's minerals. Originally most minerals were extracted and used where they could be easily found on the surface. Such deposits led to miners following the seams underground, and many early minerals were mined in labor intensive underground mines. Later, especially in the 1900s, after larger and more powerful machinery was developed, surface mining became the preferred method. In the future, however, underground mining will be the only way to get to most of our mineral wealth.

In an underground mine, a vertical shaft is dug down to the desirable seam of coal or other mineral. In some areas where the seam is exposed along the

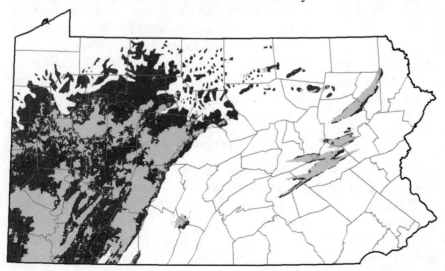

Figure 23.1 **Coal Mined Areas of Pennsylvania:** The darker and lighter gray shading is the total area underlain by coal. The areas in the lighter gray are those in which coal mining has occurred. Figure 4.1 in Chapter 4 shows the bituminous versus anthracite areas. Of the 72.8 million tons of bituminous coal mined in 2000, nearly 80 percent was extracted in underground mining operations. Of the 3.9 million tons of anthracite coal mined that same year, almost 95 percent came from surface mines (PA DEP 2001, 28-29, 97-100).

Historical Coal Production

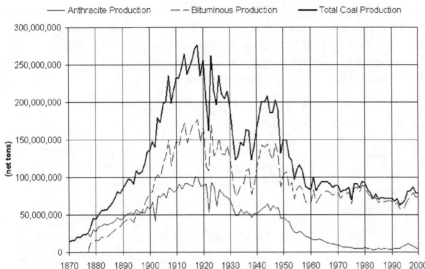

Figure 23.2 **History of Coal Production in Pennsylvania:** Coal production declined and increased during the 1900s for a mix of reasons: it declined during the Depression as the whole economy declined, increased during World War II as a domestic substitute for the oil that was so badly needed by military forces overseas, declined during the 1950s and '60s largely due to renewed competition from easier-to-use (though not necessarily cheaper) alternatives, increased during the mid to late 1970s when oil supplies were threatened, and declined in the 1980s as oil regained dominance. (Data source: PA DEP 2001, 2-4, 51-53)

side of a river valley, the shaft is not needed. The seam is then mined within the limits of the property for which the mining company has the mineral rights. In the bituminous coal region, nearly horizontal tunnels are possible from the base of the shaft. In the anthracite regions, the tunnels are often steeply inclined, making for difficult and hazardous working conditions.

The underground mining process started as a labor-intensive sequence of drilling into the seam face, setting explosive charges in these holes, detonating the charges, and then removing the coal. By the middle of the 1900s, due in part to the cost of mine labor, mining machines were developed that combined the breaking away and hauling away of the coal.

The coal removed from the mine is represented by the tunnels into the coal seam. In the early days of mining coal, using the Room and Pillar method, as much as fifty percent of the coal in that seam had to be left as pillars to support the mine roof (see the foreground portion of Figure 23.3).

Leaving the massive pillars of coal in the mine created a great financial temptation. By the middle of the 1900s, longwall mining was developed to remove all of the coal. In a longwall mine, the roof is intentionally collapsing as the mining proceeds (see Figure 23.3 back portion). While longwall mining is an effective solution in the mine, the danger is shifted to the surface. The underground collapse creates the

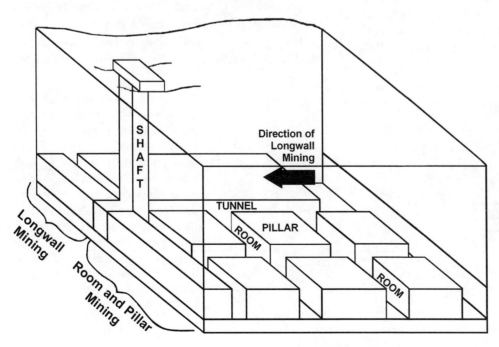

Figure 23.3 **Comparison of Room and Pillar Mining with Longwall Mining:** In Room and Pillar mining (foreground) timbers and roof bolts support the tunnel roofs. In spite of these efforts, cave-ins occasionally occurred, and could be disastrous. In longwall mining (back portion) specialized machinery starts in a tunnel perpendicular to the far end of the main tunnel, and removes all of the coal in the mine as it makes its way back toward the main shaft.

same kinds of hazard we saw in Chapter 4 with sinkhole development in areas underlain by limestone or dolomite. When a mining company owns the mineral rights for a property, and longwall mining leads to surface subsidence, the laws favor the mining company. The land owners are entitled to have their repairs paid for, but not other incidental expenses.

A surface mining operation follows a very different process, but once again results in all of the coal in that seam being extracted. An environmentally responsible surface mine (required by law today) starts by removing and storing the soil layer separately, primarily using bulldozers. Then, the non-valuable layers of bedrock (the overburden) lying over the coal seam are peeled away with repeated steps of blasting, and scooping by tremendous steam shovels, bulldozers and dump trucks. These layers are also set aside. When the coal seam is reached, it is removed for further processing using the same equipment. Once the coal is gone, the overburden and then the soil are replaced so that the area can be re-landscaped. Before such environmentally sensitive practices became required during the 1970s, most mining companies simply walked away from the mine after removing the coal seam, leaving earthen scars that persist to this day.

Not all of the coal removed is of the same quality, especially in anthracite mines. Sorting of the coal was required, and piles of the less valuable mined material were built up on the surface. In addition, there is some processing of the mined coal: a washing process removed explosive gases and dust. Some coal is mechanically broken or sorted into commercially preferred sizes. Each of these steps adds to the cost of production. They also added to its environmental impact. Given the market, safety and environmental requirements, the decision to mine a particular seam of any mineral using either the underground or surface mining process will be made only if all of the costs are less than the coal is worth.

Oil and Gas Productivity

Oil, or petroleum, is such a versatile resource because it can be made into a variety of products, from fuels to plastics to pharmaceuticals and pesticides. When it was first discovered in Venango County south of Titusville in 1859 (see Chapter 4), it was only as an illumination fuel (for oil lamps), and perhaps some "patent medicines." Some investors hired Drake, who was not a geologist or engineer (or a colonel, either), because he was adventurous and "tenacious" and had a free pass for the railroad where he sometimes worked as a train conductor (Yergin 1991, 26). They sent him to northwestern Pennsylvania (where their sample had come from) in 1857 to discover whether oil could be obtained more cheaply than by scooping it off the ground at seeps. Locals had already been boring holes into the ground and finding rock salt, so Drake found one willing to adapt the technology. After nearly two years, days before he received a letter telling him to give up, he did strike oil (Yergin 1991, 27).

Drake's well was very shallow, but proved the technical feasibility of the venture. What followed was a mad rush, similar to the California gold rush of 1849. Titusville, a logging town, was joined by such places as Oil City and Pithole as centers for oil production. Oil seekers bought tiny parcels of land sold by speculators, and built derricks so close together that once-forested valley slopes looked re-forested, but

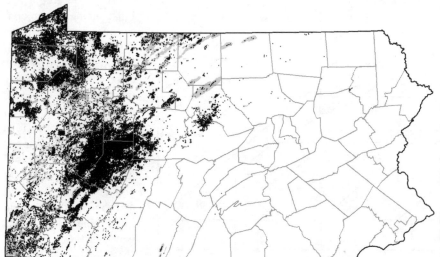

Figure 23.4 **Oil and Gas Well Locations:** All of the gray and black areas are where oil and gas deposits are known to exist. The black on this map are actually small dots, each representing a different well. Almost all of the known areas have been developed, and now the focus is on developing deeper sources. The oil and gas companies are also attempting to decide whether to explore in the Ridge and Valley landform region.

were mostly mud-baths. That was Pennsylvania's role in the oil industry for the first 40 years of its existence: proving ideas so that they could become reality.

Oil is still produced in Pennsylvania, though on a much smaller scale. A typical well today is relatively shallow (500 to a few thousand feet deep) and has a lifetime of about ten years. During its lifetime, the well will produce about 3000 barrels of oil, over half of which will come in the first few years (Harper, Tatlock and Wolfe 1999, 487).

Oil is sold by the "barrel" because that is how the production from the first wells was able to be captured and transported. Later, large oil tanks were built (initially of wood), oil tank cars were developed for train transportation, and pipelines were conceived and built. All of these happened first in Pennsylvania. Despite the early technological changes, the barrel as a unit of measure has never been replaced.

Prices and quantities of oil availability experienced a great deal of volatility in the early years. The oil rush attracted so many wells that production exceeded the market's demand, and many were losing money on their investment. After production the oil must be refined into its various products. John D. Rockefeller aggressively organized the industry by buying out the small refiners and distributors, and making them part of his Standard Oil Company, but he did not buy into the exploration and production end of the business. His influence was powerful until oil was discovered elsewhere in the country.

Initially, natural gas was nothing more than a nuisance and hazard around oil wells, although the gas flows more readily through porous rock and frequently accumulates in different bedrock strata. Natural gas became a valuable commodity in its own right in the late 1800s (Harper, Tatlock and Wolfe 1999, 486). Throughout western Pennsylvania, production companies drilled over 1500 natural gas wells in 2001, averaging about 4000 feet in depth. There were nearly 35,000 gas wells in production during 2001, more than twice the number of producing oil wells (IPAA 2004).

The ideal oil or gas well taps into a large underground reservoir, with few competing wells decreasing the natural pressure that exists there. The type of oil and/or gas produced (depending on its chemical properties) determines who has an interest in buying it. Thus, Pennsylvania crude oil is not as well suited for producing gasoline as others, but is ideal for lubricating oils.

Centralia's Fire

Coal mining and oil extraction have contributed greatly to Pennsylvania's economy over the last approximately 150 years. They helped establish our reputation as an industrial and transportation power earl on, and it still powers industrial and electricity production. On the other hand coal mining and coal use have created some of our most challenging and expensive environmental problems, as we shall see in Chapters 35 and 36.

Coal mining's impact on the economies and environments of small towns is also tremendous. Many towns were founded or at least stimulated by the opening of a nearby coal mine, and as mining has declined, those same towns have suffered economically. People in most of those towns learned to accept the occupational risks and environmental hazards of the mines. Centralia, in Columbia County has suffered from an unusual impact, however.

Centralia was founded as Centerville in the early 1800s, and was incorporated as Centralia Borough in 1866. The town sits in a valley of the Ridge and Valley landform region, under which the bedrock layers form a three-dimensional "V" like the pages of a half-opened book; several of those "pages" are coal seams. In the 1850s the Schuylkill Haven Railroad Company built a railroad to the town, and by the 1880s six different anthracite mining operations were in progress. At least two of the coal seams, the Mammoth Bed and the Buck Mountain Bed, were mined, producing nearly half a million tons in 1883. Centralia's population peaked in 1890's Census at just less than 2800 (Trifonoff 2000, 5-8).

The coal outcrop of one seam had been surface mined to about 50 feet deep, and then used as a municipal waste dump. In 1962 it appears that the trash was ignited in order to rid it of nuisance pests and odors, a common enough practice. They found themselves unable to extinguish it because the coal seam had also caught fire. Smoke and fumes became intense enough that, within a few months, federal mine officials were on the scene and closed all remaining active mines (Trifonoff 2000, 10).

Over the rest of the 1960s, strategies moved from flooding and smothering, to trenching, and then to drilling nearly two thousand "boreholes" to learn the extent of the fire and monitor its intensity. Once the fire spread into mined areas of the coal seam it gained oxygen supply, the chimney-like effect of the steeply sloped mine tunnels, and the ability to jump via mine tunnels and shafts into other coal seams. By the end of the 1970s they had given up, having spent over $5,000,000. In 1983, the federal government established a $42 million fund, partially filled with state money and contributions, to relocate all 1,200 people and over 500 homes, businesses and other buildings. By 1990 all but 30 homes had been vacated and demolished; by 1999, when relocation had been made mandatory, fourteen families obstinately remained. Many of the families that moved went to a single new development in nearby Mount Carmel (Trifonoff 2000, 10-17).

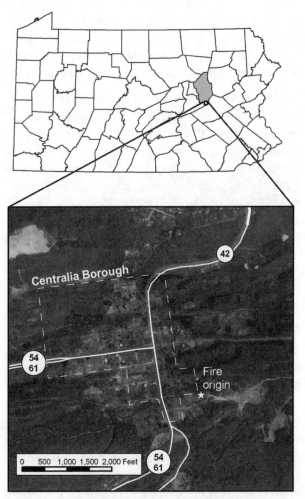

Figure 23.5 **Centralia:** Most of the burn area is in the southeast area of the borough and outside of town. The entire town was considered to be at risk, from mine subsidence as the fire burns, and also from the hazardous gases and heat. (Image source: USGS 1999)

The major road through town, PA Route 54/61 had to be relocated for a while, after parts subsided due to collapses when the mine pillars were reduced by the fire. The strategy now is simply to let the fire burn itself out (Trifonoff 2000, 20).

Conclusions

Pennsylvania coal and oil production is far below historical high points. The activities are still pursued, benefiting from the special qualities of Pennsylvania's resources and from the nearness of major eastern US markets.

Both forms of fuel production follow all of the same challenges and limitations. The resource's presence must first be surmised from surface evidence, expensive equipment must be applied to exploratory drilling, and the amount that can be extracted is influenced by the costs of production, processing and delivery.

Bibliography

Harper, John A., Derek B. Tatlock and Robert T. Wolfe, Jr. 1999. "Petroleum-Shallow Oil and Natural Gas." Chapter 38A in Shultz, Charles H. (ed.) The Geology of Pennsylvania. pp. 484-505.

IPAA 2004. Oil and Gas in Your State: Pennsylvania. Independent Petroleum Association of America, Washington, DC. Internet site: <http://www.ipaa.org/info/In Your State/default.asp?State=Pennsylvania>, visited 8/10/04.

PA DEP 2001. 2000 Annual Report on Mining Activities in the Commonwealth of Pennsylvania. Harrisburg, PA, Pennsylvania Department of Environmental Protection.

Trifonoff, Karen 2000. "The Mine Fire in Centralia, Pennsylvania." The Pennsylvania Geographer. Vol. 32, no. 2, pp. 3-24.

USGS 1999. Digital Ortho Quarter Quad (aerial photograph): ashland_pa_sw. Downloaded in TIFF format from http://www.pasda.psu.edu/access/doq99list.cgi, 9/18/04. US Geological Survey, Washington, DC.

Yergin, Daniel 1991. The Prize: The Epic Quest for Oil, Money and Power. New York, NY, Touchstone/Simon and Schuster.

Chapter 24

Early Transportation

A key part of the colonial development of America was the growth of its transportation network. Pennsylvania played a major role in that process, because of our resources and environment, and also because of our culture and location. Some of the earliest "roads" and "turnpikes" were built here during colonial times. Canals represented a major step forward in transportation, but lasted only a short while. Railroads were the main "vehicle" of late 1800s industrialization.

The greatest significance of these early transportation technologies for us today was perhaps in locations they chose for their development. Roads developed in colonial Pennsylvania have been improved many times and are still traveled today. Canals in many cases became the road beds for later railroads, and demonstrated the idea that a large engineering effort could enable the movement of goods over routes previously thought impassable. Railroads ended up being the transportation system capable of carrying an entire nation to a new future, and Pennsylvania workers played a major role. Those railroads still in use continue to hold the potential to be a primary "vehicle" of the future.

Early Highways

It is difficult for us, today, to imagine a world in which most people make most of their contacts, social or business-related, in horse-drawn wagons, by un-

Figure 24.1 **The Conestoga Wagon:** This versatile vehicle was favored by early settlers heading west. Its wheels were large and wide, enabling travel over rough terrain. Its body could double as a barge for crossing rivers. With its canvas cover in place, the Conestoga Wagon could provide a somewhat sheltered ride wihile hauling a large load. This wagon is on display at Landis Valley Farm Museum, a state museum in Lancaster County.

powered boat, or on horseback or foot. Information traveled no faster than people did. Pennsylvania, and most of the rest of the US, was still primarily rural well into the early 1800s.

Roads were the earliest modifications of nature to enable transportation, dating back to Native American times. The earliest were unpaved, but eventually stones and other means were used to improve surfaces, especially in wet weather. During colonial times, turnpikes came into use as private toll roads, an early equivalent of the modern Pennsylvania Turnpike even including the tollgates. Some continued operation into the early 1900s. By 1830 Pennsylvania had over 3000 miles of toll roads built and operated by some 200 different companies (Lapsansky 2002, 178).

One early innovation was the Conestoga Wagon (see Figure 24.1). Its was designed for hauling goods from the Conestoga River valley in Lancaster County to Philadelphia in the early 1700s, and became a major factor in the settlement of western Pennsylvania and the western US until railroads overcame its limitations (Lapsansky 2002, 167).

The roads themselves received increasing attention. Covered bridges became part of the early transportation landscape, and many have lasted until the present (see Figure 24.2). Another innovation was the use of macadam-like surfacing, first introduced on the Lancaster Turnpike out of Philadelphia, completed in 1794 (Lapsansky 2002, 178).

Canals

Canals essentially replaced rivers and streams as a mode of transportation. Their construction involved digging, lining and supplying water to a wide trench, usually parallel to a river. Where the river elevation fell, through a rapids or a falls, the canal included a system of enclosable locks. Canal boats and barges were pulled by mules or horses walking along canalside towpaths, so they were not a fast means of transportation. Even though they did not represent an increase in vehicle speed, they did improve the speed of upstream travel relative to downstream travel. More importantly, canals increased vehicle capacity relative to the effort required.

Pennsylvania's canal system began with small routes privately built in the 1790s. Major improvements came with larger ventures, 108 miles along the Schuylkill River in 1825, and along the Lehigh River in the same year. Both of these enabled the movement of goods into and out of Philadelphia (Chamberlin 2004).

When the Erie Canal was completed in New York State in 1825 and began taking transportation business away from Pennsylvania road haulers, Pennsylvania's government responded by building the Pennsylvania Canal system. In Pennsylvania construction faced the bigger challenge of surmounting the Allegheny Front. The most successful solution was a short "portage railroad," completed by 1834.

Figure 24.2 **A Covered Bridge:** Many early bridges were covered in order to protect the bridge surface and trusses (supports). It was a lot cheaper to cover the bridge than to tear down and rebuild one later. This made special sense in Pennsylvania due to the abundant rainfall. The world's longest covered bridge in the early 1800s spanned the Susquehanna River between Columbia and Wrightsville. It was two and a half lanes wide and two decks high, including a tow path for pulling barges across the wide river (Klein 1964, 92).

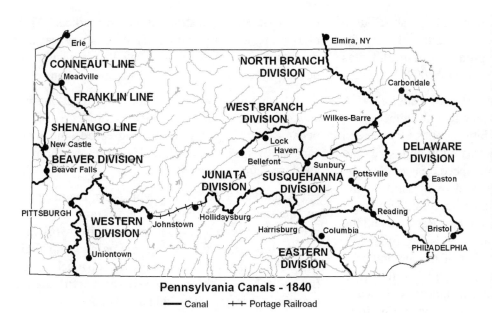

Pennsylvania Canals - 1840

— Canal   +++ Portage Railroad

Figure 24.3 **The Pennsylvania Canal System in 1840:** Canal barges improved the speed and reliability of hauling freight across Pennsylvania. The most heavily used route was the one from Philadelphia via Reading and Harrisburg to Hollidaysburg at the Allegheny Front, from there transferring the barges to the Portage Railroad for carrying up the steep climb and gentle incline to Johnstown, and then back into the canal for the final run to Pittsburgh. (Map after: PA Canal Society 2004)

The Pennsylvania Canal was organized into eleven "Divisions," not including the private canals such as the two earlier ventures on the Schuylkill and Lehigh Rivers. The divisions are shown in the map in Figure 24.3. The canal system did improve travel time between its stops, but was not a perfect solution. Many areas were still inaccessible and the canals, like the rivers and lakes, were subject to freezing over in the winter and to flooding during other seasons (Lapsansky 2002, 181-3).

By the time the canal era came to an end, Pennsylvania had had as many as 1356 miles of canals, over one third of the total for the entire US (Pennsylvania Canal Society 2004). Unfortunately for the Pennsylvania Canal system, within a few years after the completion of its main line in 1834, a number of railroad companies emerged. By the 1850s most parts had been abandoned, and in 1857 even the main line to Pittsburgh had been sold.

Several canal sections continued to operate for many years. The Schuylkill Navigation Company operated until 1875. A small piece of the original state canal at Columbia on the Susquehanna River in Lancaster County was sold, and became a successful transfer station between river goods and the railroad until 1901. The Lehigh River's canal company bought the section of the Pennsylvania Canal along the Delaware River, giving it a connection to Philadelphia which it used for anthracite coal shipments until 1932 (Chamberlin 2004).

More importantly, the canal era created several new ways of thinking in Pennsylvania. It set the state government more seriously to the task of improving transportation around the state, for the good of all of its citizens and businesses. It also created many small industrial enterprises, not just those served *by* the canals, but also those helping to create everything from locks to boats. Pennsylvania gave birth to a transportation construction industry.

Early Railroads

In 1837 a combined canal-and-railroad trip from Philadelphia to Pittsburgh three and a half days (Lapsansky 2002, 183). The early railroads required engineering on the same scale as the canals, and overcame one of their major limitations: the need for large volumes of water. Recall that the reservoir that caused the Johnstown Flood in 1889 was originally intended to supply a canal (see Chapter 7). These earlier trains were no faster than canals and probably less reliable mechanically. They were nevertheless a major step forward, and by the 1850s improved greatly in power and speed.

The emergence of the railroads coincided with the transition of fuel source from wood to coal, and the increasing productivity of factory production of goods, especially textiles out of Philadelphia in those early years. Technology improved to a level at which steel rails were required to support the weight, speed and frequency of use, especially of the main lines. The number of cars that could be pulled also increased.

By 1852, with the completion of the Horseshoe Curve near Altoona (see Chapter 2), the travel time for passengers from Philadelphia to Pittsburgh was reduced to fifteen hours (Lapsansky 2002, 185).

Fierce competition between the railroads followed. Key cargoes were heavily sought, and lumber, coal, oil, iron and steel were the biggest ones. By the end of the 1800s engines had increased in both size (and therefore pulling power) and speed. Success in the freight competition game, and ownership of the Baldwin and Eddystone engine building facilities in Philadelphia, gave the Pennsylvania Railroad and the Reading Railroad early upper hands. Another key development was the invention of the air brake by George Westinghouse of Pittsburgh, which gave greater control over stopping distances with heavy loads. Once again, Pennsylvania was benefiting from both the transportation improvements and the additional industrial opportunities involved in building the system.

The Railroads' Heyday

The railroads expanded immensely during the rest of the 1800s and into the early 1900s. A peak total of over 12,000 miles of track crossed Pennsylvania in the 1920s. As early as 1865, but especially by 1920, the Pennsylvania Railroad was one of the largest, most profitable corporations in the world (Licht 2002, 232). Railroads were almost everywhere in the state (see Figure 24.4). At least fifteen other railroads also had significant connecting operations in the state. They were an integral part of the US transportation system, because Pennsylvania is a key crossroads state for the major east coast industrial cities and markets. They also were a significant part of Pennsylvania's manufacturing employment, a role that Pennsylvania never played in the later automobile and airplane eras.

The engines powering the trains were coal-burning steam generators on wheels (see Figure 24.5). They became ever larger as time progressed, and were capable of speeds exceeding 100 miles per hour in level, straight sections of track.

After the 1930s' Great Depression, the railroads and the steel and coal industries never recovered their previous dominance. Competition from oil and automobiles, and the burden of maintaining their expensive equipment and tracks were too much. Even the switch to lighter and still-powerful diesel locomotives in the 1940s and 1950s could not make them cost-efficient.

Pennsylvania's Three Railroad Museums

Railroads were such an important part of Pennsylvania's (and America's) rise to industrial prominence in the late 1800s and early 1900s that it makes sense that many want to preserve reminders of that past. Pennsylvania holds many places of distinction in railroad history. We have already seen the importance of Altoona and the Horseshoe Curve (Chapter 2), which is now a and railroads in lumbering (Chapter 10). Without railroads, Pennsylvania may never have dominated so strongly (if temporarily) in the markets for bituminous and anthracite coal production, for steel making in Pittsburgh and Bethlehem, for oil production, and even for railroading itself.

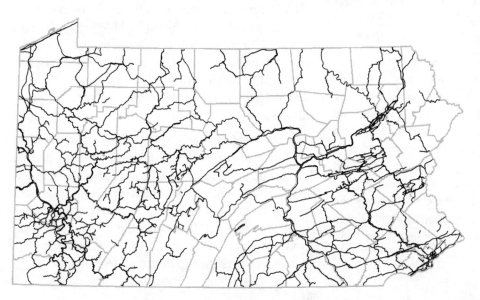

Figure 24.4 **Pennsylvania's Many Railroads in the 1920s:** At or near its peak in 1920 was both the number of railroads and the amount of use of those tracks and equipment. Although both have declined in magnitude today (see Figure 24.5), railroads still form an important part of the state's and country's freight transportation system.

Figure 24.5 **A Coal-Burning Steam Engine:** This steam engine still pulls trains almost daily at the Strasburg Railroad in Lancaster County. This tourist railroad, adjacent to the Railroad Museum of Pennsylvania, features vintage engines and passenger cars. Pennsylvania has many such excursion railroads, as well as many museums, historical societies and other organizations, and commercial establishments who own and display historic railroad vehicles and equipment.

When the question of how to commemorate the railroad industry's past arose, it was easy enough to decide to collect examples of many different kinds of railroad equipment, even the train locomotives and cars themselves. At least three places across Pennsylvania are making names for themselves among railroad buffs: Strasburg, Scranton and Altoona (see Figure 24.6). The first two are state- and federal-government sponsored, respectively, and the third is a locally organized entrepreneurial venture.

The site with the longest railroad history is Altoona's "Railroaders Memorial Museum." The Pennsylvania Railroad's Altoona railroad construction and repair shops were the world's largest in 1945. This museum grew from concept in 1966 to its first opening in 1980. By the middle 1980s it had acquired one of the former Pennsylvania Railroad's premier locomotives and restored it for excursion runs. In the early 1990s it acquired one of the former Pennsylvania Railroad shop buildings, and moved the entire museum to its new site in 1998.

Altoona benefits from the fact that the Horseshoe Curve has continued in use, even when the Pennsylvania Railroad became part of the Penn Central Railroad, then Amtrak and Conrail, and most recently Norfolk Southern and CSX. The Curve has also been

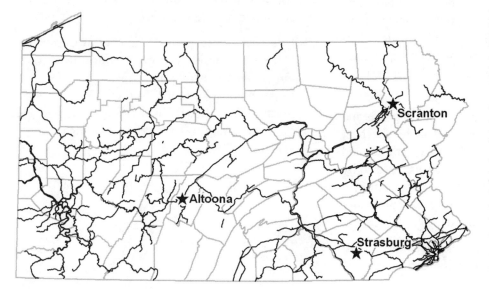

Figure 24.6 **Railroad Museums in Pennsylvania:** The "Railroaders Memorial Museum" in Altoona, "the Steamtown National Historic Site" in Scranton, and the "Railroad Museum of Pennsylvania" in Strasburg are three places to see railroad equipment up close. They are joined by many other railroad-oriented societies, and even a trolley museum in Washington, Pennsylvania. This map also shows the state's actively-used rail lines as of 1996.

named a National Historic Landmark. Nearby also are the Allegheny Portage National Historic Site and America's first railroad tunnel (RHC 2005).

Altoona was edged out by Strasburg Township for hosting the Railroad Museum of Pennsylvania. Part of the reason was the presence, literally across the street, of the privately owned, steam powered Strasburg Railroad, a significant tourist attraction in a larger prime tourist area. Because the tourist railroad provides visitors a short excursion trip, the museum can focus on its mission of collecting and restoring railroad equipment. The Museum's collection of over 100 train cars and locomotives includes some from the earliest years of railroading, as well as diesel locomotives from the middle 1900s (RRM PA 2005).

Scranton's entry in this trio was established as a federal government venture: Steamtown National Historic Site. Like Altoona, Steamtown's location in downtown Scranton and its mission most reflect the city's early history serving the mining, manufacturing and passenger railroad activities. This museum also offers a steam engine pulling an excursion train through the local countryside. Its regional location also is a benefit, as it draws most of its visitors from New York and New Jersey, as well as Pennsylvania (US NPS 2003).

Conclusions

Early transportation was more a set of solutions than a system. Settlers and their needs moved west, and some of their products made the return trips to the east during the Colonial and Early American periods. Canals and railroads forced a larger perspective because they were expensive to build and operate. Similarly, industry was smaller scale and its products were more often meant for more local markets until mass production and large-scale transportation systems were even more greatly needed after the Civil War.

Meanwhile, for the local movement of goods and people, animal power still proved to be the best alternative. As the Amish still do today, families and goods produced locally for local markets traveled by horse-drawn wagon. Public roads, mostly unimproved, became increasingly common and competed with the private turnpikes.

The industrial revolution changed all of that. Canals, and then railroads, had industrial America operating on huge scales. Population growth, especially due to immigration, corresponded. But, in the 1930s the big picture changed.

Bibliography

Chamberlin, Clint 2004. Canals in Pennsylvania. Northeast Rails. Internet site: <http://www.northeast.railfan.net/canal.html>, visited 11/7/04.

Klein, Frederic Shriver 1964. Old Lancaster: Historic Pennsylvania Community from its Beginnings to 1865. Lancaster, PA, Early America Series, Inc.

Lapsansky 2002. "Building Democratic Communities: 1800-1850." Chapter 4 in Miller and Pencak (eds.) Pennsylvania: A History of the Commonwealth. pp. 151-202.

Licht, Walter 2002. "Civil Wars: 1850-1900." Chapter 5 in Miller and Pencak (eds.) Pennsylvania: A History of the Commonwealth. pp. 203-256.

PA Canal Society 2004. Pennsylvania's Canal Era 1792-1931. Easton, PA, Pennsylvania Canal Society. Internet site: <http://www.pa-canal-society.org/>, visited 7/25/04.

RHC 2005. About the Railroaders Memorial Museum. Altoona, PA, Railroaders Heritage Corporation. Internet site: <http://www.railroadcity.com/museum.htm>, visited 2/8/05.

RRM PA 2005. About Us. Strasburg, PA, Railroad Museum of Pennsylvania. Internet site: <http://www.rrmuseumpa.org/about/welcome/aboutus.htm>, visited 2/8/05.

US NPS 2003. Steamtown/Chamber of Commerce Release Annual Report: Economic Impact Favorable Upon Region. Washington, DC, US Department of Interior, National Park Service, Steamtown National Historic Site. Internet site: <http://www.nps.gov/stea/economicimpact.htm>, visited 2/8/05.

Chapter 25

Modern Transportation

Modern transportation is identified most strongly with the automobile and the airplane. Both were developed in the late 1800s, with the airplane's success delayed by the greater scientific obstacles it faced. Both used the same basic technology, the internal combustion engine, and similar fuels refined from petroleum.

However, what really captures the importance of these two modes of transportation as agents of change in society is the length of time it took them to become part of everyday life. The automobile was developed in Europe and America in the 1880s, was still being only reluctantly acknowledged in the first decade after 1900, and did not become widely demanded until it became affordable in the 1910s. Even then it took another decade before programs to pave rural roads began to make it practical for driving longer distances. Airplanes didn't begin taking passengers until after they finally had commercial success as part of the US mail delivery system in the 1920s. However, it was not until around the time of World War II that regularly scheduled commercial flights began.

Once the automobile and the airplane caught on, both changed American culture, and Pennsylvania's economy, profoundly. The changes have been mixed, however. On one hand, our adaptations have been timely, and modern transportation reflects Pennsylvania's crossroads location. On the other hand, the success of these forms of transportation dealt tremendous blows to the railroad industry, and coincided with the shift of much manufacturing away from Pennsylvania. They also represent a significant economic challenge as we attempt to pay for their upkeep and improvement.

Pennsylvania's Road System

Road building in Pennsylvania faced different challenges in different parts of the state. In the east, so much of the land was already owned and settled that most paved roads simply took the place of the previously unpaved "unimproved" (dirt) and "improved" (gravel) roads. In the Ridge and Valley region transportation was easy as long as it followed the long narrow valleys, but crossing the ridges was a challenge for many cars. On the Appalachian Plateau, the upper plateau areas were easy to pave and drive, as were the flood plains of the river valleys where most of the towns and cities are located. The difficult routes were the ones that climbed out of the valleys, creating many very steep slopes, and that crossed the valleys from one upland area to the next. The solutions were longer routes to climb the valley slopes more gradually (recall the railroad version of this solution, the Horseshoe Curve, in Chapter 2), and many bridges to cross the valleys. Today we have one of the highest ratios of miles of highway per square miles of land, and of bridges per square mile of land.

Along with the rest of the country, the first paved roads were usually local projects. Then, in 1916, the federal government passed legislation to create the US Highway system. As it was built and modified over the years, it became a series of US Routes with two-digit route numbers for major roads and three-digit route numbers for a second tier of roads. The two-digit route numbers began with lower even numbers

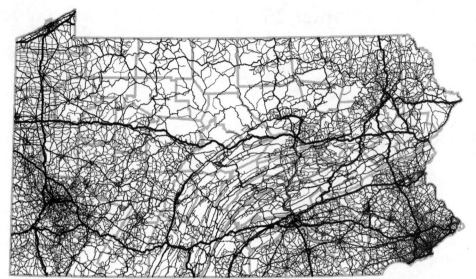

Figure 25.1 **Highways and Other Major Roads of Pennsylvania:** This map shows the major connecting roads of Pennsylvania. Heavier lines represent the Interstate routes, medium lines show US routes, and the lightest lines show other roads under state management. Not shown are roads built and maintained by municipalities or as parts of housing, business or other developments.

for east-west roads in the northern US, and larger route numbers further south. Within Pennsylvania these include US Route 6 across the northern part of the state near the New York border, US Route 22, and US Route 30 across the southern part of the state near the Maryland border. The two-digit route numbers began with lower odd numbers for north-south roads in the eastern US, and larger route numbers further west. Within Pennsylvania these include US Route 1 in the eastern part of the state connecting Philadelphia to New York City and Washington, D.C., the unusual crossing of US Routes 11 and 15 around Harrisburg, and US Route 19 near the Ohio and West Virginia borders. The strategy of these routes was to connect most cities at their most central crossroads; in later years, though, many of the more heavily used US Routes were shifted away from these downtown alignments when bypass routes were built and given the original US Route number.

In 1944 the federal government decided to proceed with a new generation of highways, called the Interstate Highway System. The interstate highways can trace their roots in large part to the Pennsylvania Turnpike. The Turnpike, in turn, can trace its roots to a failed attempt to build a railroad in the late 1800s. The owners of the New York Central Railroad created a subsidiary called the South Penn Railroad and attempted to build a line to compete with the Pennsylvania Railroad; the move was in retaliation for the Pennsylvania Railroad's construction of a line parallel to the New York Central's up the Hudson River valley. The South Penn Railroad had completed much

of the roadbed, including several tunnels, when the two companies called a truce and mutually abandoned their projects in 1885. The work sat idle for more than 50 years, when construction of the Turnpike began (Lewis 1997, 57). Even though the original route still needed work to be adapted for automobile requirements and tunnels and other areas had suffered significantly, workers were able to complete the 160 miles of new road in less than three years, opening in 1940 (Lewis 1997, 67).

The original route ran from Middlesex, 15 miles west of Harrisburg, to Irwin, about 18 miles east of Pittsburgh (current exits 226 to 67). It was successful enough that by 1950 it was extended all the way to Philadelphia, and by 1957 the Northeast Extension was added from Norristown to Scranton. The Pennsylvania Turnpike was the first large-scale application of the limited access design, with no road-level intersections. That meant that every road perpendicular to the route (even exits) required a bridge crossing: 307 of them on the original route. Almost every aspect of the Turnpike was an innovation at this large scale: the tunnel ventilation systems, the tollbooths, the service plazas, the roadbed design, and especially the level of organization and coordination to accomplish it all so quickly (Lewis 1997, 59-67).

The success of the Turnpike encouraged the federal government to adopt the legislation creating the Interstate Highway system. It was constructed mostly during the 1960s, but new routes are still being completed (as I-99 in Pennsylvania demonstrates). In several cases roads originally constructed by the state became parts of Interstate highways, including the Pennsylvania Turnpike, which became Interstate 76. The numbering pattern for the Interstates was similar to that for the US Routes, with two-digit route numbers for major roads and three-digit route numbers for connecting highways. The two-digit route numbers began with higher even numbers for east-west roads in the northern US, and lower route numbers further south. Within Pennsylvania these include Interstate 90 along the Lake Erie coastal plane, Interstates 80, 78, 76, and 70 in southwestern Pennsylvania near the Maryland border. The two-digit route numbers range from higher odd numbers for north-south roads in the eastern US, toward lower route numbers further west. Within Pennsylvania these include Interstate 95 along the Atlantic Coastal Plain and through downtown Philadelphia in the eastern part of the state Interstates 83 and 81 through Harrisburg and I-79 connecting Erie, Pittsburgh and points south. With few exceptions, the Interstate Highway system was designed to connect to the outskirts of the major cities, just as the building of bypasses was designed to bring higher-speed and larger-volume traffic to the outskirts of the city and let local roads do the rest.

The highways had major impacts wherever they were, and were not, built. First and foremost, they attracted the most mobile, automobile-owning segments of the population. In the early years (1900s and 1910s), cars were a nuisance to pedestrians and horses in the towns and cities. By the time the US Highway system was built, the increased accessibility made city centers into magnets for mobile activity and shopping-related investment.

By the 1950s, downtowns were getting clogged with cars because there were too few places to put them (see Schuyler 2002, 37, for example). "Bypasses" were the order of the day, and they ended up back-firing on the cities by pulling development and economic investment (not to mention

population) out into the newly forming suburbs. The building of Interstate highways only exaggerated the problems; their ramp areas became the new magnets for truck- and car-oriented development.

As it now stands, Pennsylvania has major challenges with its road system. Table 25.1 provides several statistical measures of transportation accessibility here.

Air Travel

Airplanes came into their own as part of the larger transportation system in the 1920s. They primarily delivered high-priced goods, especially ones such as the mail for which timing was critical. However, reliability was a major problem. By the late 1930s improvements in the planes' construction, durability (through more common use of aluminum), and

Table 25.1 Pennsylvania Transportation Statistics (US DOT 2003)

	Quantity	Ratio or Density	US Rank
Public roads	120,298 mi.	. . .	6
"	"	2.61 per sq. mi.	6
Interstate Highways	1,757 mi.	1.46 per 100 mi. of roads	19
"	"	3.82 mi. per 100 sq. mi. area	6
Road Bridges	21,873	18.2 per 100 mi. of roads	18
"	"	47.5 per 100 sq. mi. area	9
Deficient/Obsolete Road Bridges	9,407	42% of total	5
Railroad Trackage (Class 1 lines)	3651 mi.	. . .	8
"	"	7.9 mi. per 100 sq. mi. area	8
Cars, Motorcycles, Buses & Trucks	9,746,000	. . .	7
"	"	81 per mi. of public roads	14
Airports (total public & private)	467	. . .	8
Airports Certificated for air carriers	17		
Enplanements	22,560,674		7

151

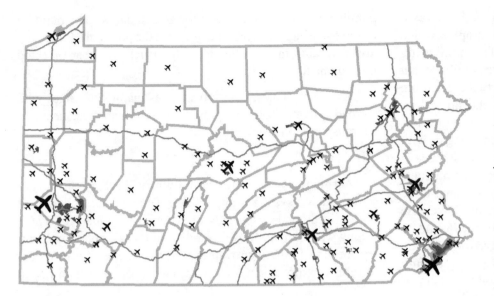

Figure 25.2 **Pennsylvania's Airports:** In this map the sizes of the airplane symbols correspond to the relative sizes of the airports. Like the larger cities they generally serve, Pennsylvania's larger airports are widely spaced. Larger airports serve larger jets, which require longer runways, and many more passengers, which require greater transportation access and parking and other facilities.

navigation systems, but not comfort, were accomplished. They started taking occasional passengers along with their cargo.

After World War II, the first commercial airline operations started. By the 1950s most cities were building ever larger airports with paved runways, control towers and other facilities in order to attract air traffic. The introduction of jet engines and "jumbo" sized planes meant larger passenger capacities and lower fares, but also required longer runways. In some cases the existing airports could be expanded, but for Philadelphia and Pittsburgh entirely new airports were built to meet the large increase in volume of freight and passenger traffic.

The airlines changed significantly, as well. From many small companies emerged a smaller field of much larger companies by the 1970s. For example, "All American Aviation," born in the 1930s as a mail delivery system serving western Pennsylvania, became "Allegheny Airlines" in 1953. Allegheny acquired several other companies and became "US Air" in 1979, and then US Airways in 1996 (PA HMC 2004; US Airways 2004a).

Airline deregulation, intended to eliminate extensive government regulation of air routes and timetables, also changed the ways airlines operated by allowing them to move to a "hub-oriented" model. In this system most of an airline's flights are channeled through its primary airports, some of which also serve as hubs for other airlines. The hub-based system allowed passengers greater numbers of connecting

flights, and reduced maintenance and other operating costs for the airlines. For US Airways both Pittsburgh, the airline's original home base and its window on the nation's Midwest, and Philadelphia, its center for east coast and international flights, serve as primary hubs within Pennsylvania (Getis et al. 2001). A recent study for US Airways found that its operations at its Pittsburgh hub contributes over $3 billion into the region's economy and directly and indirectly supports over 33,000 jobs (US Airways 2004b).

Breezewood

Breezewood is a prime, perhaps extreme, example of highway-oriented development. The town is very small and some distance from the intersection of Interstates 76 and 70 (the two coincide from New Stanton to Breezewood) and US 30. The activity started in the early 1940s after the completion of the Turnpike exit onto US 30. In the early 1960s the Interstate 70 interchange was added, though the entire length of Interstate 70 was not completed for several more years. The odd thing about this interconnection of the two Interstates is that they do not connect directly. Instead, traffic traveling east to Baltimore exits Interstate 70/76 and travels a short distance on US 30 with all of its development and two traffic lights, before continuing on Interstate 70.

Development grew along US 30 with the popularity of such Pop Culture amenities as fast-food restaurants (Breezewood has fifteen), gasoline stations (there are nine) and motels (there are six). All of this development is within a short stretch of less than a mile. More recently, plans for reconstruction of the Turnpike were supposed to solve the disconnection. A lawsuit prevented a significant realignment, because it would have severely altered the local economy.

Conclusions

Modern transportation options have changed the cultural and economic landscape of Pennsylvania. They reflect modern concepts of time and space, both of which appear to be shrinking.

One question that remains is whether significant changes in passenger or freight transportation will again result from structural changes in the transportation industry. Will new technologies create the degree of change that railroads, highways and air travel have made in the twentieth century? What will Pennsylvania's reputations for level of service in the transportation sector be in ten to twenty years' time?

Bibliography

Getis, Arthur, Judith Getis and I.E. Quastler 2001. The United States and Canada: The Land and the People. Boston, McGraw-Hill.

Lewis, Tom 1997. Divided Highways: Building the Interstate Highways, Transforming American Life. New York, NY, Penguin Putnam.

Miller, E. Willard 1995. "Transportation." Chapter 14 in Miller, E. Willard (ed.) A Geography of Pennsylvania. pp. 234-251.

PA Highways 2004. Breezewood Services. Pennsylvania Highways. Internet site: <http://www.pahighways.com/interstates/bwoodservices_files/bwoodservices_vml_1.htm>, visited 11/9/04.

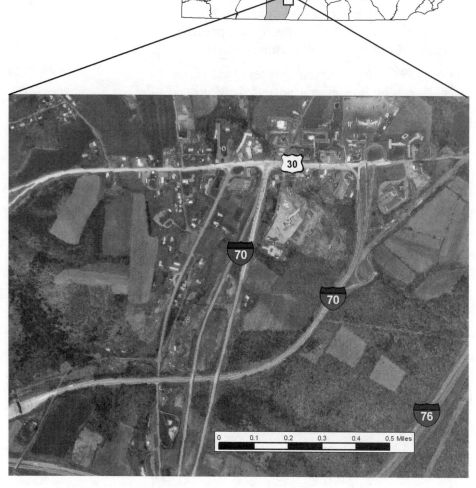

Figure 25.3 **Breezewood, Bedford County:** Breezewood's odd configuration was because of the sequence of events in the development of the Breezewood Exit of the Pennsylvania Turnpike (Interstate 76), and the decision to let the Turnpike serve as both Interstate 76 and Interstate 70 in western Pennsylvania. Now, travelers on Interstate 70 have to follow a short stretch that also serves as US Route 30. This stretch includes traffic lights, the only ones in the whole Interstate Highway System.

153

PA HMC 2004. Pennsylvania State History: Maturity 1945-1995. Harrisburg, PA, Pennsylvania Historical and Museum Commission. Internet site: <http://www.phmc.state.pa.us/bah/pahist/mature.asp?secid=31>, visited 11/9/04.

Schuyler, David 2002. A City Transformed: Redevelopment, Race, and Suburbanization in Lancaster, Pennsylvania 1940-1980. University Park, PA, The Pennsylvania State University Press.

US Airways 2004a. US Airways History. Arlington, VA, US Airways, Inc. Internet site: <www.usairways.com/about/corporate/profile/history/company_history.htm>, visited 2/11/05.

US Airways 2004b. Economic Impact Study Finds Pittsburgh Air Hub Generates $3.1 Billion for Region, Supports 33,300 Jobs. Arlington, VA, US Airways, Inc. Internet site: <www.usairways.com/about/press_2003/nw_03_0917.htm>, visited 2/11/05.

US DOT 2003. Pennsylvania Transportation Profile. Washington, DC, US Department of Transportation, Bureau of Transportation Statistics. Internet site: <http://www.bts.gov/publications/state_transportation_profiles/pennsylvania/index.html>, visited 8/23/04

USGS 1999. Digital Ortho Quarter Quad (aerial photograph): breezewood_pa_nw. Downloaded in TIFF format from http://www.pasda.psu.edu/access/doq99list.cgi, 11/9/04. US Geological Survey, Washington, DC.

Chapter 26

Energy in Pennsylvania

Making the best use of raw materials requires energy, and Pennsylvania has been one of the country's best-endowed states, with both the energy resources and the technical ability to develop them. Energy production has gone through stages of manual and draft animal labor, biomass (mostly wood) energy, coal and petroleum fossil fuels, and the electricity era. We may be on the cusp of a new energy era, but the production industry is holding us to our existing course, and it is unclear whether the next stage will be based on renewable energy resources at lower intensity and greater efficiency, or on technologies beyond nuclear fission that allow us to increase production and continue to push our scientific technological limits.

Today our energy systems can be said to use all of the earlier technologies, especially if you include the

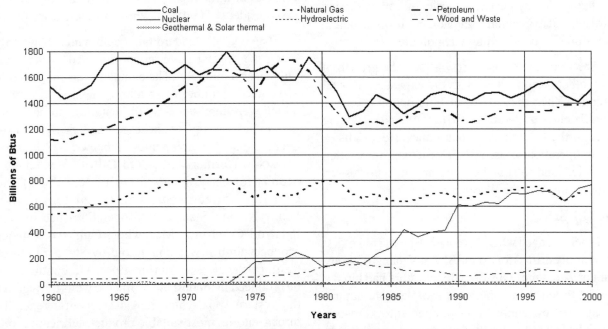

Figure 26.1 **History of Energy Consumption in Pennsylvania:** This graph shows the changes in consumption totals for our major energy sources. It reports the quantities in Btus (British thermal units, a standard unit of measure for comparing different forms of energy). Using Btu's allows us to show comparisons of oil (usually measured in barrels) with coal (measured in tons), natural gas (measured in cubic feet) and other sources using other units. Since 1960, two trends stand out: the decrease in oil and coal consumption from the middle 1970s to the early 1980s, and the rising contribution of nuclear power from the early 1970s until 1990. (Data source: US DOE 2004)

Amish reliance on draft animals in that picture. But while all of the energy resources are present in-state, we have become net importers of oil products and of nuclear fuels.

Early Times

During colonial and early American times the main energy needs could be accomplished with simple resources and technologies. The needs included transportation by wagon or boat, farm fieldwork, home heating, blacksmithing and the grinding of grain in grist mills. The transportation and fieldwork chores relied on horses, mules and cattle, heat was provided by burning wood and charcoal, and mills ran on water power.

As industrialization began and the first steam engines were put to work in both stationary and railroad applications, wood was still the fuel. Lighting was a greater challenge, accomplished using candles or, if you were wealthier, lanterns burning whale oil. By the middle 1800s some cities were installing gas street lights, for which the gas came from coal. Isolated uses of coal in blacksmith forges had occurred in both the bituminous and anthracite regions during the 1700s. By the early to middle 1800s, coal was finding increased use as a general purpose heating fuel, eventually fuelling the vast majority of iron works and railroads, not to mention many homes. It was a natural extension of wood-burning technologies.

The founding of many towns, and the growth of many others, came about with the expansion of coal mining. When the coking of bituminous coal was added, beehive ovens and spoils (or tailings) piles were added to the landscape. When mines, steel mills or coking operations closed, their structures were often simply left, and still dot the landscape of southwestern Pennsylvania and the Wyoming Valley.

The Oil Era

Much changed when "Colonel" Edwin Drake first successfully drilled for oil in 1859 in Venango County south of Titusville. Out-of-state investors had hired a chemist to tell them whether a sample from natural seepages of oil could be developed into kerosene (which had already been made from coal), a substitute for whale oil as a lantern fuel. When the chemist said that *better* kerosene could be made from the crude petroleum, they hired Drake, who found an economical

way to extract the oil from the ground (Yergin 1991, 26; see Chapter 23).

Pennsylvania was the number one producer of oil until 1900 (Harper, Tatlock and Wolfe 1999, 487), and oil, or at least kerosene, was America's fourth most valuable export in the 1870s and 1880s. The market for oil was based mostly on kerosene (and only a few by-products, such as Vaseline), and the geologists were busy figuring out more reliable ways to predict oil. When oil was discovered in Ohio in the 1880s, and then in larger quantities in California in the 1890s, and finally in tremendous discoveries in Texas and Oklahoma starting in 1901, Pennsylvania's dominance ended (Yergin 1991, 52-85).

The oil industry of the late 1800s featured the rise to dominance of John D. Rockefeller's Standard Oil Company, controlling over 90 percent of the market at its peak. An additional highlight was the attempt by independent producers to challenge Standard Oil. In one early incident, a group of small companies, looking for ways to beat Standard Oil's preferred rates for shipping by railroad, built the first long distance oil pipeline. Standard Oil soon followed with a bigger and better one (Yergin 1991, 43). One of the most successful challengers was William Mellon, of Pittsburgh, a member of the Mellon banking family and president of Gulf Oil, whose oil came from Texas.

The end of Standard Oil's dominance came only when the federal government objected to Standard Oil's near monopoly. The company was dissolved in 1911 into a number of smaller companies, many of which were the Standard Oil operations in one state or another. Many of those companies went on to become major oil companies in their own right, such as Exxon (formerly Standard Oil of New Jersey), Mobil (formerly Standard Oil of New York) and Chevron (formerly Standard Oil of California). Pennsylvania's Standard Oil operations became the Pennzoil Company. Meanwhile, Gulf was a major producer until it was bought out by Standard Oil of California (now Chevron) in 1984. Oil continues to be produced in modest but profitable wells throughout the western Appalachian Plateau area, but is generally turned into its most suitable product, lubricating oil, under such brand names as Quaker State and Pennzoil (see Chapter 4).

The other oil story that had an impact in Pennsylvania and everywhere else, was the development that saved the oil industry from its certain demise. In the late 1880s a new technology threatened to eliminate the main reason for the existence of the oil industry: electricity represented an easier and cheaper way to create light. It was the birth and growth of the automobile industry that saved Standard, Gulf and the other oil companies. Despite its importance in the steel industry, and the steel industry's importance in Pennsylvania, automobile manufacturing was drawn to the Detroit area. However, a huge impact was felt when road construction became a necessity in order to serve the industry. Pennsylvania now has one of the densest networks of roads in the nation, measured as miles of road per square miles of territory (see Figure 25.2 in Chapter 25).

As we have seen, natural gas is a byproduct of the processes that created oil. Today, natural gas is one of the cheapest and cleanest alternatives for home space and water heating, and for cooking. It is also the fuel most likely to be selected in heat-intensive industrial and electricity-generating applications. In 2001, natural gas wells earned $535 million, nearly ten times the value of Pennsylvania crude oil production (IPAA 2004).

Quickly Followed by Electricity

Electricity was recognized as a phenomenon through experiments with static electricity and Benjamin Franklin's famous kite and key. Scientists even knew how to generate electricity. But it took Edison's invention of the light bulb to prove electricity's practical value. He did so in 1879, and by 1882 he had built the first large (for its time) electric generator and installed it in New York City. Edison illuminated buildings in a few blocks of Wall Street in order to attract investors. He realized, though, that it could have a tremendous impact in smaller cities and towns, and so began the process of setting up small franchises of Edison Electric Illuminating Companies. His first ventures were in a group of towns in central Pennsylvania: Sunbury, Williamsport, Shamokin, Mount Carmel and Hazleton. Sunbury's was the first to start operation (Beck 1995, 49). His generators burned coal and created direct current (DC) electricity. Within a few years many small towns, especially in northern Pennsylvania, had Edison generators and lights.

By the middle 1880s George Westinghouse's company had perfected the generation of alternating current (AC) electricity, which was better suited to traveling longer distances over wires and to running motors. There was fierce competition between Westinghouse and Edison's General Electric, and by the early 1890s alternating current became the standard (Beck 1995, 66-68).

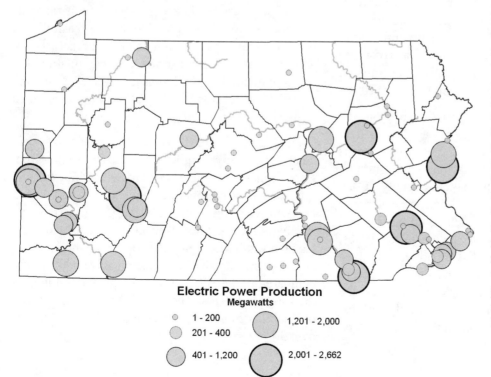

Electric Power Production
Megawatts

○ 1 - 200
◔ 201 - 400
⬤ 401 - 1,200
◯ 1,201 - 2,000
⬤ 2,001 - 2,662

Figure 26.2 **Relative Sizes of Electric Power Stations in Pennsylvania:** Notice that the largest power stations are located further from the major cities than smaller (mostly older) stations are. There are two reasons: coal-burning stations are located closer to the mines in order to minimize coal transportation costs; and all large power stations (mostly coal and nuclear) are located along major rivers for cooling water and further from population centers for safety reasons. (Data source: US DOE 1991)

157

Soon electric trolleys or streetcars were operating in many cities, and many of the trolley companies also operated as local electricity suppliers. There were relatively few home appliances to spur residential demand until well into the 1900s. In the 1930s a number of occurrences coincided. Before the Depression many smaller electric companies were being bought by larger ones, allowing the latter to expand their territories and to justify installing larger generators in their power plants. These electric utilities came to monopolize production in their regions, and the Pennsylvania Public Utility Commission, formed years before to watch over the railroads, was given the job of regulating them. Finally, a host of new appliances, including washing machines, fans and early refrigerators, began appearing, which greatly stimulated the market (Beck 1995, 213-217).

Early generators were small, probably 20,000 watts (20 kilowatts or Kw) or less, because the only work they did was to light a couple hundred light bulbs each. By the 1920s they were capable of as much as 20,000,000 watts (20 megawatts or Mw), and by the 1960s the electric utility companies were installing generators that produced over 1000 Mw each in their power plants. Such large-scale production was the result of a complex decision. One alternative was to build many smaller plants serving smaller numbers of customers and located closer to the communities where the customers lived, with the fuel delivery and air pollution costs also spread out. Their decision was to maintain large centralized power plants, often with two or more of the very large generators, each of which would serve hundreds of thousands of customers, along with large distribution networks (Geiger 2004).

The power sources were varied right from the beginning, since the technology is based on boiling water into pressurized steam. The blast of steam is directed at the fan-blade-like turbine whose turning axle drives the generator to make electricity. Coal, oil or natural gas can be used to boil the water, and the fuel used depended on which could be obtained most cheaply at that location. Later, in the 1950s, uranium was added to the list of boiler fuels. In addition, some of the earliest and the most recent power plants were hydroelectric, which eliminate the boiler and let the flowing river water turn the turbine. The hydroelectric facilities generally produce much less electricity, up to a few hundred megawatts per site.

Nuclear Power in Pennsylvania

Pennsylvania has three claims to fame in the world of nuclear-generated electricity, two of which were significant in the birth of the industry, and the third of which has been a factor in its struggles. First of all, realize that nuclear power plants operate on a different principle than those burning fossil fuels. In a nuclear power plant the boiling process is accomplished by concentrating atoms of uranium in the fuel rods of the nuclear reactor. Doing so makes each uranium atom more likely to be hit by a free neutron. If the uranium atom is unstable, as about 3 percent of them are, the splitting does three things: it creates by-product elements, it produces heat, and it releases a few more free neutrons to continue the chain reaction. The heat is the ultimate goal of the process, so the reactor is connected to a system of pipes feeding the water which the heat turns to steam. The uranium fuel makes it possible to produce that heat with much smaller quantities of fuel and virtually no air pollution. The risk is that a loss of control of the chain reaction could lead to a release of radiation beyond the power plant.

Saxton, in Bedford County, was the site of a research reactor that operated from 1962 to 1972. It was owned by a consortium of companies in the nuclear industry, headed by GPU Nuclear, which was itself the result of a merger of several electric utilities with operations in Pennsylvania and New Jersey. The reactor was very small, generating electricity at a rate of only 23 Mw, but was used to test the feasibility of many technologies later used in nuclear reactors. For 26 years after shutting down, decontamination and removal of auxiliary buildings took place. Beginning in 1998 the reactor building itself was dismantled (NukeWorker.com 2004).

Shippingport, in Beaver County was a relatively small community 25 miles down the Ohio River from Pittsburgh near the state boundary with Ohio. In 1957 the US's first commercial nuclear power plant, a cooperative venture between the US government's Atomic Energy Commission and the Duquesne Light Company, of Pittsburgh, began operating there. It was small by later standards, producing only about 60 Mw compared to the 800 to 1200 Mw giants designed in the 1960s and 1970s. The Shippingport station proved the viability of nuclear power and operated safely for over 30 years.

Figure 26.3 **The Three Mile Island Nuclear Power Plant:** The active Unit 1 is shown, while the buildings for Unit 2 are to the left of the photo area. The accident to Unit 2 in March 1979 caused Unit 1 to have a six-year break from operating. It has been back online since 1985.

Pennsylvania also is home to nuclear power stations at (arranged in the order that they started producing electricity): Peach Bottom (York County, an early unit started in the 1960s, but the two currently operating started in 1974), Three Mile Island (Middletown, Dauphin County, 1974), Beaver Valley (adjacent to the Shippingport reactor, Beaver County, 1976) and Susquehanna (, Luzerne County, 1982), Limerick (Montgomery County, 1985). By the time the last two came on line, Three Mile Island had suffered its famous accident (see below), and the cost of constructing, powering up and operating a nuclear power plant had skyrocketed. The increase was so significant that a Pennsylvania Power and Light utility company administrator was quoted as saying that "If management had known in advance what the ultimate costs would be, the units would never have been built" (Beck 1995, 384).

The Accident at Three Mile Island

On the other hand, Pennsylvania is also the home of the Three Mile Island nuclear power station; one of the giant reactors there never got a chance to operate commercially. The first generator, Unit 1, had been operating for five years in March 1979, but was shut down for routine maintenance. Unit 2 had been built and tested, and was in the final phase of starting up to full power before getting permission to connect to the electricity distribution grid. In the late hours of March 29, a pump feeding water to the reactor failed, and a valve automatically opened to relieve the pressure of the stalled water system. However, the valve failed to

close when the pressure was reduced and remained open for over two hours, while the water loss continued. Confusion in the control room ensued. The reactor overheated, causing a partial melting of metal parts and the fuel in the reactor. It was several days before Unit 2 was under complete control and was shut down, never to operate again (GPU Nuclear 1991, 2).

Due to the nature of the accident, Unit 1 was not restarted for six years, and the safety systems of nuclear power plants around the world were reexamined. Costs of nuclear power stations had already been rising at rates unexpected by the utility companies. The costs of meeting stricter safety standards compounded that problem. Many plants that were already under construction, including Pennsylvania's Limerick station outside Pottstown and Susquehanna station outside Berwick, were completed and started. However, many other nuclear power plants at varying stages of progress around the US were cancelled.

Alternative Energy Use

In 1998 Pennsylvania passed legislation changing the structure of the electricity industry. What had evolved into a set of regional monopolies was opened to any number of competitors. Any company can generate electricity, and any company can market and sell electricity to consumers. One result has been the sale of Three Mile Island and Peach Bottom from their former utility owners (GPU-Nuclear and the Philadelphia Electric Company) to a company,

159

Amergen, who owns over a dozen nuclear plants around the US.

Another result has been the emergence of several production and sales companies that specialize in "green energy". These companies have added several alternative energy facilities for generating electricity. Wind power is currently produced for sale to the public in several locations in Somerset County. In fact, a plan to increase the role played by renewable energy in Pennsylvania was recently signed by Governor Rendell. The law, entitled Advanced Energy Portfolio Standard, requires that 18 percent of Pennsylvania's electric power come from a list of renewable sources by 2019.

Conclusions

Energy production has benefited from our natural resources in the past, and once again has given Pennsylvania leadership roles in many ways. However, Pennsylvania's energy resources and technologies are in danger of causing more resource and environmental harm than good.

Bibliography

Beck, Bill 1995. PP&L 75 Years of Powering the Future: An Illustrated History of Pennsylvania Power and Light Co. Allentown, PA, PP&L.

Geiger, Charles 2004. "Historical Electricity Production in Pennsylvania." The Pennsylvania Geographer, Vol. 42, no. 2 (Fall/Winter), forthcoming.

GPU Nuclear 1991. The TMI-2 Story. Middletown, PA, GPU (originally General Public Utilities) Nuclear, Public Affairs Department.

Harper, John A., Derek B. Tatlock and Robert T. Wolfe, Jr. 1999. "Petroleum-Shallow Oil and Natural Gas." Chapter 38A in Schultz, Charles H. (ed.) The Geology of Pennsylvania. pp. 484-505.

IPAA 2004. Oil and Gas in Your State: Pennsylvania. Independent Petroleum Association of America, Washington, DC. Internet site: <http://www.ipaa.org/info/In Your State/default.asp?State=Pennsylvania>, visited 8/10/04.

NukeWorker.com 2005. Saxton Nuclear Experimental Facility. Powell, TN, NukeWorker.com. Internet site: <http://nukeworker.com/nuke_facilities/ North_America/usa/NRC_Facilities/Region_1/ saxton/index.shtml>, visited 1/13/05.

US DOE 1991. Inventory of Power Plants in the United States 1990. Washington, DC, US Department of Energy, Energy Information Administration.

US DOE 2004. Table 7. Energy Consumption Estimates by Source, Selected Years, 1960-2000, Pennsylvania. Internet site: <http:// www.eia.doe.gov/emeu/states/sep_use/total/ use_tot_pa.html>, visited 11/11/2004

Yergin, Daniel 1991. The Prize: The Epic Quest for Oil, Money and Power. New York, Simon and Schuster.

Chapter 27

Manufacturing

At the secondary level of economic activity, raw materials are processed in some way into intermediate or finished products. Intermediate products are inputs for further manufacturing, while finished products are headed for their end users in the residential or commercial markets. Some manufacturing is heavy energy-intensive work, while other "manufacturers" simply clean and package their goods.

Table 27.1 shows a list of 21 main NAICS (see the Section C introduction and Table C.2) categories of manufacturing, with their three-digit codes, as used by the US Bureau of the Census for their Economic Census every five years. Even though the range of inputs and outputs (from raw materials to produced or processed goods) for these categories is great, they all share some common characteristics.

First, all of the raw materials produced in primary level activities in Pennsylvania are potential inputs for manufacturing. Only a few of those primary level outputs, mostly agricultural, are consumed in the form that they are produced. Pennsylvania's history of producing all of those primary level products (except perhaps fish) has positioned us well to be leaders in the manufacturing sector.

Secondly, in addition to the component raw materials, Pennsylvania's rich supplies of energy played an equally important role. As Chapter 26 showed, we played significant roles in all types of fossil and nuclear fuel development, and even in the biomass period that preceded those fuels. All of those fuels and technologies were significant in the progress of our manufacturing sector. The fact that our energy production has declined relative to other parts of the US and world is related to the declined we have experienced in our manufacturing sector.

Thirdly, more than for the primary and tertiary sectors, a very important geographical decision is part of every manufacturing deployment, and that is the question of where to put it. The location of manufacturing inputs is dictated by environmental factors, and the destination of the outputs is not likely to be a single place. The factory location must balance

Table 27.1 North American Industry Classification System: Manufacturing Sector
(US Census Bureau 2004)

311	Food mfg.
312	Beverage and tobacco product mfg.
313	Textile mills
314	Textile product mills
315	Apparel mfg.
316	Leather and allied product mfg.
321	Wood product mfg.
322	Paper mfg.
323	Printing and related support activities
324	Petroleum and coal products mfg.
325	Chemical mfg.
326	Plastics and rubber products mfg.
327	Nonmetallic mineral product mfg.
331	Primary metal mfg.
332	Fabricated metal product mfg.
333	Machinery mfg.
334	Computer and electronic product mfg.
335	Electrical equip., appliance, & component mfg.
336	Transportation equipment mfg.
337	Furniture and related product mfg.
339	Miscellaneous mfg.

Figure 27.1 **Total Value of Manufacturing Around Pennsylvania:** The circles represent the total value of manufacturied goods shipped from factories in each county. In only six counties are there so few manufacturing facilities that the US Census Bureau is unwilling to publish financial totals so as not to reveal proprietary information. What this map does not show is that the factories are most often in urban locations. Counties that do not have large cities will still have factories, usually located in smaller towns, and usually not in rural areas. (Data source: US Census Bureau 2001)

Manufacturing Shipments
Total $ Value

- $81,470,000 - $500,000,000
- $500,001,000 - $2,000,000,000
- $2,000,001,000 - $5,000,000,000
- $5,000,001,000 - $20,666,642,000
- ND Too Few Establishments

those input and output locations, and also minimize energy, transportation and other local costs. It is a complex issue, as we will discuss below.

Pennsylvania's Range of Production

In 1997, Pennsylvania companies in the manufacturing sector earned $172.2 billion, good for sixth among US states. Significant production occurred in every NAICS manufacturing category (see Table 27.2). Food manufacturing (NAICS code 311), chemical manufacturing (code 325) and primary metal manufacturing (code 331) lead the list with over $17 billion each in total value shipped. The manufacturing of leather goods (code 316), in last place, still generates over $675 million in total value shipped. In terms of employment, fabricated metal products (code 332), food (code 311) and machinery (code 333) manufacturing employed over 73,000 each, and again leather products manufacturing was last with about 5,200 employees (US Census Bureau 2001).

It is well known that Pennsylvania's iron and steel industry has declined precipitously from its peaks in the first half of the twentieth century. As a group, though, the production categories most directly related to that industry still lead the state's manufacturing economy. NAICS categories (codes) 331, 332, 333 and 336 totaled over $55 billion in value shipped in 1997. The second closest cluster of activities is the

set of companies producing mineral and chemical goods (NAICS codes 324 to 327) at $43 billion total value. In this same sense, food products (codes 311

Table 27.2 Pennsylvania Manufacturing Productivity (US Census Bureau 2002)

NAICS code	Number of establishments	Total value of shipments (x $1,000)
311	1,368	20,374,271
312	132	2,688,042
313	223	1,704,867
314	290	1,028,084
315	751	3,717,401
316	73	676,577
321	997	3,292,725
322	328	8,901,062
323	1,795	6,260,876
324	165	9,447,025
325	616	19,295,648
326	779	8,414,514
327	820	6,095,887
331	421	17,057,715
332	3,185	14,593,912
333	1,531	12,632,173
334	815	12,840,103
335	378	5,186,779
336	400	10,843,428
337	854	2,819,727
339	1,207	4,322,400

162

and 312) contribute another $23 billion, and wood products (codes 321 to 323) add another $21 billion. The smaller categories of electronic and electrical equipment manufacturing (codes 334 and 335, contributing $18 billion) and textile production (codes 313 to 316, at $7 billion) comprised the final broader categories (US Census Bureau 2001).

The Location Factor

A key decision, if you are a manufacturer, is where to locate your manufacturing facility. You must take into account the locations of: the sources for your inputs, the type of energy needed, availability of water for cooling (or any other locally determined environmental need, including air pollution limits and waste disposal options), quantity and quality of labor available, and distances to your customers. As we have seen, many goods are hauled (both to and from the factory) by trucks nowadays, but some processes might require railroad, ship, pipeline or even airport access.

If your manufacturing process takes a very bulky input and reduces that bulk considerably, like a food processing factory that uses a small part of the harvested plant, then you might locate closer to the source of that input in order to reduce the larger transportation cost. For example, the town of North East, Erie County, hosts a production plant for Welch's,

the marketing division of The National Grape Cooperative Association, Inc. headquartered in Concord, Massachusetts. Welch's is known for its grape juices and jellies. Northeastern Erie County, along with nearby areas of western New York State, are one of the area's where the company's Concord and Niagara grapes are grown (Welch's 2005).

Another process might turn a variety of relatively small or scattered inputs into large heavy products, like the companies who manufacture high-quality paper do. Inputs for making paper include wood or cotton rags, water, lime, sizing agents that affect the papers "weight," and bleaching or coloring agents. Then, especially if you are limited to truck transport to your printing company customers, you will more likely locate nearer to those customers.

Every factory and product is different. The challenge is to learn enough about a manufacturing process to be able to look at a map of its factory locations or higher-producing areas and explain or criticize them.

In Pennsylvania, most towns and cities have had at least one period of their history in which manufacturing played a key role in the place's development. Perhaps it is a company town, located near the primary input resource and originally employing a relatively pure stock of ethnic immigrants,

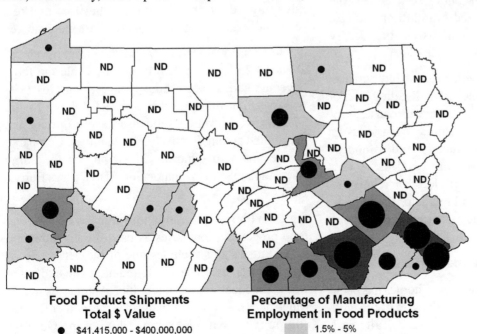

Figure 27.2 **Food Processing Around Pennsylvania:** This map shows both the total value of processed food products shipped from each Pennsylvania county, and the relative importance of food processing as a percentage of total manufacturing employment in each county. Note the much stronger concentration in the most productive agricultural lands and near the largest population centers. (Data source: US Census Bureau 2001)

Food Product Shipments Total $ Value

- • $41,415,000 - $400,000,000
- ● $400,001,000 - $1,000,000,000
- ● $1,000,001,000 - $2,616,818,000

Percentage of Manufacturing Employment in Food Products

- 1.5% - 5%
- 5.1% - 15%
- 15.1% - 27.4%
- ND Too Few Establishments

163

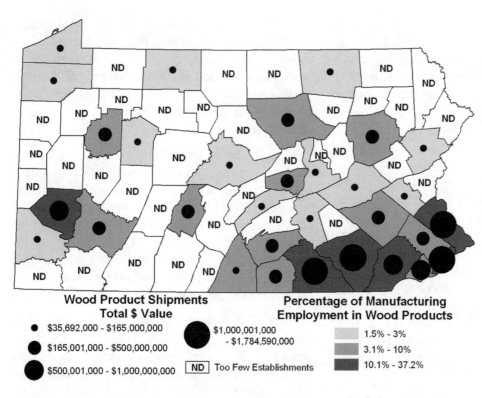

Figure 27.3 **Wood Products Around Pennsylvania:** Again, fewer counties have large numbers of factories making paper, furniture and other wood products, but those that do benefit greatly in economic terms. It is quite possible that similar relative values of production are recorded in many of those counties labeled "ND" just because they have fewer factories. (Data Cource: US Census Bureau 2001)

Wood Product Shipments Total $ Value

- $35,692,000 - $165,000,000
- $165,001,000 - $500,000,000
- $500,001,000 - $1,000,000,000
- $1,000,001,000 - $1,784,590,000
- ND Too Few Establishments

Percentage of Manufacturing Employment in Wood Products

- 1.5% - 3%
- 3.1% - 10%
- 10.1% - 37.2%

or located near the primary market for its product. An example of the latter is the Woolrich Company, located in Woolrich, Clinton County (the town was named after the company). The company made its start by buying wool to make wool socks and other articles of clothing, and selling them to the lumber men in north-central Pennsylvania beginning in 1830. It has become an internationally recognized brand, and added a distribution center in nearby, curiously-named Jersey Shore, Lycoming County (Woolrich 2004).

In other cases, the larger cities tend to have greater varieties of manufacturing activity. Reading, Berks County, for example, was a center for railroad activity and also developed a strong manufacturing industry focused on iron and related metals, and a modest secondary industry based on textiles. The combination, as well as Reading's location relative to major east-coast cities such as Philadelphia and New York, led to some cost advantages for product distribution.

Some Additional Sample Industries

Much of the food that reaches our homes has been processed in some way. Compare the size of the area of your grocery store devoted to fresh fruits and vegetables to the store area holding packaged goods. Food processing lengthens the time the foods remain edible and reduces the time consumers spend preparing meals.

As an agricultural state, Pennsylvania's main competition is the cereal grains, livestock and dairy product region of the Midwest and the fruit and vegetable growing in central California. Our advantage over those areas is that our markets (especially our out-of-state markets) are much nearer.

In Pennsylvania in 1997 and 2002 farms produced about $4.25 billion worth of goods, while the food, beverage and tobacco industries' earnings for 1997 was over four times greater, at $23.1 billion. Figure 27.2 shows how that food processing output was distributed among Pennsylvania counties. Compared to the maps in Chapter 21, it shows the same orientation toward the major cities (US Census Bureau 2001).

Both Philadelphia and Pittsburgh have had strong food processing industries as part of their industrial past. Pittsburgh's includes one of the better known brand names, as it was the founding home and is the world headquarters of Heinz, Inc. Another example, which does not show up on the map, is another famous company town, Hershey, Lebanon County. Strategically located near a strong source for milk, Hershey has to import its cocoa beans, sugar and other ingredients. As a Piedmont city, it can also take strategic advantage of markets and shipping opportunities via the major east coast cities.

A second example of manufacturing strongly tied to raw material production in Pennsylvania is the set of categories related to our wood resources: wood products, paper, printing and furniture manufacturing. The Economic Census shows total manufacturing productivity in this area within Pennsylvania to be $21.3 billion, ranking the state second in the US after California. The separate category ranks are: fourth for wood products, second for paper, and first for both printing and furniture making (US Census Bureau 2001).

In Pennsylvania our strengths are in the paper and furniture trades. Many other states have productive forests, and those of the southeastern US Appalachian region and the Pacific Northwest tend to produce much more structural lumber. Pennsylvania's hardwood-rich forests are better suited to furniture and other highly finished products such as wood flooring.

Industrial Decline

Pennsylvania's manufacturing economy is still one of its stronger economic sectors, but has declined significantly since the early 1900s. Many factors led to the decline: the shifting locations of the American public (as both customers and laborers) as the western and southern states grew in population, the discovery of new raw materials and energy sources in those other parts of the country, the relative aging of our factories as newer factories elsewhere adopted more modern technologies, the higher expectations of Pennsylvania's professional workforce as workers' experience and union organization grew in prominence, and the snowball effect of one factory's closing negatively affecting another that used its products or affecting the costs of running the railroad that used to haul its products. The most devastating declines tended to be in the iron and steel industry, as we shall see in the next chapter.

As a result, the towns and cities that once were leaders in certain industries have frequently had to rebuild their economies to make up for the loss of many manufacturing jobs. Some interesting adaptations have been created, which will be studied in greater detail later, since they require the town to develop a stronger retail- or service-oriented economy. One example is how Reading's expertise in textile factories and distribution centers led to some of the early

"outlet" stores there in the 1980s. In other places, they replaced heavy industry with a tourism economy by re-opening and advertising the same closed factories as museums of past glory.

York: Still an Industrial City

York, the county seat of York County was among the earliest of Pennsylvania cities to take advantage of its location, relative to inputs such as minerals and farmland, and to markets including Philadelphia and Baltimore, Maryland, to develop a strong manufacturing economy (see Figure 27.4).

Straddling the Codorus Creek, York was the first Pennsylvania town west of the Susquehanna River when it was founded in 1741. York County was formed

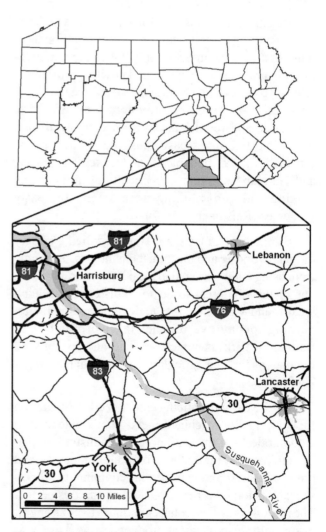

Figure 27.4 **York, Pennsylvania:** York's location relative to Harrisburg, Lancaster and Baltimore (not shown, south of York) gave it significant advantages. The heaviest lines, which indicate Interstate and US route highways today, were the approximate routes of railroad lines in the 1800s.

165

out of Lancaster County soon afterwards (Smith 1987, 8-9). It grew first as an agricultural market town, and later as the county's center for tools and machinery manufacturing. Its industrial growth for wider markets took place after railroads connected it, first eastward with Lancaster via the Columbia-Wrightsville covered bridge in the middle 1830s (see caption for Figure 24.2 in Chapter 24). Within a few years it was also connected with Baltimore, and later connections were made northward toward Harrisburg along the west side of the Susquehanna and southwestward to western Maryland and Pennsylvania via Hanover, York County (Boas 1987, 12). York's location advantage was brought into the twentieth century when Interstate 83 established a highway connection to Harrisburg and Baltimore.

Reisinger explains that for a place to develop a manufacturing base, in addition to transportation there must first be a manufacturing culture. This includes raw materials, people with the knowledge to create relevant products, investors and a workforce that value production-oriented employment, and effective wholesale and retail links to local or more distant markets. In York's case these developed early, especially in the merging of the Quaker, Scots-Irish and German immigrants. Agricultural mills, barrel making and gun-smithing were among the early successes (Reisinger 2003, 70-76).

By the time of the Civil War, York had a variety of metalworking and machinist businesses, and some were using anthracite coal in local forges and furnaces, one of the first areas to do so. This expertise led a local foundry to be the first to develop an anthracite-burning locomotive for the Baltimore and Ohio Railroad. York manufacturers had also established reputations for their agricultural equipment (Reisinger 2003, 80-83).

By the early twentieth century, York's reputation for high-quality machined metal goods extended to automobiles and trucks, before the industry came to be dominated by the major automobile companies, and other industrial and commercial equipment. York's specialties today were established during this period. The central core of the county continues to be strongest in specialty manufactured products, as well as food- and wood-related production. It is the home of several manufacturers who are very well known within certain special interest groups if not the general public: Pfaltzgraff Pottery dinnerware, York Barbell fitness

equipment, York International air conditioners and refrigeration equipment, and Harley-Davidson motorcycles (Reisinger 2003, 67).

Conclusions

Manufacturing has always been a key component of Pennsylvania's economy, and probably will continue to be so well into the future. The York example can be found repeated in numerous towns and cities, on both smaller and larger scales.

Bibliography

Boas, Charles W. 1987. "The Railroads of York County, PA: Rail Center to Recreation." The Pennsylvania Geographer. Vol. 25, nos. 1&2 (Spring/Summer), pp. 12-17.

Reisinger, Mark E. 2003. "The Origins and Development of the York Industrial District in South Central Pennsylvania: 1700-1920." The Pennsylvania Geographer. Vol. 41, no. 2 (Fall/Winter), pp. 67-94.

Smith, Joe Bart 1987. "The Growth and Development of York City from 1740 to 1890." The Pennsylvania Geographer. Vol. 25, nos. 1&2 (Spring/Summer), pp. 8-11.

US Census Bureau 2001. 1997 Economic Census: Manufacturing. Washington, DC, US Department of Commerce, Bureau of the Census. Internet site: <http://www.census.gov/epcd/www/97EC_US.HTM>, visited multiple times.

US Census Bureau 2004. North American Industry Classification System (NAICS). Washington, DC, US Department of Commerce, Bureau of the Census. Internet site: <http://www.census.gov/epcd/www/naics.html>, visited multiple times.

Welch's 2005. Welch's Company Information. Concord, MA, Welch Foods, Inc. Internet site: <http://www.welchs.com/company/general_co_info.html>, visited 1/23/05.

Woolrich 2004. About Woolrich. Woolrich, PA, Woolrich, Inc. Internet site: <http://www.woolrich.com>, visited 8/23/04.

Chapter 28

Iron and Steel

Iron and steel was once the industry that epitomized manufacturing, and among Pennsylvania's many nicknames was one which showed its role in that industry: The Steel State. The story is one of a strong rise to early dominance, followed by a gradual decline in stature.

It will help to understand how iron ore is turned into iron and steel products. First the iron must be mined as ore containing a significant percentage of iron mixed with impurities. Then the ore must be "smelted," the function of a "furnace," in order to extract nearly pure iron. Finally, the iron must be "worked," the function of a forge or mill, into finished products. Working the iron further in a process that adds carbon (creating an iron-and-carbon alloy) helps to turn iron into steel.

Pennsylvania's early importance was an outcome of two levels of location decisions. At one level, decisions favored some parts of Pennsylvania above other parts; at the other level our state was boosted past others. Because both the smelting and working processes require intense heat, the furnaces and forges demanded the best fuel sources. The availability of iron ore became a factor only later in that history. Even limestone, water as an ingredient, and water and railroads for transportation became important location factors.

In the earlier times there were many small operations, serving their local towns and rural areas. By the 1870s these were replaced by much larger operations, combining most of the processing steps into very large *integrated* steel mills. Many innovations in the industry came out of Pennsylvania as it took production from crude beginnings to high-volume efficiency. The decline came later, when other producers continued that trend while Pennsylvania companies felt compelled to stand pat.

The Steel-Making Process

The ore that comes to the furnace is up to 60 percent iron. The intense heat in the furnace melts the iron, but the other impurities react differently (turning to gases or remaining solid), allowing the iron to be separated. The furnace produced an intermediate product called pig iron, which became the input to other processes.

In the early years molten pig iron was poured into carefully designed molds or castings, to become cast iron. The pig iron could also be reheated and beat into shape at iron forges by blacksmiths, or reheated and repeatedly forced between rollers to change its shape in rolling mills. With enough beating or rolling in the presence of carbon-rich fuels the iron became steel.

In order to accomplish such work, very high temperatures are needed. In colonial times this was accomplished by charcoal-fueled fires. Put simply, charcoal is wood which has been baked to remove water, trapped gases and some impurities, leaving behind a more carbon-rich fuel. A similar feat is accomplished by nature in the creation of anthracite coal, which has a higher heat potential than charcoal, and requires mining but no processing. Even greater heating was accomplished by applying the charcoal-preparation process to bituminous coal to create "coke." In the middle 1800s coke was produced in "beehive ovens," named because of their shape, built near the early coal mines in western Pennsylvania (Rodgers 1995, 287). Later the coking process became part of the large integrated mills.

By the middle 1800s blast furnaces made the smelting process more efficient by forcing more oxygen into the furnace. In the resulting pig iron up to half of the impurities is carbon, which makes it much easier to work into steel. With that little extra processing, another intermediate product of many mills was steel ingots, ready to be worked into finished steel products. In time, both the blast furnace and the production of raw steel became stages in integrated steel mills. The Bessemer process, which again improved the blast furnace method of making steel ingots, was introduced in the 1860s but had its greatest impact in integrated mills of the 1870-80s.

When the steel ingots are reheated they can be worked into an enormous variety of shapes and qualities. For example, rolling mills (and the rolling mill portion of the integrated mills) created such products as bars, railroad rails, sheets, structural shapes (such as girders), tubes (such as pipes) and wire products of varying dimensions.

Steel has much greater tensile strength than pure iron. Other specialized steel products come from melting the pig iron with other metals, or adding the other ingredients at later stages of processing. These alloys, usually metals, add such qualities as flexibility, rigidity, strength or rust resistance. The results include stainless steel, galvanized steel, chrome-plated steel, tin-plate (from which "tin" cans are made) and others.

Larger operations became necessary to keep up with demand and to try to reduce production costs. The open hearth furnace, also first introduced in the 1860s, represented yet another improvement, and came into its own in the 1890s. As part of an integrated mill, the open hearth furnace made it easier to add those other ingredients to make a host of new alloys. Unfortunately, as we shall see below, it was also part of the reason for the decline of the Pennsylvania steel industry in the twentieth century.

The Locations of Steel Mills

A steel mill is located, as we saw in the last chapter, in order to minimize the costs of production and of transportation, both *from* its raw material suppliers and *to* its customers. Transportation access is an important consideration, since the inputs and outputs are all heavy and bulky materials. Changes in iron ore availability had the greatest impacts in location decisions as the sizes of the mills increased. Small deposits had been sufficient for small furnaces and forges, but larger operations and integrated mills required greater quantities and reliability of supplies.

The first large steel mills were in Pottsville and Bethlehem, in eastern Pennsylvania. These locations were favored because anthracite coal was available from the Southern Field anthracite deposits, up the Schuylkill and Lehigh Rivers, and larger iron deposits were available in Lebanon County and northern New Jersey. In addition to those inputs, limestone was available in the Lehigh Valley. A big factor in their early success was the fact that the region's biggest market, Philadelphia, was downriver (Rodgers 1995, 286). Bethlehem Steel was the longest-lived product of that legacy.

Later, of course, Pittsburgh gained the upper hand in the industry. First, coke's advantages over anthracite were discovered, and several bituminous deposits were found to be ideal for coking. The best deposits were located in the Connellsville area on the Youghiogheny River. Then, when local iron deposits proved to be unreliable and insufficient, much higher grade ores were discovered along the shores of Lake Superior and Lake Michigan (Rodgers 1995, 287-288). Pittsburgh's advantage was the quality and quantity of coking coal in the region. Twice as much coke as iron ore was required in the smelting phase of production (Miller 1981, 9). Pittsburgh already was known to be ideally positioned for trade down the Ohio River, and was well connected by railroads. By the early 1890s all the pieces were in place for Pittsburgh's rise.

Later in the 1900s, the Pittsburgh region's steel mills closed in increasing numbers along with mills in Johnstown, Bethlehem and other locations. Once again, location was one of the factors in that decline (see below). However, Pennsylvania remains a significant, if not dominant, producer of finished products from the iron and steel industry. These factories, some still referred to as mills, purchase steel ingots or other unfinished steel, and produce finished goods from fabricated metal parts, machinery and transportation equipment to smaller items like nails, wire and springs.

The map in Figure 28.1 shows where many of the Pennsylvania steel mills were located in the 1950s. Notice the two magnet areas at either end of the state, but also the other locations, such as Johnstown and Steelton (just south of Harrisburg). Others had existed before this period, but were shut down when they couldn't compete with the newer and larger integrated mills.

Figure 28.2 shows that primary metals manufacturing is still active in Pennsylvania. Primary metals is the Bureau of the Census NAICS category that would include steel mills, but also includes operations that work with aluminum and other metal resources. The strength of the Pittsburgh region is still evident despite its lower level of activity compared to even a few decades ago.

The Industry's Rise and Decline

Pittsburgh's success came from a combination of large quantities of local experience, and leaders who were willing to take some risks in order to make their operations grow. As early as 1856, there were nine rolling mills and at least eighteen foundries in the Pittsburgh area (Ingham 1991, 27). The various improvements in the steel-making process increased outputs and encouraged the addition of new mills. The Pittsburgh area grew to 36 rolling mills and other operations in 1874, and to 58 total operations in 1894 (Ingham 1991, 203-209). Almost all of those

operations were operated by independent owners; few owned more than one facility.

Andrew Carnegie was perhaps the best known of these leaders, who invested not only in steel, coal and railroad operations, but also in Minnesota iron mines and Great Lakes shipping operations. His Carnegie Steel Company was re-incorporated in 1901 as US Steel, headquartered (then and now) in Pittsburgh. Carnegie Steel was the largest company in the area in 1900. With the formation of US Steel it added 19 operations and brought its capacity up to 2.7 million tons of steel per year. Jones and Laughlin, which had been around since 1852, was the second largest company with 1.4 million tons capacity. Union Steel, a new company formed in 1900, was the third largest at 850,000 tons. By 1901, 36 other independent firms had the capacity to produce another 1.7 million tons of steel ingots and primary steel products (Ingham 1991, 141-149). US Steel was aggressive about expanding its operations, and soon had much greater capacity.

Pennsylvania steel was used to construct the cables supporting the Brooklyn Bridge, the entire Golden Gate Bridge, and the Empire State Building, to name a few high-profile projects. More importantly, it produced much of the rail that carried the large number of trains, as well as building the trains themselves, during the height of the railroad era.

Among steel companies, US Steel held the largest share of the nation's production capacity in the first years following its founding, and propelled the US to be the world leader in steel production as well. In

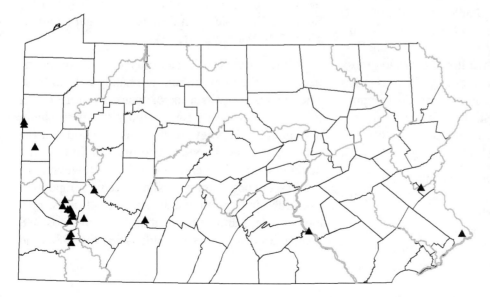

Figure 28.1 **Some of the Iron and Steel Manufacturing Facilities in Pennsylvania:** These locations represent most of the better-known iron and steel mills around Pennsylvania. It excludes some of the large steel mills, and all of the companies who manufacture that iron and steel into more finished consumer goods. The clustering of these mills in the Pittsburgh region would not change if those additional operations were shown.

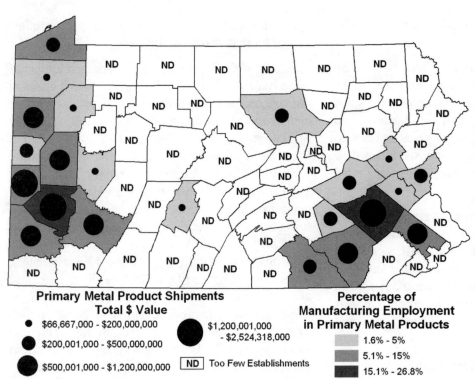

Primary Metal Product Shipments
Total $ Value

● $66,667,000 - $200,000,000

● $200,001,000 - $500,000,000

● $500,001,000 - $1,200,000,000

● $1,200,001,000 - $2,524,318,000

ND Too Few Establishments

Percentage of
Manufacturing Employment
in Primary Metal Products

1.6% - 5%

5.1% - 15%

15.1% - 26.8%

Figure 28.2 **Primary Metals Manufacturing around Pennsylvania:** A number of metals are mined in Pennsylvania (see Table 4.1 in Chapter 4). Most will go though at least initial processing here, since processed metals will be cheaper to transport and the fuel in abundant in Pennsylvania. Allegheny County is still a relatively strong center of activity and, in the east, Berks County is strategically located for anthracite coal and major transportation access. (Data source: US Census Bureau 2001)

fact it was responsible for much of the production leaving Pennsylvania, as it opened mills in such areas as Cleveland, Youngstown and Lorain, Ohio, Gary, Indiana and Joliet, Illinois. Those states also had bituminous coal resources, were closer to the iron mines in the upper Great Lakes region, and were closer to developing markets in such growing metropolitan areas as Chicago and St. Louis.

US Steel's share of US production slipped, too, as more competitors entered the scene. In 1902, US Steel produced 65 percent of all US rolled iron and steel products; by 1920 their share fell to 43 percent (Warren 2001, 138). Bethlehem Steel, also a relatively small and local company in Bethlehem at the turn of the 1900s, under the leadership of Charles M. Schwab, also followed the practice of buying out smaller companies and building new mills in order to grow. This pattern continued during the early 1900s as mills opened in many parts of the country to satisfy emerging markets in the US Midwest and West. When the automobile industry developed in Detroit, Pennsylvania took another blow.

US Steel was able to keep the competition from getting the upper hand for many years, by working out an exclusive deal with the other steel producers and with the railroads. Known as "Pittsburgh Plus," the scheme involved the steel companies charging the same prices for steel produced anywhere in the country

as if it was being produced in Pittsburgh (as the "basing point"), and the railroads charging shipping costs as if the steel was being delivered from Pittsburgh. US Steel, as the largest single producer, set those prices. It meant that Pittsburgh steel could always complete in pricing as long as their cost and quality were comparable. It also meant that other locations profited from higher profits on local projects. The overall effect was to encourage steel companies to expand their operations into new parts of the country, without US Steel losing its profitability. In 1924 it was declared illegal and a similar system with several more production centers was adopted. In 1948 the "multiple basing point" system was also stopped when a US Supreme Court decision declared it illegal (Rodgers 1995, 289; Warren 2001, 43).

"Pittsburgh Plus" was only one factor in the decline of the industry, and ultimately a minor one. The Depression was a major factor, as were newer more competitive resource supply locations, labor unions and technology changes. After World War II a renewed surge in steel production led to the development of several new mills. However, the success and optimism were short-lived. Competition from rebuilt German and Japanese steel mills, now more modern and efficient than US mills, cost US companies many overseas markets. By the 1980s and

1990s, the US was importing significant amounts of steel products.

The latest technology was the electric-arc furnaces, in which electricity became the means for heating the pig iron or other steel-making inputs, allowing for much more precise control over the entire finishing operation. American firms in general, and Pennsylvania mills in particular, could not afford to cast aside their existing technologies and replace them with electric furnaces. Most mills closed rather than make the change. The 1970s and 1980s saw the greatest numbers of closings, putting tens of thousands of steel mill workers out of their jobs. Many related industries, which take steel mill outputs and make finished products such as machinery and other consumer goods, have survived the transition by purchasing their steel from other, often overseas, suppliers.

Another development has been the mini-mill, a return in some ways to the early specialized rolling mills. These mills generally use scrap steel as their input, maintain electric furnaces to work the steel, employ non-union labor, and produce a small range of products each. Their numbers and locations (especially outside the traditional producing states such as Pennsylvania) are out-maneuvering the big steel mills, which still produce most of the volume (Warren 2001, 286-7).

Homestead

Homestead, Allegheny County was a small town along the Monongahela River just south of Pittsburgh (see Figure 28.2) when a competitor to Carnegie Steel decided to build a mill to make steel rails and other products in 1878. Within a few years, the mill was operating and competing effectively with Carnegie's mills. However, in 1883, the tides changed. An economic depression cut into the company's profits, and the workers, who had joined the Amalgamated Association of Iron and Steel Workers union, were making increasing demands of the owners. The company president was forced out because his anti-union attempts were only making matters worse. The remaining owners sold the whole operation to Andrew Carnegie.

One of the most famous events in steel-making history occurred at Homestead under Carnegie's ownership. In 1889, the workers were attempting to negotiate a new contract with Carnegie Steel. Carnegie

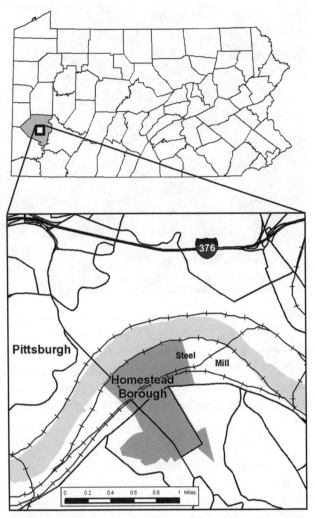

Figure 28.3 **Homestead, PA:** Homestead is located at the southeastern edge of the City of Pittsburgh.

tried to implement a new wage system with longer hours, in order to force the union to dissolve, and the workers at the Homestead Works went on strike. The plant president negotiated a compromise with the workers which cost him his job but kept the union intact for three more years (Ingham 1991, 132). In 1892, Carnegie locked out the union workers again, and most of the non-union workers went on strike. This time owners Henry Frick and Andrew Carnegie did not budge, and called in the state militia. The outcome this time went in the company's favor, and the union was essentially removed. This started a trend that spread to the other steel mills. Steel production remained essentially non-union work until the late 1930s, when the United Steel Workers union was formed.

The Homestead mill was expanded in 1902. It produced primarily structural steel girders and beams,

171

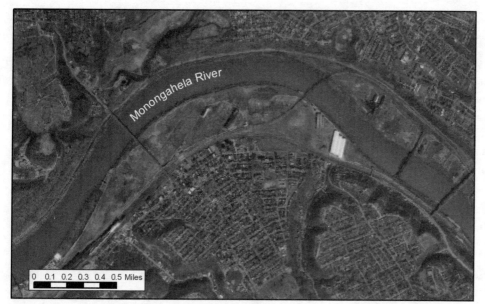

Figure 28.4 **Aerial Photograph of the Homestead Area:** This photo shows that the area once occupied by US Steel's Homestead Works has been almost completely cleared since the steel mill closed in 1986. Part of the site is being redeveloped as a major shopping center. Most of the steel mill was on the south side of the Monongahela River, but the two pareserved blast furnaces are on the north side, just northeast of the white buildings. (Image source: USGS 1999)

and steel plates. The latter operation was especially important to the US military efforts in World War I. Its work was again in demand for aiding the war efforts during World War II. A major expansion of the mill was carried out at the expense of over a thousand low income homes housing over 10,000 people (Warren 2001, 168).

Production was significant during the post-War years, but gradually declined as US Steel failed to compete effectively. In 1983 parts of Homestead's mill were shut down as un-needed. By 1986 the entire mill, which was five years earlier US Steel's third largest plant, was closed.

Since 1986 the mills occupying Homestead's riverside site have been removed. A set of about a dozen smokestacks, two of the blast furnaces, and a pump house that was a focus of the 1892 battle between striking steelworkers and US Steel's hired militia, are all that remains of the mill. In the mill's place has developed a major shopping center, with space to grow. The mill landmarks have been adopted into the Rivers of Steel National Heritage Area, affiliated with the National Park Service, but maintained by a private non-profit agency, the Steel Industry Heritage Corporation (Pitz 2003).

Conclusions

In addition to Homestead, US Steel closed many operations in the 1970s and 1980s. Other steel companies were similarly impacted. Warren (2001, 336) reports that whereas steel companies in the Pittsburgh metropolitan area employed 100,000 in the middle 1960s, the corresponding number in 1994 was 28,000. US Steel remains a strong and efficient company, operating two mills in the Pittsburgh area at Irvin and Clairton, the Fairless Works in Bucks County near Levittown, and many others around the world.

Bibliography

Ingham, John N. 1991. Making Iron and Steel: Independent Mills in Pittsburgh, 1820-1920. Columbus, Ohio State University Press.

Miller, E. Willard 1981. "Pittsburgh: Patterns of Evolution." The Pennsylvania Geographer. Vol. 19, no.3 (October), pp. 6-20.

Rodgers, Allan L. 1995. "The Rise and Decline of Pennsylvania's Steel Industry." Chapter 16 in Miller, E. Willard (ed.) A Geography of Pennsylvania. pp. 285-295.

US Census Bureau 2001. 1997 Economic Census: Manufacturing. Washington, DC, US Department of Commerce, Bureau of the Census. Internet site: <http://www.census.gov/epcd/www/97EC_US.HTM>, visited multiple times.

USGS 1999. Digital Ortho Quarter Quad (aerial photograph): pittsburgh_east_pa_se. Downloaded in TIFF format from http://www.pasda.psu.edu/access/doq99list.cgi, 11/9/04. US Geological Survey, Washington, DC.

Warren, Kenneth 2001. Big Steel: The First Century of the United States Steel Corporation 1902-2001. Pittsburgh, University of Pittsburgh Press.

Chapter 29

Retail Economy

The retail portion of the tertiary level of economic activity is essentially the insertion of "middlemen" into the relationship between producers and consumers: one merchant buys goods from the producer and transports them to regionally central locations, and another merchant buys those goods and sells them to the customers. The first merchant is the wholesaler, and the second is the retailer.

The development of the retail sector made several profound changes in the ways people lived. It proved the viability of the cash economy, and the end of complete self-sufficiency. It showed that value and cost could be added to a product without changing the product. It created the realm of advertising, in which customers who have choices are enticed toward one alternative. It helped to establish the business district of cities. It also helped to create the middle class, a large portion of whom were those merchants.

In this chapter we will look at the function of retailing in the economy, its effects on the making of Pennsylvania's landscape, and the geographical distribution of factors related to retailing. Almost everything considered in this discussion is common to the rest of the US, although, as before, Pennsylvania contributed several firsts and was ideally located in earlier economic times.

Wholesaling and Retailing

The wholesale/retail economy developed along with manufacturing in order to provide an efficient system for delivering manufactured goods into the hands of customers. In adding links in the chain from producer to consumer, the prices paid for goods increased. However, the value that those costs represented, the convenience of having a ready supply outside the home, is considered worthwhile for consumers.

Both activities developed first in the cities, and later spread into the rural parts of the state (see Figure 29.1). Philadelphia developed retail activities from its very beginnings as a city. One type consisted of peddlers roaming the residential streets. Another kind of retailing was market buildings in which producers of goods and traders from Europe or the Caribbean could set up stands. The same thing occurred in market sheds set up mostly along East High Street, which later became East Market Street. Finally a small number of specialty shops sold goods to the wealthy. That urban landscape persisted until the 1860s (Rees 2004, 32).

Pennsylvania can lay claim to several firsts in retailing. The John Wanamaker's store in center city Philadelphia was one of the first large department stores. His initial retail effort, a clothing shop on Market Street opened in 1861, was joined later that decade by Strawbridge and Clothiers. In 1876 he moved his operations to a huge abandoned Pennsylvania Railroad station closer to the center of the city, expanded his merchandise offerings, organized them into departments and called it a "department store." He is credited with "inventing" the price tag and the white sale, and with being among the first to have electric lights, an elevator and a restaurant in his store later in the 1800s (PBS 2004). The names Wanamaker and Strawbridge, as well as others such as Lit Brothers and Gimbels who opened

Figure 29.1 **Early Retail Establishment:** The "general store" at Landis Valley Farm Museum "stocks" a variety of goods that might have been tyupical of a country store in the 1800s. From locally ground flour to imported fabrics, it served all of its regular customers. Notice the prices, the selection and the packaging. Most containers or wrapping were expected to be re-used.

large downtown Philadelphia stores selling clothing and household goods in the late 1800s, may not be familiar to all Pennsylvanians, but those from the Philadelphia area recognize them as retailing institutions (Rees 2004, 32).

The first store aimed at selling low-cost goods only, F.W. Woolworth's "Five and Ten" store, began in downtown Lancaster in 1879. Other stores had offered bargain tables, but Frank Woolworth came up with the idea of an entire store full of bargains. Woolworth's stores became common in almost every town in America, and even spread its franchises overseas. It declared bankruptcy, and closed its American stores in 1998, though stores in the United Kingdom and other countries still operate under the Woolworths name (Superbrands 2005).

Retailing includes many other kinds of selling, including automobile dealers, gasoline stations, grocery stores and convenience stores. The large-scale department store and the bargain store remain part of the retail system today, but have been joined by a number of other types of large retailers. The small neighborhood specialty shop never went away, and survives in small towns and big cities alike, but is increasingly being replaced by franchised chains, such as video rental stores and pharmacies.

Behind the scenes in such retailing situations is the wholesaler, the matchmaker who purchases and transports goods from the manufacturers for sale to

the retailers. The Census Bureau classifies wholesalers as dealing in either durable goods (expected to last) or non-durable goods (expected to degrade in quality over a relatively short time). In past times the wholesaler was completely dependent on warehouse space to hold goods in transit, because buying in large quantities while selling in smaller lots helped to increase their profits.

The lines between retailing and wholesaling have been blurred in recent years. In certain product areas today, communication and computer technologies allow manufacturers and wholesalers to deal in "just in time" manufacturing and delivery, reducing storage needs. Another manufacturer's innovation is the outlet store. Outlets are owned or franchised by the manufacturer in order to remove the wholesaler and reduce consumer prices. A similar aim is implemented by "buyers clubs" (formerly referred to as cooperatives) who buy and sell in bulk sizes and quantities, like buying directly from the wholesaler. Still other retailers buy manufacturers' remainders and goods at reduced prices formerly held by stores going out of business, and sell them to consumers at retail prices lower than those suggested by the manufacturers.

The Retailing Landscape

Retailing began as an urban function, and farmers and other rural folk traveled into towns and cities to purchase their clothing and other supplies. As cities grew, they tended to have central business districts which featured large and higher-priced retailers in addition to government, financial and other services (see Chapter 30). The smaller retailers usually specialized and were located along more heavily traveled streets in residential areas of the city. Their selections of goods often reflected the ethnicity of the neighborhood.

Meanwhile, wholesaling was located in the industrial district of the cities. In this district, space is least expensive and transportation for bulk quantities is most accessible. In early cities, especially Philadelphia and Pittsburgh, that location was most often along a river. Later, railroads served the same function.

The development of the automobile and the suburban landscape had far-reaching implications for retail sales. Beginning in the 1920s, but especially by the early 1950s, the shopping plaza or shopping center was a common feature of the suburbs. Today, interstate-type highways augment and sometimes replace the railroads that served inner city industry and wholesalers. Wholesaling is now attracted to suburban locations with the best highway accessibility.

Beginning in the 1950s, but especially by the early 1970s, the shopping mall played a similar role. Advertising itself as "the largest mall on the east coast," the King of Prussia Mall northwest of

Philadelphia was also among the first. It opened in 1959 as a large shopping center, the King of Prussia Plaza, but by the middle 1960s had added indoor walkways connecting many stores. The Court at King of Prussia was added in 1981 adjacent to the original mall, forming a complex that includes over 400 retailers.

These developments often drew retail business away from larger downtown or center city stores, just as the wholesalers were being attracted to suburban industrial parks, and contributed to the declining economies of our larger and older cities. Today, the shopping center concept is making a comeback, with the difference that the stores included in the development are mostly very large retailers that are parts of nation-wide chains.

Geographic Patterns

The counties highest in total retail sales are those with the highest populations, including Philadelphia, Allegheny and the more suburban counties surrounding Philadelphia. If we divide those total sales by the number of people resident in each county, we get another measure of relative sales activity.

Notice first of all in Figure 29.2 that Philadelphia has a very low level of retailing jobs per capita, but that suburban Montgomery County is in the highest category. The same will be true of the other larger and older cities, which lost their bigger spenders when

Total Retail Sales

• $24,668,000 - $900,000,000

● $900,001,000 - $4,000,000,000

● $4,000,001,000 - $12,929,651,000

Retail Employment Per Capita

☐ 27.8 - 43

▨ 43.1 - 56

▨ 56.1 - 68

▨ 68.1 - 76.9

Figure 29.2 **Retailing around Pennsylvania:** Retailing contributes over $100 billion to Pennsylvania's economy. The values depicted in this map reflect both the relative wealth of each county's consumers and the number of establishments relative to the population served. Larger cities together with their suburbs show up in higher-value counties. The Philadelphia area is the pprimary case in which that combination does not occur within a single county. Cumberland, Centre and Lycoming Counties include significant regional cities and are surrounded by very rural counties. (Data source: US Census Bureau 1999)

175

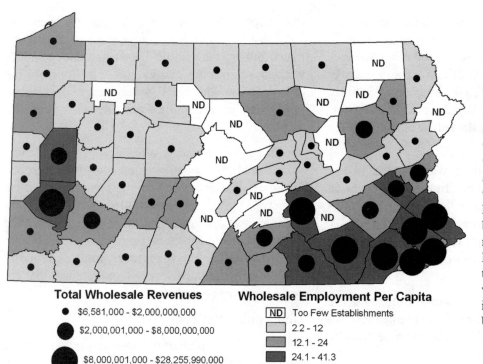

Total Wholesale Revenues

- ● $6,581,000 - $2,000,000,000
- ● $2,000,001,000 - $8,000,000,000
- ● $8,000,001,000 - $28,255,990,000

Wholesale Employment Per Capita

- ND Too Few Establishments
- ☐ 2.2 - 12
- ▨ 12.1 - 24
- ▨ 24.1 - 41.3

Figure 29.3 **Wholesaling around Pennsylvania:** Wholesaling contributes nearly $160 billion to the Pennsylvania economy, which is more than retailing contributes. The larger metropolitan areas again show up the strongest, especially where Interstate highway access is present. As before, the counties labeled "ND" have too few wholesaling businesses. Non-disclosure rules prevent the Census Bureau from releasing county totals because it would virtually reveal proprietary information. (Data source: US Census Bureau 2000)

the suburbs seemingly vacuumed away many of the middle and upper class. Pittsburgh, with the highest total sales, does not show this as clearly only because its suburbs are largely within Allegheny County. In Harrisburg's case the suburban destinations are largely to the west side of the Susquehanna River, in Cumberland County.

Wholesaling leads to similar, but not identical, patterns (see Figure 29.3). Philadelphia, Pittsburgh and Harrisburg are again the focus for wholesale sales and jobs. Their suburban areas reflect the loss of dominance of inner city industrial districts and of railroads as transportation, in favor of rural sites on the fringe of suburban development and with maximum access to Interstate highways.

Conclusions

Retailing is a visible part of the American lifestyle and of Popular Culture. Wholesaling is more hidden from most of the buying public, but still plays an important role in the retailing economy. The nature of retailing has changed, but location factors such as access to customers and transportation facilities are still some of the most important considerations. They are important for the companies deciding where to locate operations, and for the business geographer attempting to explain the success or failure of such ventures. The Internet is just beginning to have an

impact on traditional retail channels, so the future could evolve very differently.

Bibliography

Public Broadcast System 2004. They Made America: John Wanamaker. Internet site: <http://www.pbs.org/wgbh/theymadeamerica/whomade/wanamaker_hi.html>, visited 11/19/04.

Rees, Peter 2004. The Spatial Evolution of Center City, Philadelphia. In Dougherty, Percy H. (ed.) Geography of the Philadelphia Region: Cradle of Democracy. pp. 25-44.

Superbrands 2005. Woolworth's. London, UK, Superbrands, Ltd. Internet site: <http://www.superbrands.org/21843>, visited 3/2/05.

US Census Bureau 1999. 1997 Economic Census: Retailing. Washington, DC, US Department of Commerce, Bureau of the Census. Internet site: <http://www.census.gov/epcd/www/97EC_US.HTM>, visited multiple times.

US Census Bureau 2000. 1997 Economic Census: Wholesaling. Washington, DC, US Department of Commerce, Bureau of the Census. Internet site: <http://www.census.gov/epcd/www/97EC_US.HTM>, visited multiple times.

Chapter 30

Services

Our discussion of the economy has so far focused on activities which produce and exchange tangible products. The "services" part of the tertiary level of economic activity deals with the exchange of such intangibles as information, wealth, health and entertainment. Services include activities in which the means to an end (such as food preparation and "atmosphere" at a restaurant) is considered worth paying for in the same sense as the end itself (the food) is. Restaurants are not considered manufacturing establishments for that reason; neither are they simply retailers. Many sources would include transportation companies and facilities as service activities, contrary to the approach taken in this text.

Table 30.1 lists the Census Bureau's 2-digit and 3-digit categories for service activities. It almost ends up being the "Other" category in analyzing the economy because it seems to consist of activities that certainly don't fit at the primary and secondary levels of economic activity, or in the retail/wholesale activities of the tertiary level. Even though many of the activities listed in Table 30.1 seem impossible to directly compare with each other, they all have a connotation of quality of service being more important than simply quantity of output.

Early Services

The earliest of the services were financial. Banks were built in Philadelphia as early as 1781. Insurance companies soon followed, and Philadelphia had its first financial district (Rees 2004, 31).

Schools were also built at an early stage in Pennsylvania history. When the Free School Act of 1834 was passed, free public education became mandatory and became an important means for the collection and transfer of ideas and values, not to mention language (Lapsansky 2002, 176). Pennsylvania was the first state to officially require students to attend school, starting in 1848. However, the law had no teeth, the government did not enforce it, and many children were to be found working in mines and factories as late as the early 1900s (Licht 2002, 239).

Many services that had very small beginnings in colonial times became much more significant as industrialization developed, and as the shift progressed from an economy based on self-sufficiency to one based on mutual interdependence. Almost all service activities started in Philadelphia, where the larger numbers of people, the people who were more likely not to be self-sufficient, and the people who were more likely to have a formal education all lived. The service activities moved from Philadelphia to smaller cities and towns down the urban place hierarchy.

The Range of Services

Because services consist more of human effort than of material goods, there is greater variability of availability than for most extracted or manufactured goods. The urban focus that characterized their early development still persists, and you are likely to find much greater accessibility and selection for them in the larger cities and their suburbs.

All levels of government hire a large number of workers who will generally be categorized as service personnel. Police, social workers, museum curators

Table 30.1 NAICS Categories for Service Activities (US Census Bureau 2004)

51   Information
  511   Publishing Industries (except Internet)
  512   Motion Picture and Sound Recording Industries
  515   Broadcasting (except Internet)
  516   Internet Publishing and Broadcasting
  517   Telecommunications
  518   Internet Service Providers, Web Search Portals, and Data Processing Services
  519   Other Information Services
52   Finance and Insurance
  521   Monetary Authorities - Central Bank
  522   Credit Intermediation and Related Activities
  523   Securities, Commodity Contracts, and Other Financial Investments and Related Activities
  524   Insurance Carriers and Related Activities
  525   Funds, Trusts, and Other Financial Vehicles
53   Real Estate and Rental and Leasing
  531   Real Estate
  532   Rental and Leasing Services
  533   Lessors of Nonfinancial Intangible Assets (except Copyrighted Works)
54   Professional, Scientific, and Technical Services
55   Management of Companies and Enterprises
56   Administrative and Support and Waste Management and Remediation Services
  561   Administrative and Support Services
  562   Waste Management and Remediation Services
61   Educational Services
62   Health Care and Social Assistance
  621   Ambulatory Health Care Services
  622   Hospitals
  623   Nursing and Residential Care Facilities
  624   Social Assistance
71   Arts, Entertainment, and Recreation
  711   Performing Arts, Spectator Sports, and Related Industries
  712   Museums, Historical Sites, and Similar Institutions
  713   Amusement, Gambling, and Recreation Industries
72   Accommodation and Food Services
  721   Accommodation
  722   Food Services and Drinking Places
81   Other Services (except Public Administration)
  811   Repair and Maintenance
  812   Personal and Laundry Services
  813   Religious, Grantmaking, Civic, Professional, and Similar Organizations
  814   Private Households
92   Public Administration
  921   Executive, Legislative, and Other General Government Support
  922   Justice, Public Order, and Safety Activities
  923   Administration of Human Resource Programs
  924   Administration of Environmental Quality Programs
  925   Administration of Housing Programs, Urban Planning, and Community Development
  926   Administration of Economic Programs
  927   Space Research and Technology
  928   National Security and International Affairs

and librarians (although some museums and libraries are private enterprises or the public face of a foundation or other organization), elected officials and tax office secretaries are all employed in the service sector of the economy. There are both public and private schools and universities, of course; all of their employees, not just the educators, would be considered service providers (Abler 1995, 298.

The availability of many services is distributed in direct proportion to the population. Others will choose locations based on specific site characteristics; thus, military bases require isolation and accessibility, and sports stadiums maximize their audience reach, but ski slopes require both colder temperatures and steeper slopes. In some cases, as is true with retailing, there is a minimum population to support the service; this is why hospitals, for example, are usually located in the larger cities and boroughs.

Finally, many services feed off of other services. You seldom find a motel without a restaurant and gas station nearby. Lawyers' offices are more frequently located near courthouses than near customers, and health care specialists are usually located nearer to the hospitals. The central business district of any town or city hosts more restaurants, copy shops, banks and delivery services than it does automobile repair shops or doctors' offices.

One special category of the services sector is often identified as a separate level of economic activity: the "high"-technology firms associated with the Information Age. These are the companies who specialize in computer technologies and software, telecommunications, medical and other scientific research and other types of technology-oriented businesses that have primarily developed since World War II.

Service Patterns

While some high-tech firms (or at least their headquarters) may be located in the central business districts of major cities, many more of them are to be found on the suburban fringe of major urban areas such as Philadelphia. Recall that Interstate Highways were built around cities' downtown cores. It is those same peripheral Interstates, or roads with similar design and capacities, that attract the high-tech businesses. They fight with shopping malls and outlet centers for prime real estate near the highway ramps. These firms are attracted to locations closer to universities in many cases and, as we saw with inner city businesses, to areas which can provide the auxiliary services they require.

Hospitals and health care employment distribution rates provide an interesting illustration of these location generalizations. Initially most hospitals were

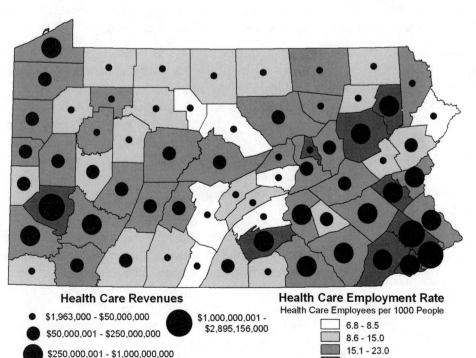

Health Care Revenues

- $1,963,000 - $50,000,000
- $50,000,001 - $250,000,000
- $250,000,001 - $1,000,000,000
- $1,000,000,001 - $2,895,156,000

Health Care Employment Rate
Health Care Employees per 1000 People

- 6.8 - 8.5
- 8.6 - 15.0
- 15.1 - 23.0
- 23.1 - 39.3

Figure 30.1 **Health Care Services around Pennsylvania:** This map shows the revenues from health care services, as well as the proportion of people who work in health care. Health care contributes over $17.5 billion to the Pennsylvania economy. Health care employees include as doctors, nurses, and other workers in hospitals, clinics and offices of medical, dental, and mental health professionals. The urban orientation of health care is clear, including the shift from central city locations to the suburbs. (Data source: US Census Bureau 1999)

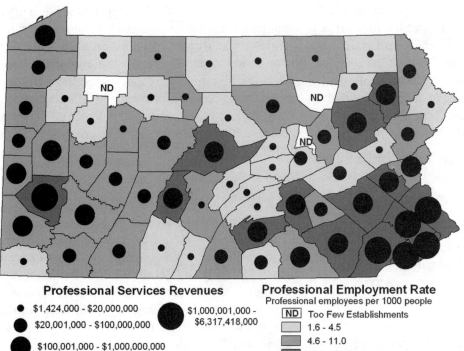

Figure 30.2 **Professional Employment across Pennsylvania:** Professionals provide advice and assistance in business management, scientific research and other technical services. These jobs tend to be attracted to urban areas and places where universities and similar institutions provide expertise, facilities and a pool of potential employees. Such professional services contribute over $26 billion to the Pennsylvania economy. (Data source: US Census Bureau 1999)

Professional Services Revenues

- $1,424,000 - $20,000,000
- $20,001,000 - $100,000,000
- $100,001,000 - $1,000,000,000
- $1,000,001,000 - $6,317,418,000

Professional Employment Rate
Professional employees per 1000 people
- ND Too Few Establishments
- 1.6 - 4.5
- 4.6 - 11.0
- 11.1 - 43.8

built near the central business districts of cities because these were central locations in all directions relative to the area's population. Public transportation routes were likely to include them as special nodes. As larger numbers of people moved out of the cities, and as we became more automobile-oriented, city locations presented limitations. The hospital's ability to expand as the urban area's population grew was limited, and the need to provide parking used valuable real estate. By now many major hospitals have developed satellite campuses.

At the scale shown in Figure 30.1 a few of these patterns are visible. The higher health care employment rates in Allegheny, Lackawanna, Luzerne and Lehigh counties reflect each urban area's population and the fact that the suburbs are in the same county as the city. Philadelphia, on the other hand, is in a lower category than its neighboring suburban counties. The interesting exception is Montour County. Danville is the home of the Geisinger Medical Center, part of the Geisinger Health System. The Medical Center started as Geisinger Memorial Hospital in 1915 and grew over the years as a well-endowed regional hospital able to serve a largely rural population. It has since expanded to a system of facilities in Danville, Wilkes-Barre and Scranton, along with a health insurance program (Geisinger Health System 2004).

Professional employment (Figure 30.2), however, is more consistently urban or suburban in its focus. This category only includes businesses whose service is professional management, either for other businesses or for government projects, or is scientific or technical expertise. These are firms that are hired as consultants by other firms or government agencies. Even though every manufacturer and every retailer has a professional staff leading the organization, they are not counted in this Census category.

The interesting case visible in Figure 30.2 is the relatively high placement of Centre County. State College, Centre County's largest population center, is one of the faster growing communities in Pennsylvania. It is the presence of the Pennsylvania State University which has both attracted and spun off many research-related businesses in Centre County. The number of such firms there employ more professionals than in other counties with comparable populations.

Real estate is another services category which tends to be urban-oriented, with its main focus in urban and suburban residential and business properties. Real estate salespeople earn more money on sales of more expensive properties. Others who own real estate for rent to businesses or residents will also be attracted to sites whose location makes them very desirable.

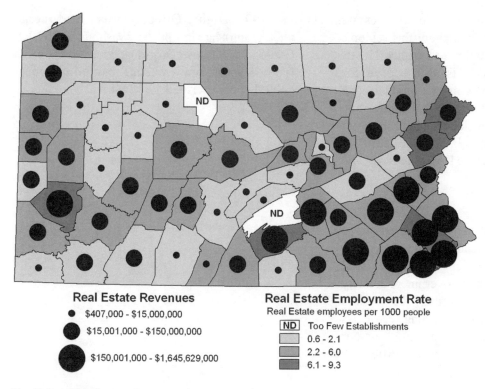

Figure 30.3 **Real Estate Employment across Pennsylvania:** Real estate developers, sales agents and rental or leasing services are attracted to where there are large numbers of porperties changing hands very rapidly. Real estate revenues for the entire state add up to over $7.5 billion. The usual major metropolitan areas are joined here by the northeastern counties of Wayne and Pike, near the New Jersey/New York border. (Data source: US Census Bureau 1999)

Real Estate Revenues

- ● $407,000 - $15,000,000
- ● $15,001,000 - $150,000,000
- ● $150,001,000 - $1,645,629,000

Real Estate Employment Rate
Real Estate employees per 1000 people

ND	Too Few Establishments
	0.6 - 2.1
	2.2 - 6.0
	6.1 - 9.3

Traditionally the most expensive properties were in the central business districts of larger cities, as demonstrated by the multi-story buildings constructed there. Such high-rise towers now occasionally are visible along urban perimeter routes and near Interstate exits. The more such buildings house multiple tenants, the more likely real estate professionals are involved.

Figure 30.3 shows the attraction of real estate activity to places with the best transportation access, with one major exception. Monroe and Pike Counties

Figure 30.4 **Pike County Development:** Part of the Appalachian Plateau glaciated area, which has left many natural lakes and made it relatively easy to create man-made lakes. These lakes are the focus of many new developments, including these which still have many lots left to sell. (Image source: USGS 1999)

are home to an alternative focus of real estate development. More will be said about this region's recreation and tourism reputation in Chapter 31. Its newest attraction, however, is as a region for expatriates from the New York City metropolitan area. The aerial photograph in Figure 30.4 illustrates the style of development that developers and realtors are promoting. The area shown is in northwestern Pike County, a short distance from the Delaware River and Interstate 84.

Conclusions

The service functions of our economy had roots in very old economic or financial activities, but it was a matter of specialization in particular services, as well as the increasing involvement of government in many aspects of everyday life, that has made the services sector of our economy the fastest growing sector. As industrial enterprises continue to fail, more of our citizens are turning to the service activities for employment.

Bibliography

Abler, Ronald F. 1995. "Services." Chapter 17 in Miller, E. Willard (ed.) A Geography of Pennsylvania. pp. 296-311.

Geisinger Health System 2004. About Geisinger. Danville, PA, Geisinger Health System. Internet site: <http://www.geisinger.edu/about/history.shtml>, visited 12/13/04.

Lapsansky, Emma 2002. Building Democratic Communities: 1800-1850. Chapter 4 in Miller, Randall M. and William Pencak (eds.) Pennsylvania: A History of the Common-wealth. pp. 153-202.

Licht, Walter 2002. Civil Wars: 1850-1900. Chapter 5 in Miller, Randall M. and William Pencak (eds.) Pennsylvania: A History of the Commonwealth. pp. 203-256.

Rees, Peter 2004. "The Spatial Evolution of Center City, Philadelphia." in Dougherty, Percy H. (ed.) Geography of the Philadelphia Region: Cradle of Democracy. pp. 25-44.

US Census Bureau 1999. 1997 Economic Census: Services. Washington, DC, US Department of Commerce, Bureau of the Census. Internet site: <http://www.census.gov/epcd/www/97EC_US.HTM>, visited multiple times.

USGS 1999. Digital Ortho Quarter Quad (aerial photograph): hawley_pa_ne. Washington, DC, US Geological Survey. Downloaded in TIFF format from <http://www.pasda. psu.edu/access/doq99list.cgi>, visited 12/12/04.

Chapter 31

Recreation and Tourism

One part of the tertiary sector of the economy that is particularly strong in Pennsylvania now, and is expected to gain strength in the future, is recreation and tourism.

"Recreation" is a vague word that encompasses many outdoor activities and land uses, not to mention many at-home indoor activities (for example: card, board, video, and computer games). Recreation on a scale that makes an interesting subject for geographical study requires sufficient economic presence that it either affects a significant portion of the population economically, or has a major impact on land use. Generally, this extent of recreation was achieved only when technology and other factors enabled sufficiently large numbers of people the luxury of leisure time away from work and housekeeping. The economic impact of recreation is measured by examining the amount of land dedicated to its pursuit, the amount of income generated in businesses and organizations hosting such activities, and the amount of money spent by recreation participants.

The word "tourism" is also vague. Even though it refers to travel-related activities, it encompasses both the businesses that represent the tourists' destinations and those that provide more incidental services. We could try to categorize recreation and tourism activities by the types of activities, by the types of facilities, by the locations of the facilities, or by the types of owners or operators of the facilities. We will take a geographical approach to looking at them according to the types of businesses or governments that own the places.

Parks, Forests and Other Preservation Efforts

A list of federally protected sites includes Allegheny National Forest, which is huge, and many smaller areas. They consist of National Historical Parks (Independence and Valley Forge), a National Military Park (Gettysburg), a National Battlefield (Fort Necessity), National Historic Sites (Allegheny Portage Railroad, Eisenhower, Friendship Hill, Hopewell Furnace, and Steamtown), two National Memorials (Johnstown Flood and Thaddeus Kosciuszko), two National Wild and Scenic Rivers (the Delaware and the Upper Delaware), a National Recreation Area (Delaware Water Gap), and parts of two National Scenic Trails (the Appalachian Trail and the Potomac Heritage Trail), and many are shown on Figure 31.1. All of these federal sites, except for the Allegheny National Forest, are administered by the National Park Service. The various categories have been created to designate different goals, management strategies, funding mechanisms, and of course types of focal point. The preservation of natural beauty or significance, and historical importance on the national stage are the most common focal points.

The federal park system got its start in 1872 when Yellowstone National Park was preserved in Wyoming. Once the federal government established the concept and the mode of implementing government-owned parks, the state and local governments got into the act as well. Most park creation efforts by federal, state and county governments, thus, happened in the 1900s. State-owned lands include categories similar to the federal designations, though presumably of greater

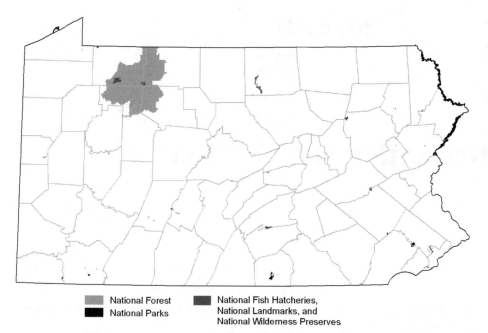

Figure 31.1 **Federal Preserved Lands in Pennsylvania:** The US government has recognized Pennyslvania's important, if not unique, landscapes and historical sites in its selection of places and areas administered by the National Park Service and National Forest Service. Most such places are mere dots on this map. Independence National Historical Park, which includes the Liberty Bell and Independence Hall, is so small at this map's scale, that it is not even visible.

National Forest
National Parks

National Fish Hatcheries,
National Landmarks, and
National Wilderness Preserves

interest to residents of Pennsylvania than of the rest of the country. There are approximately 111 State Parks hosting a wide range of recreational activities, and 28 State Forests featuring recreation facilities such as picnicking and fishing (see Figure 31.2). Pennsylvania also created an additional category, the State Game Lands to provide land for regulated hunting. The state also includes museums, battlefields, famous homes, former industrial sites and other Historic Sites in a category that straddles the line between park and educational facility (not shown in Figure 31.2).

Lower levels of government are also in the preservation and recreation business. Counties are relative late-comers in acquiring lands for such purposes. Pennsylvania cities and other communities have recognized its value ever since William Penn created five public parcels in his original plan for Philadelphia (see Figure 15.1 in Chapter 15). Such plots were not necessarily for recreation purposes; some were for public gathering areas and simply for city beautification. Community parks usually include playground, ball fields and courts, and other amenities. Schools also include such play areas in their ever-increasing land areas.

The latest trend in land acquisition and preservation is referred to as "rails to trails;" this effort involves the purchase of abandoned railroad rights of way, removing the tracks but improving the railroad bed for use by bicyclists, hikers, cross-country skiers, and in many cases riders of motorcycles and other powered vehicles. Counties and municipalities (or regional groups of municipalities) are often involved in such purchases, with funding often coming from state or federal sources (Boas 1987, 17).

As we saw with farmland preservation efforts, non-profit organizations also purchase land for public benefit. Their income sources range from organization members to private corporations. These non-profits include the Nature Conservancy, a national group, as well as local organizations. A well-known example active in southeastern Pennsylvania is the Brandywine Conservancy.

Private companies are also in the business of buying land in order to provide settings for such active pursuits as paintball, skiing, golf and, in more urban locations, skateboarding. The public's willingness to pay for such activities, as well as for the preserved history and nature sites of government and other organizations, demonstrates the degree to which leisure time and mobility have become part of American popular culture.

Tourism

In addition to pursuing recreational activities and their active interest in preserved pieces of history and nature, the public also are willing to dedicate their leisure time and spare money to other non-essential travels. Theme parks and other amusement parks are among the most highly visited travel destinations. The fact that they are considered "parks" shows that they

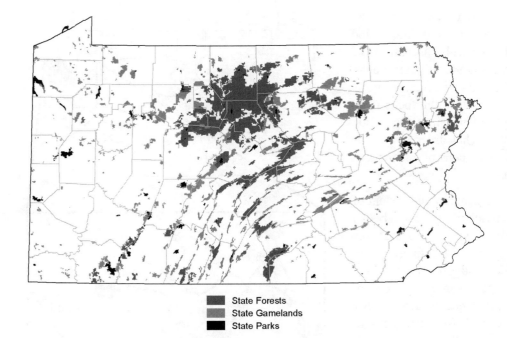

Figure 31.2 **State-Owned Recreation and Land Preservation Tracts in Pennsylvania:** Following the lead of the federal government, the state has acquired extensive land areas dedicated to preserving nature and providing for outdoor activities. Note the full extent of the natural area in north central Pennsylvania that includes our managed elk habitat (see Figure 12.1 in Chapter 12). The state also manages dozens of smaller sites as museums and similar historical facilities

State Forests
State Gamelands
State Parks

are descended from their more nature- and history-related forbears.

The concept of tourism captures the idea that the population is attracted to passive as well as active leisure activities. Even though facility owners reap significant income from the attractions they host, tourism spending goes much farther. Hotels, motels and restaurants, for example, rely to a great extent on money spent by tourists, in addition to business travelers and others. Add to that the specialty shops appealing directly to the tourists, and the impact is clearer. This is why we have referred several times to former industrial towns looking to tourism to make up for lost revenues.

Spectator Sports

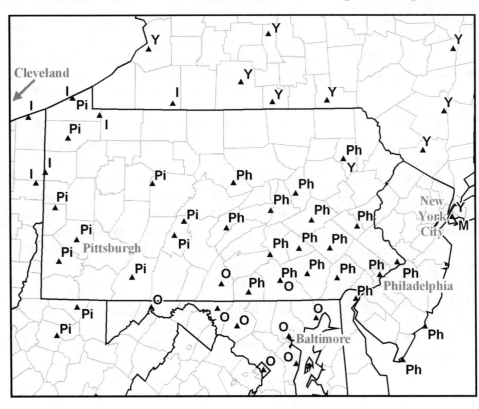

Figure 31.3 **Radio Stations that Broadcast Major League Baseball in Pennsylvania:** The Philadelphia Phillies (Ph), Pittsburgh Pirates (Pi), Baltimore Orioles (O), New York Yankees (Y), and Cleveland Indians (I) all have sufficient followers in Pennsylvania to justify allowing radio stations here to broadcast their games. The New York Mets (M) do not appear to have any. (Data source: Marvel 2005)

185

Professional sports also represent recreation/tourism draws, and have their roots in the same time period and social changes. Beginning with baseball in the middle to late 1800s, professional athletes began performing before large crowds. Spectator sports combine aspects of recreation (since the focus is on athletic performance) and tourism (since the public are viewers, not participants). They are major points of economic activity, as well as icons of our modern American Popular Culture.

Figure 31.3 illustrates the geographical extent of interest in one such sport, baseball, in Pennsylvania as indicated by the existence of radio stations broadcasting major league games. Such interest is an expression of culture. It shows that Pennsylvanians will follow teams that are closer, even if they are from surrounding states. Similar patterns exist for professional ice hockey and for college and professional football and basketball.

Maps of major professional sports also represent economic patterns. First, of course, the team owners and players benefit economically from spectator and fan spending. Figure 31.3 is showing that such spending is not just found at the team's home location. Broadcast sponsors and sporting goods stores in the radio stations' towns will also profit from the team's fans. Within the team's home city, restaurants and bars, sporting goods stores and transportation-related businesses will all get increased income from game-day crowds. In addition, these reasons can then be used by the team owners to justify appeals to the state and home city for funding assistance for stadium construction projects.

The Pocono Mountains Resorts Area

The Poconos area is a good example of a travel destination that combines activities and land uses focused on nature, history, participatory sports and tourist destinations. North of Philadelphia and Allentown, east of Scranton and Wilkes-Barre, and west of New York City and Newark, New Jersey (see Figure 31.4), the modern resort reputation of the Poconos started from very non-resort beginnings. Its most positive attribute was that it was not far from New York and Philadelphia, and upstream from the latter (it was eventually connected by canal and railroad to New York City). Otherwise it began as a forested and formerly glaciated piece of the Appalachian Plateau.

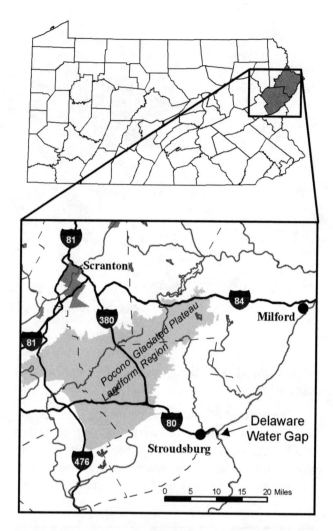

Figure 31.4 **The Poconos Resort Area:** Hundreds of tourist hotels, ski slopes, golf courses and second-home develoments dot the landscape of this area of northeastern Pennsylvania. The area's boundaries are vague, but relate strongly to the landform region (in gray) described earlier.

By the middle 1800s trees were being harvested to the same devastating effect that eventually affected all of Pennsylvania. Tanning of leather also became an important local industry. Its higher elevation, abundant shallow glacial lakes and relative location (north of Philadelphia and inland from New York and Newark) also created the ice-cutting industry of the late 1800s, before mechanical-chemical refrigeration became widely available in the post-World-War-I early 1900s (Ackroyd-Kelly 1987, 19).

Early in the 1900s, beginning around the Delaware Water Gap and gradually spreading westward, it gained its first reputation as a respite from city life. Several luxury hotels attracted upper class visitors from New York and Philadelphia. Soon, as a leisure activity for the rich, some second-growth forest land was being cleared again for a new phenomenon: golf. The upper class resorts (owned by wealthy outsiders or their clubs) were built on large privately-owned tracts, usually in a lake-side setting (Ackroyd-Kelly 1987, 20).

In the 1930s a second tier of visitors began arriving as part of a new trend: the locals were building smaller, less ornate hotels and renting rooms to affluent, but not upper-crust, visitors. This soon developed into a honeymoon-oriented clientele, a reputation the Poconos enjoyed, and advertised, well into the 1960s (long after the wealthiest abandoned the area). Changing times have relabeled the honeymoon resorts as "couples" and even "family" resorts (Ackroyd-Kelly 1987, 21-22).

In the 1970s another identity of the Poconos developed, when ski slopes began to dot the sides of glacial hills. Ackroyd-Kelly attributes this development to two necessary precursors: the Interstate highway (especially I-80, I-84 and the Northeast Extension of the Pennsylvania Turnpike now labeled I-476), and artificial snow-making (1987, 22).

Finally, the latest trend also developed as a response to improved highway accessibility. "Second home" communities have become big business to area developers. The typical practice is to buy large tracts of land facing a natural lake or an easily-created reservoir, subdivide them into small properties, and then build hundreds of "cabins" or other small homes (Ackroyd-Kelly 1987, 22-23). Many of the communities have some year-round residents, and home-owners associations create rules and "neighborhood watches" to help protect their investments. Whether this is the ultimate destiny of the Poconos, or another identity surfaces, remains to be seen.

Conclusions

Recreation and tourism represent a late development in the tertiary economic sector. Almost entirely a trend of the 1900s, they reflect society's willingness to spend money on leisure activities. Automobiles and highways, the five-day work week and 50-week work year helped to create the demand. Once again Pennsylvania's location as a crossroads for the country has helped to strengthen its tourism-related activities. Our many key events in the early history of the country, our extensive territory with many large, beautiful, sparsely populated areas, have all created the economic opportunities. The decline of several key manufacturing industries has led many parts of Pennsylvania to advertise its service and retail sector, and recreation and tourism are key parts of that.

Bibliography

Ackroyd-Kelly, Ian 1987. "The Near Country: The Historical Geography of the Pocono Resorts." The Pennsylvania Geographer. Vol. 25, nos. 1 & 2 (Spring/Summer), pp. 18-23.

Boas, Charles W. 1987. "The Railroads of York County, PA: Rail Center to Recreation." The Pennsylvania Geographer. Vol. 25, nos. 1 & 2 (Spring/Summer), pp. 12-17.

Marvel 2005. Radio Roadtrip.com – Baseball. Marvel Company, LLC. Internet site: <http://radioroadtrip.com/baseball.htm>, visited 3/5/05.

Section D: Chapters 32 to 37

Urban and Environmental Issues

In the coming final chapters we shall see just how the concepts studied so far come together in bigger, more complex contexts. The smaller place studies and the discussions within each chapter have established the importance of historical precedents. The landscape and other natural conditions are important in demonstrating that there is always a natural setting or process at work that must be considered. Our studies of the people and their cultural and economic activities show that decisions made, and even any lack of decisiveness, have implications in day-to-day life. These chapters will focus on urban and environmental *issues*, meaning that we will pay special attention to aspects of cities and the environment that are uncertain or that stimulate conflicting opinions.

Urban Issues

We will start by examining cities and other urban areas and by pointing out that, despite their common features, which we called their "city-ness" in Chapter 17, each is unique. Our larger cities have grown more by design and leadership than by luck of location and situation. The issues that have emerged as most visible and salient in each place, then, are the products of many conditions, events and decisions. One generation's chosen solution might be later generations' problem or blessing.

Pittsburgh and Philadelphia are obvious choices for more in-depth study. Pittsburgh's shorter history and clearer identity as the center of steel production in the late 1800s and early 1900s make it easier to consider first. Its collective journey since the 1960s has also been more clearly defined. Philadelphia, on the other hand, has been faced with similar challenges in a different physical, cultural and structural setting. Progress has been made, but in a less focused way.

The industrial past of Philadelphia and Pittsburgh certainly made many fortunes and employed many workers, feeding their families and creating opportunities. Now, those same industrial sites have to be cleared and cleaned, sometimes at public expense. The landscapes are landscapes of challenge rather than of opportunity. A major issue, therefore, is "Who should bear the costs of meeting the challenges and creating new opportunities?"

Cities reflect much larger regions than just the space within their boundaries. As we saw in Chapter 18, if it were not for key decisions made over 150 years ago, our cities might *be* those larger regions. Could there be a point, today, at which the larger urban region or the state gains when, for example, Philadelphia receives financial support from its neighboring counties? That larger population and tax base might have provided more resources for the cities to resolve their issues. On the other hand, as comparisons between all of Pennsylvania's cities suggest, perhaps such larger sizes would create even more conflicts of interest, and trying to find widely accepted solutions would become even more difficult.

Interestingly, though, state government has no agency whose purpose is to study, coordinate and provide one source for state resources to help resolve urban issues. The implication seems to be that each city must resolve its own issues, or at least request state assistance on a case-by-case basis. Cities *are* both universally similar and universally unique.

Environmental Issues

We will conclude the book by studying a selection of environmental issues. The three types of environmental issues chosen, air pollution, water pollution and waste disposal, are very large and complex. They are felt most strongly at or near specific sites, such as landfills, electric power stations and large factories. At the same time, they are repeated in many or all areas of the state.

Just like our urban studies, most environmental issues result from past decisions impacting natural systems in different locations and may have been made with the best of intentions. Most of these issues are common throughout America and in many places across Pennsylvania. A few are unique, either to specific places or to Pennsylvania.

Cleanup and prevention of air and water pollution in Pennsylvania are the responsibility of two government agencies: the US Environmental Protection Agency and the Pennsylvania Department of Environmental Protection. Federal environmental regulations were first imposed long before The Environmental Protection Agency was created in 1971. Just like we learned in the time following the terrorist disaster of September 2001, when environmental issues became a focus of national concern in the 1960s it was realized that the federal government's efforts were neither strong enough nor well coordinated.

The Pennsylvania Department of Environmental Protection and the Pennsylvania Department of Conservation of Natural Resources were originally under one agency, the Department of Environmental Resources, formed in 1971. Today's agencies were separated by Governor Ridge in 1995. Much effort at the state level is dictated by the need to meet federal guidelines, with the threat that not doing so may cause the state to lose some federal funding for other needs.

Like our urban issues the environmental problems can raise difficult questions. What kinds of problems are blamed on specific companies or land owners, and which are blamed on "society" as a whole? Who should pay to clean up or prevent the continuation of the problem? Who is impacted by the different problems? How should we prioritize the problems? What if solving the environmental problem hurts the economy, such as when the only solution is to shut down an employer?

Issue Convergence

In several ways, urban issues and environmental issues are the same types of problems. The first way this becomes obvious is the relevance of geography as the discipline that best understands them. Geography, more than other disciplines, looks at how historic, economic, cultural and technological aspects of any issue blend, or conflict, with its setting and location. Geographers understand that the nature of a problem changes depending on the scale at which it is examined. A city with crime problems has to make at least two decisions: how many new police officers it can afford to hire; and where among the city's neighborhoods to deploy them. Similarly, acknowledgement that Pennsylvania has a problem with mercury residues in water and in the fish taken form these waters forces two decisions: how much money and other resources the state will muster to address the issue, and where the critical watersheds in which to deploy them are.

A second way in which urban and environmental issues are similar is that they both have emerged strongly from our industrial past. The maximum rates of city growth, whether measured by numbers of people, size (area) dimensions or possibly even dollars invested, were achieved during the late 1800s and early 1900s. Population growth led to issues of economic class, ethnic differences and political engagement. The growth of early industry created an acceptance of dirt, smoke, noise and congestion, at least in some parts of the cities. Waste was created in ever-mounting quantities, especially as we became consumers and retailers. Even though these are the problems with which we must now deal, we may never have accomplished the lifestyles of affluence and leisure many of us now enjoy if we had not created them.

Third, thinking politically, the uncertainty and conflict play into the differences between the two primary political parties. The Democrats favor strong governmental involvement, leadership and financial support toward attempting solutions. The Republican perspective sees government involvement as a last resort, to be pursued only if the other attempts fail. These perspectives are frequently voiced in discussions of urban and environmental issues alike.

Chapter 32

Uniqueness in Urban Places

There are plenty of similarities between urban places, as we saw in Chapter 17. There we looked at their common development, structure and functions. Pennsylvania's urban places include incorporated cities and boroughs (and town), and many of its unincorporated towns and villages. In this chapter we will look at factors that make any urban place unique compared to others.

The common structural and functional characteristics of urban places include street patterns, governmental systems, and physical settings (a relatively limited range of variability in climate, landforms, major soil types and natural vegetation characteristics). For example, the rectangular grid of streets is generally quite independent of any natural landscape features (other than large water bodies) such as hills, valleys or smaller streams. Common development included founding and preconceived ideas about what cities should look like.

On the other hand, an urban place's character depends on many factors. Its identity is a reflection of the geographical patterns of features and areas within it, of the individual people who have been leaders, designed pieces of it and lived there, and of the particular resources (including variations in quality and accessibility) nearby. The list in Table 32.1 presents several questions which are likely to produce unique answers for any place examined.

Several of these questions will form the core of our examination of the unique identities of several Pennsylvania places in this chapter. The places vary in size, from the small borough of Montrose in

Table 32.1 Place Identity Questions

When was _____ founded?

By whom was _____ founded?

What resources are nearby?

What ethnic groups came to _____, and when?

What is _____'s population, and what has it been at different stages of its growth?

Where is _____ in relation to other places?

What highways, railroads, airports or ports serve _____?

Which companies or other organizations are major employers in _____?

Is _____ a significant tourist destination?

What is _____'s economic condition?

Susquehanna County to the bustling city of Erie. We will consider information about their development and current conditions, including what their specializations have been.

Erie

On the American shore of Lake Erie is the nearly triangular Erie County, a late addition to Pennsylvania's territory that gave Pennsylvania three successful port locations (after Philadelphia/Chester and Pittsburgh). This triangle was purchased from Native American tribes, and has proven to be invaluable. In particular the site of the City of Erie, with its harbor protected by the "sand spit" peninsula known as Presque Isle, came to be an ideal port (see Figure 32.1).

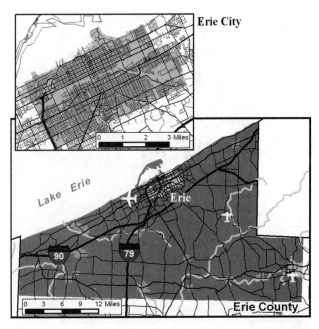

Figure 32.1 **Erie, Erie County:** Erie's lakeside location has been a major factor in its military and economic success.

In the late 1700s a series of forts were built by French and English colonists and the area changed hands between them and Native Americans. One tribe of Native Americans was the Eries. The English finally established the stronger claim by starting a permanent town in 1795.

Erie's most famous contribution to American history came during the War of 1812, when Commander Oliver Hazard Perry led a small fleet of American ships against a British invasion from Canada. His victory is heralded in several preserved sites around the city, including his flagship the US Brig Niagara. The Niagara has been restored and offers trips out onto Lake Erie.

Later, Erie played a key role in the rise to dominance of the Pittsburgh iron and steel industry. Its port facility handled large volumes of iron ore from the upper Great Lakes, as well as iron and steel products headed westward. Its strategic location was strong because of the confluence of lake ships, rail lines to Pittsburgh, and rail lines parallel to the Lake Erie coast. Today, Interstates 90 and 79, and Erie International Airport, add other transportation options.

Erie's location was strong enough that Edison's General Electric (GE) Company chose to locate one of its major manufacturing facilities there in the late 1800s. It became GE's center for transportation

research, development and manufacturing, and still builds a number of large transportation-oriented industrial products. Its best-known product is the GE diesel-electric railroad locomotive. General Electric-Erie was developing its design as early as the second decade of the 1900s. It also produces railroad signaling equipment, large "off-highway" vehicles for mining and construction industries, and well-drilling motors (GE Transportation 2005). In an age when many manufacturers are closing such plants in many parts of the US, Erie's GE plant still employs over 4,000 workers. Erie's manufacturing sector also has strong production and reputations in the plastics and paper industries, too.

Lancaster

Lancaster was the American colonies' first significant inland town, and remained the largest one from soon after its founding in 1718 for about a century. It was even the capital of the US for a day in 1777 while the Continental Congress was en route from Philadelphia to neighboring York (see Figure 32.2). It was also the state capitol briefly around 1810. Lancaster was designated a city in 1818.

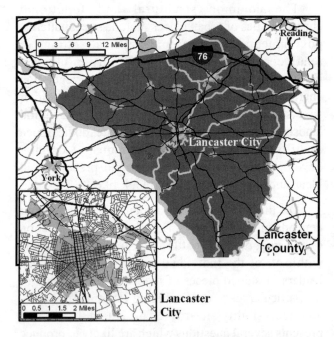

Figure 32.2 **Lancaster, Lancaster County:** The city of Lancaster is the center of a very productive agricultural region, but also serves as a concentration of manufacturing employment. Lancaster's industrial economy mixes older manufacturing based on primary metals and floor and ceiling products with more modern technologies.

192

Most of Lancaster's economic significance derives from its function as a market center for one of the country's most productive agricultural counties (see Chapter 21). It has also been an important regional retail center, spawning not only F.W. Woolworth's economic career (see Chapter 29), but Milton Hershey's first candy-making efforts as well. Since around 1909, it has also been the headquarters for Armstrong World Industries, manufacturers of ceiling and floor tiles as well as other building construction products.

Lancaster today is one of the fastest growing urban areas in Pennsylvania. Its location on the outer fringe of urbanization from Philadelphia and Wilmington, DE, and not far from the state capitol in Harrisburg, gives it a desirable economic location. It has accomplished all of this with rail connections to Philadelphia, Harrisburg and (in the past) Baltimore, and moderately strong highway connections, but no access via water and a very modest airport. Its agricultural landscape makes it attractive to suburbanites escaping those cities, creating the problems with disappearing farmland noted in Chapter 21.

Like most cities, Lancaster saw its population and its tax base undergo major changes beginning in the early 1950s (Schuyler 2002, 52-53). While the county prospers with tourists, outlet shoppers and suburban residential and light manufacturing development, the city economy struggles. Recent efforts have focused on building a convention center in Lancaster City's central business district and attracting an independent-league baseball team to locate in the Lancaster area and to build its stadium in the city's struggling northwestern industrial/warehousing district.

Sharon

Literally on the western (Ohio) boundary of the state, and on the Shenango River that flows into the Ohio River further south, lies Sharon, in Mercer County (see Figure 32.3). Sharon was originally settled in 1795 and chartered as a borough in 1841. Like many other communities in Pittsburgh's sphere of influence, it became a center for coal mining and steel production, industries which carried its economy well into the 1900s.

In fact, it was not until the 1970s that Sharon saw significant impacts of industrial decline. Several

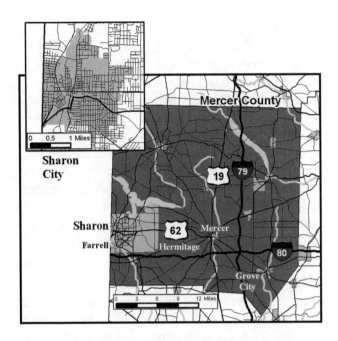

Figure 32.3 **Sharon, Mercer County:** The city of Sharon is representative of many cities in western Pennsylvania that were formerly dependent on the coal and iron and steel industries. Sharon's uniqueness, though, stems from Sharon Steel having remained independent of US Steel and other major steel companies, and having outlasted many of its competitors, into the 1990s.

important factories closed between 1979 and 1983, producing a decline in population seen in the 1980 Census. The biggest single blow was the closing of Sharon Steel in 1992. Both periods created significant peaks in the town's and county's unemployment history. As a result of the changing economy, Sharon's population has aged significantly, its median age rising from 29.2 in 1970 to 36.4 in 1990. Per capita income and median house values are also well below the levels for all of Pennsylvania (MC RPC 2004).

A significant local issue for years after the closing of Sharon Steel was the clean-up of contaminated soils and groundwater, and the removal of industrial waste known as slag, from its site. The US Environmental Protection Agency decided in 2000 that the slag would be able to be reused as fill material elsewhere, saving millions of dollars in clean-up costs (US EPA Region III 2000).

Within Mercer County, the population is shifting away from the cities of Sharon, Hermitage and Farrell toward the boroughs of Mercer (the county seat) and Grove City, located closer to the intersection of Interstates 79 and 80.

Montrose

Susquehanna County is along the eastern end of Pennsylvania's northern border with New York State (see Figure 32.4), an area of the Appalachian Plateau known as the Endless Mountains. Montrose's founding dates to a time when control over this area of Pennsylvania was disputed by Connecticut. The layout of the town's streets is nothing like the standard grid of most Pennsylvania towns. Since it is the county seat, the county courthouse is the most prominent building in town. Typical of New England towns, the courthouse is adjacent to a "village green," along with the library/historical society and the fire company.

Montrose is somewhat unusual in that it is located on the Appalachian Plateau, miles from any significant river. Access requires a drive of approximately ten miles from Interstate 81, and railroads do not serve the town. Its main industrial products are wooden coat hangers and concrete pipes used for water diversions in construction projects.

Montrose's claim to fame outside Susquehanna County is that it is the home to about a dozen churches, in a town whose population is only 1,664 (US Census Bureau 2002). The religious presence in this small town is partly due to the annual Montrose Bible

Conference. This conference is a well-known event every August that swells the size of the borough significantly.

Conclusions

Given the history of the development of each of these places, or any place you examine, is their fate now cast forever? Consider that the railroad and the tall building period, around the turn of the 1900s, gave a virtual permanence to transportation accessibility (and routes) or inaccessibility, and that suburban development in the 1950s have forced many other changes in city management of its land and building resources. Consider that the resource basis for founding most towns is seldom the economic basis of that town today.

Given these realities, could a town like Montrose develop into another Philadelphia, or even a Scranton?

Bibliography

GE Transportation 2005. Facts and History. Erie, PA, General Electric Transportation Rail. Internet site: <http://www.getransportation.com/general/locomotives/fast_facts.asp>, visited 3/15/05.

MC RPC 2004. Mercer County's Comprehensive Plan – Trends and Data. Hermitage, PA, Mercer County Regional Planning Commission. Internet site: <http://www.mcrpc.com/trends.htm>, visited 11/30/04.

Schuyler, David 2002. A City Transformed: Redevelopment, Race, and Suburbanization in Lancaster, Pennsylvania 1940-1980. University Park, Penn State Press.

US EPA Region III 2000. Slag from Sharon Steel to be Reused Under Innovative Agreement. Environmental News, September 20, 2000. Philadelphia, PA, US Environmental Protection Agency Region III, Mid-Atlantic Hazardous Site Cleanup. Internet site: <http://www.epa.gov/reg3hwmd/super/sites/PAD001933175/pr/2000-09-20.htm>, visited 12/1/04.

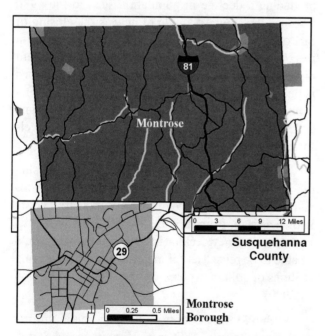

Figure 32.4 **Montrose, Susquehanna County:** The borough of Montrose is a small town in northeastern Pennsylvania. Its conservative nature is represented in its strong religious community, over a dozen churches in the borough and its surrounding townships.

Chapter 33

Pittsburgh

Pittsburgh's roles in the westward expansion of the state and in our dominance of the iron and steel industry around the turn of the twentieth century have been portrayed already. Those accomplishments were more than just the luck of Pittsburgh's location. At the same time, there is more to Pittsburgh's location than its importance for industry and trade. Pittsburgh has put the quality of its setting and of its people to great use in overcoming some of the negative consequences of its past.

Early Growth

Pittsburgh began as a Virginia frontier fort in 1754 at the junction where the Allegheny and Monongahela Rivers meet to form the Ohio River. It was taken by the French (as Fort Du Quesne) soon after and then retaken by the British forces (and rebuilt as Fort Pitt) during the French and Indian War. Once its position was secured by the English, it grew as a trading center, especially after settlers continued down the Ohio River valley toward the country's new territories in early American times. Finally, by the 1820s it was building its reputation as an important regional commercial and industrial center.

Pittsburgh gained population steadily but slowly during its first hundred years (Toker 1986, 9). The earliest settlers were mostly English, Scottish and Scots-Irish. By the 1850s, with its rail connection to Philadelphia established, it was growing much more quickly, especially with Irish, German and British immigrants (Miller 1995, 391). Iron furnaces, forges, rolling mills and machinery works were an important part of that development, as were glass factories,

potteries and, later, oil refineries (Toker 1986, 10). Once bituminous coal became preferred as a fuel over wood and charcoal, and once the coking process was introduced, the iron industry took off.

Many early entrepreneurs laid the groundwork for later tycoons such as Andre Carnegie, Andrew Mellon and H.J. Heinz, who proved to be shrewd businessmen. They built up their corporations, and their personal wealth, on the backs of a largely immigrant work force. Immigrants in the late 1800s and early 1900s came largely from Russia, the Ukraine, Poland, Italy and other countries of eastern and southern Europe (Miller 1995, 392). Pittsburgh also gained substantial Jewish and African-American populations in the 1900s.

The steel mills dominated American production for a long enough time to establish Pittsburgh's reputation as the "Steel City" and the "Smoky City" (Miller 1995, 376). Steel producers such as US Steel and Jones and Laughlin Steel attracted other manufacturers of finished goods that were made using the mills' steel as inputs, including Westinghouse. Other industries also were attracted to Pittsburgh's location and other resources. Metal working surfaced also in the early aluminum industry, when the Aluminum Company of America (now ALCOA) made its start. Heinz combined glass production, and later the "tin" can (a steel alloy), in its food processing and packaging plant. Pittsburgh Plate Glass (now just PPG) also took advantage of local resources and expertise in glass production (Kline 1981, 2).

At its peak of industrial productivity, Pittsburgh was well connected to areas down the Ohio and Mississippi Rivers, the Great Lakes, and the major east coast cities (see Figure 33.1). Railroads were the most important of those connections, and some of Pittsburgh's economic muscle was flexed in its control of them, as we saw in the discussion of the Pittsburgh Plus pricing scheme (see Chapter 28).

Pittsburgh was already well established when, in 1881, a federal ruling attempting to end spelling inconsistencies and confusion about place names, decreed that any place whose name ended in "burgh" and was pronounced "berg" must drop the final "h." On appeal by local leaders, who proved that Pittsburgh had always officially been spelled with the "h," an exception was granted. Only Pittsburgh was exempted (Van Trump 2004).

Pittsburgh's Valleys and Hills

Pittsburgh's natural landscape has presented both opportunities and challenges (see Figure 33.2). First, of course, is the fact that the two large rivers, the Allegheny and Monongahela Rivers, met to form the even larger Ohio River. The transportation opportunities were tremendous, as all three were far more navigable (by barges) than the Susquehanna and

northern Delaware Rivers. In addition, at their junction, all three had substantial floodplains.

The floodplains were mixed blessings. Just as we saw with Johnstown, periodic flooding came with the territory. The flooding limited development of the floodplains to industrial and transportation uses, and to poorer housing. Later in the middle 1900s the problem was essentially solved by the construction of many dams on the Allegheny and Monongahela Rivers and on upriver tributaries.

As the coal and iron and steel industries grew, the flood plains became long narrow strips of industrial factories and mills, warehouses and transportation facilities for both barges and railroads. These floodplain sites were connected by multiple rail lines with many sidings on both sides of each river. The factories and locomotives belched the smoke of incompletely-burned coal. The impurities, especially sulfur, created a pall that obstructed visibility, stuck to buildings, and was unhealthy to breathe. The valley slopes helped to trap much of the heavy smoke in lower elevations.

The rivers themselves also proved to be mixed blessings. Despite being major transportation arteries, helping to create Pittsburgh's fortunes, they also

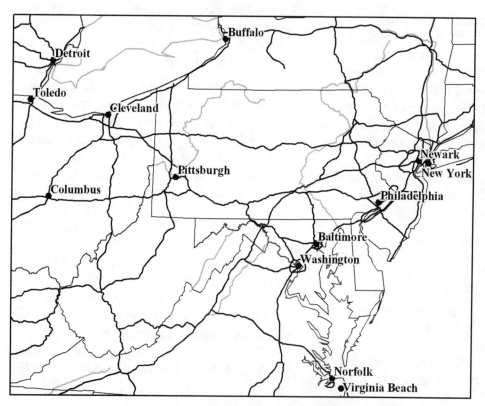

Figure 33.1 **Pittsburgh's Larger Region:** In its early days Pittsburgh benefitted from its ability to receive resources from a large area and to distribute steel and other products to an even larger area. Access to both the Great Lakes and the Atlantic seaboard with more modern forms of transportation continue to benefit Pennsylvania's second largest city.

196

Figure 33.2 **A City of Rivers and Bridges:** This scene looks up the Allegheny River from Pittsburgh's Roberto Clemente Bridge looking west, with several additional bridges visible beyond the 7th Street Bridge. Some of the bridges carry rail or mass transit traffic only, while most allow automobiles and pedestrians to cross. The tallest building in the photograph is the US Steel Tower. (Photo courtesy of Jonathan Egger.)

separated the urban area into three disjointed parts. Much effort has gone into connecting those parts, via many bridges over the rivers, as well as a series of tunnels under the rivers.

Better residential areas were built higher up the valley slopes to the Appalachian Plateau (the "mountains") and, between the two in-flowing rivers, on "the hills." The steep slopes of the valley, rivaling San Francisco's landscape, created transportation challenges which were partially met with a number of "incline railroads," two of which are still in use today.

The Renaissances

In the 1940s, concerned about the beginning of what would become a substantial decline in steel and related industries, several leaders undertook to change Pittsburgh's image and character. Government and corporate leaders agreed that major redevelopment could erase Pittsburgh's image as a dirty industrial city, and proclaim it to be modern and successful, thereby continuing to attract people, companies and investments. The project became known as "the Renaissance."

The district between the Allegheny and Monongahela Rivers nearest their meeting was known as "The Point" and was largely devoted to the Pennsylvania Railroad and a substantial but old central business district. Most of the Renaissance projects efforts were focused on removing railroad and industrial land uses from The Point, and reforming it into a "Golden Triangle." In the process, the entire railway yards were removed, and many blocks of

downtown were demolished to make way for new construction. Thirty six acres at the tip of The Point became Point State Park, featuring resurrected parts of Fort Pitt and Fort Duquesne and a huge water pool and fountain. Next came a project coordinated by the redevelopment authority to build a cohesive set of office towers, hotel and apartment buildings and plenty of parking. This complex stretched from the Allegheny to the Monongahela and became known as Gateway Center. Additionally, some of the major corporations built their own office towers in the district now collectively referred to as "The Golden Triangle." These corporations included Mellon Bank (Richard K. Mellon, the son of Andrew Mellon, was the project leader on the corporate side), ALCOA and US Steel. In addition, a multi-function performance-and-sports facility known as the Civic Arena was built to attract an arts and recreation-oriented public (Muller 2000, 20-21).

The Renaissance was a substantial success. The exception was the Civic Arena's hoped-for impact, which failed to materialize due to the fact that the Arena became separated from the rest of the Golden Triangle by an expressway. The highway construction did bring the additional tourist and business traffic into the city, though. Even though Pittsburgh continued to lose steel mills during the 1950s and 1960s, it maintained a strong corporate presence, and its downtown, or "dahntahn" (see Chapter 17), thrived.

Drawing on the success of the Renaissance, a second wave of demolition and rebuilding centered on the Golden Triangle took place in the 1970s and early 1980s. During this period US Steel moved into its new taller headquarters (made almost entirely of steel), PPG built a high-rise headquarters (covered almost entirely in glass), a new performance center, Heinz Hall, was born, and the David L. Lawrence Convention Center (named after the city's mayor who also helped guide the first Renaissance and later became Pennsylvania's governor) and hotel complex was built.

Pittsburgh's success can be attributed to two major factors: wealth and planning. Bank and other investment-based firms, and the accumulated wealth of the Carnegie, Heinz, Mellon and other families, foundations and corporations were brought together for the first Renaissance before much federal funding for urban renewal projects was available (Muller 2000, 19). In addition to buildings, much attention was paid to transportation needs, including highways and mass transit.

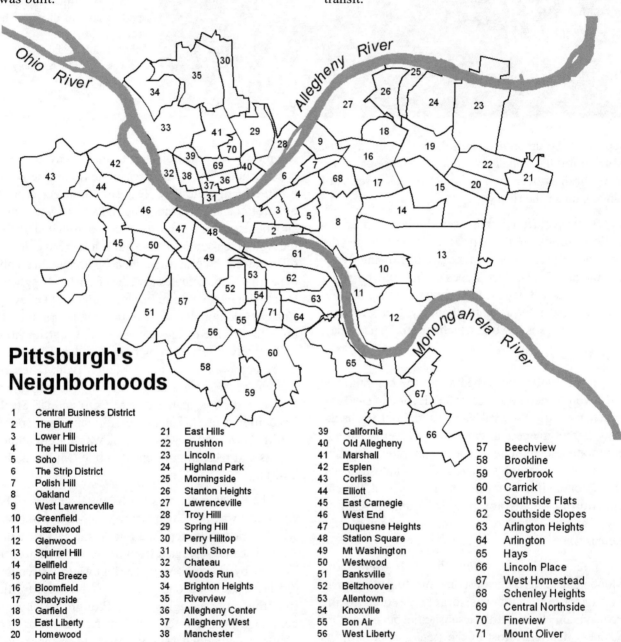

Pittsburgh's Neighborhoods

1	Central Business District	21	East Hills	39	California
2	The Bluff	22	Brushton	40	Old Allegheny
3	Lower Hill	23	Lincoln	41	Marshall
4	The Hill District	24	Highland Park	42	Esplen
5	Soho	25	Morningside	43	Corliss
6	The Strip District	26	Stanton Heights	44	Elliott
7	Polish Hill	27	Lawrenceville	45	East Carnegie
8	Oakland	28	Troy Hilll	46	West End
9	West Lawrenceville	29	Spring Hill	47	Duquesne Heights
10	Greenfield	30	Perry Hilltop	48	Station Square
11	Hazelwood	31	North Shore	49	Mt Washington
12	Glenwood	32	Chateau	50	Westwood
13	Squirrel Hill	33	Woods Run	51	Banksville
14	Bellfield	34	Brighton Heights	52	Beltzhoover
15	Point Breeze	35	Riverview	53	Allentown
16	Bloomfield	36	Allegheny Center	54	Knoxville
17	Shadyside	37	Allegheny West	55	Bon Air
18	Garfield	38	Manchester	56	West Liberty
19	East Liberty				
20	Homewood				

57	Beechview
58	Brookline
59	Overbrook
60	Carrick
61	Southside Flats
62	Southside Slopes
63	Arlington Heights
64	Arlington
65	Hays
66	Lincoln Place
67	West Homestead
68	Schenley Heights
69	Central Northside
70	Fineview
71	Mount Oliver

Figure 33.3 **The Neighborhoods of Pittsburgh:** These neighborhoods are defined by natural features such as streams and rivers, by political boundaries of the city (note that #71 on the map, Mount Oliver, is a separate borough, and is not part of the city), and by built features such as roads and rail lines. The pattern effectively continues into the suburbs, as each additional borough and each town within the townships will have its own identity.

Figure 33.4 University of Pittsburgh Cathedral of Learning: This building is on the U Pitt campus in Oakland and was built with money contributed largely by Andrew Carnegie. (Photo courtesy of Jonathan Egger.)

Not all of the projects succeeded. For example, the attempts to upgrade slum areas just east of the Golden Triangle by building the Civic Arena had little cooperation from the local neighborhood. The Civic Arena and the expressway were built on many city blocks of demolished housing and neighborhood businesses. Once the fate of the Civic Arena was cast and the surrounding cultural center failed to materialize, the remaining demolished areas were turned into parking facilities, hardly an uplifting land use for the community (Muller 2000, 21).

Neighborhoods

Over time, with the changes the city went through, new urban developments were built on the outskirts of the city, and once-outlying communities were absorbed into the city. Each usually grew largely with one strong ethnic group. In addition, some of its older areas absorbed different ethnic groups of immigrants as previous residents moved out. In the process, all of these areas took on their identities as Pittsburgh's neighborhoods (see Figure 33.3).

The area directly east of The Point became known as "The Hill," and was the initial settling area for most immigrants, the poorest elements of the population. Today it has a largely African-American population.

Other neighborhoods have equally colorful histories. The Strip was largely industrial development and lower class housing. Industrial buildings were removed, as were some of the lower class housing. However, the remaining residents formed a tightly knit group that resisted further demolition. Some of the factories have been turned into farmers market and boutiques, and have become quite trendy.

Oakland also has a distinctive character. Dominating its land area and its skyline is the University of Pittsburgh. Towering above the university is the landmark building "The Cathedral of Learning (see Figure 33.4).

Metropolitan Connections

The ultimate success of Pittsburgh's redevelopment is best seen as a transition from a manufacturing-based economy to a services-based economy. Pittsburgh used its new downtown, improved transportation system and strong university resources to attract new outside investment in the region. Even though the city continued to struggle outside of the central business district with job losses and declining population, its metropolitan area grew.

The Pennsylvania Turnpike, the downtown expressways, a new enlarged airport (built around 1970 and expanded in 1992) and the regional transit system allowed the growing suburbs to maintain a strong connection with the city. All of this came despite a very challenging landscape.

The poorest connections today are with the communities south of Pittsburgh, many of which were once thriving steel and coal towns in their own rights. Planning agencies and state government are well aware of the situation. Curiously, it has been the Pennsylvania Turnpike Commission that has stepped forward with the most ambitious solution. It is building a new Turnpike extension, the Mon/Fayette Expressway and Southern Beltway, circling the southern part of the metropolitan area (PA Turnpike Commission 2004). That highway is expected to connect to another new highway project, connecting the airport to US Route 22 on the west side of the city. The Mon/Fayette Expressway and Southern Beltway will be a toll road, like the rest of the Turnpike system, and much controversy surrounds whether drivers will be willing to afford its fees.

Conclusions

Pittsburgh has come a long way since its days of heavy industry. Perhaps because that legacy lives on in the names of its leading corporations (and in the name of its professional football team), many from outside the city have not adjusted their image of it. It will probably be another generation or so before the transition is complete; much of the local workforce has not made the transition to jobs that newcomers to the region have been taking. Rebuilding the fabric of the city certainly requires the types of working class skills that the earlier steel mills took advantage of. A services-oriented economy values both higher education and people skills. Pittsburgh continues to appear to be on track to succeed, though.

Bibliography

Kline, Hibberd V.B. 1981. "Welcome to Pittsburgh." The Pennsylvania Geographer. Vol. 19, no. 3 (October), pp. 1-6.

Miller, E. Willard 1995. "Pittsburgh: An Urban Region in Transition." Chapter 21 in Miller, E. Willard (ed.) A Geography of Pennsylvania, pp. 374-395.

Muller, Edward K. 2000. "Downtown Pittsburgh Renaissance and Renewal." in Patrick and Scarpaci (eds.) A Geographic Perspective of Pittsburgh and the Alleghenies: From Precambrian to Post-Industrial, pp. 17-27.

PA Turnpike Commission 2004. Mon/Fayette Expressway and Southern Beltway. Harrisburg, PA, Pennsylvania Department of Transportation, Pennsylvania Turnpike Commission. Internet site: <http://www.paturnpike.com/monfaysb/>, visited 12/5/04.

Toker, Franklin 1986. Pittsburgh: An Urban Portrait. University Park, PA, Pennsylvania State University Press.

Van Trump, James 2004. The Controversial Spelling of "Pittsburgh," or Why The "H?" Pittsburgh, PA, Pittsburgh History and Landmarks Foundation. Internet site: <http://www.phlf.org/phlfnews/essays/pittsburg.html>, visited 12/3/04.

Chapter 34

Philadelphia

Philadelphia is still a top-tier city in the US, even though it has slipped from former higher ranks in everything from population to economic measures. The core economic functions that generate its wealth and the role that it plays regionally in relation to money matters are still important, though perhaps different than in earlier times.

Also changing is Philadelphia's racial and ethnic mix, its very cultural identity. The city's cultural base is changing largely by subtraction, with comparatively fewer of the ethnically diverse European immigrants or their descendants who established its early identity. The suburbs, the destination of much of that "white flight," flourish in their independence, while the city struggles to maintain cohesiveness.

Philadelphia's challenge is to rebuild its infrastructure and economy with an assortment of residents and businesses that are committed to making it happen. This task is challenging because there are high expectations based on Philadelphia's and Pennsylvania's former glory. In fact the rebuilding is *more* difficult than the initial building because it is starting from a problematic situation, not from a clean slate. It requires the demolition and removal of materials with little resource value; the solutions need to fit an old street pattern and infrastructure; and there are modern (more expensive) expectations.

Two questions that need to be asked are:

- What will it take to return Philadelphia to sustainability?

- How much should the state, the region and the city contribute towards and take responsibility for its rebuilding?

Physical Setting

Philadelphia was first accessed via the Delaware and Schuylkill Rivers. The Atlantic Coastal Plain ends just upriver along both the Delaware and Schuylkill from the site chosen by William Penn for his model city. The area proved an excellent early location with productive soils in all directions to support a rapidly growing city (Dougherty 2004, 5). Its harbor was adequate for the smaller colonial-era ships.

By the time the United States of America was born, the deficiencies of Philadelphia's location relative to New York City's were beginning to show. The river's length and depth placed some limits on the harbor, despite inspiring some technologically innovative solutions. Philadelphia was not so easily connected to the Appalachian Plateau or to the nation's interior. Baltimore had better access to Pennsylvania's Ridge and Valley region, not by navigable river, but by the Susquehanna's river valley. Industrial Philadelphia in the late 1800s was not even able to match New York's awe-inspiring skyline because its geological base could not so easily support tall skyscrapers (Dougherty 2004, 5).

The elevation drop from Piedmont to Coastal Plain is logically called the fall line. One early industrial benefit to Philadelphia's location was immediate access to good natural sites for water-wheel and, later, hydroelectric dams. Again, at early levels of the use of such technologies, Philadelphia was well enough positioned to become an industrial leader; later those resources were not adequate themselves and had to be supplemented. The best news for Philadelphia was that it had much more manageable routes to receive delivery of anthracite coal.

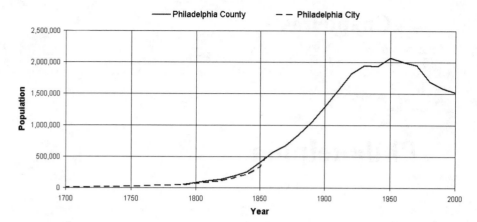

Figure 34.1 **Philadelphia's Population History:** Philadelphia grew consistently and steadily until the 1920s when growth potential became limited by its boundaries. The growth of the suburbs after World War II led to a population decline which is still occurring. (Data source: UV GSDC 1998)

Background

William Penn's original "vision" of Philadelphia was as a garden city of large lots surrounded by open space (whole city blocks in extent) and broad avenues. The grid street scheme of the original city plan, from Vine Street to South Street and from river to river (see Figure 15.1 in Chapter 15), was a radical break from older European cities. The use of the perfect right-angled intersections everywhere was meant to make no part of the city better than any other, so that people from all religions and ethnic backgrounds (as long as they were strong economic contributors) would feel welcome (Rees 2004).

The city that actually developed required many modifications to the original plan. For example the city center was supposed to be defined by an open square at the intersection of the two principle streets, Broad and Market (originally High) Streets, halfway between the rivers and on the highest ground. Since the highest ground was not at the halfway point, the road positions were shifted, and the city plan became slightly unbalanced. Penn and his planner also wanted the streets' Delaware River ends to be at low points of the river bank to enable better landing sites. This also required shifting the positions of several streets, so they became unevenly spaced. Later, riverfront lots were made narrower than the original square block so as to accommodate more purchasers who wanted river access. Many city blocks were made smaller and access was given to their centers by extending alleys through them. Many of those alleys eventually, in the early 1800s, became streets. By late colonial times, the now variably-sized lots and the development of many different professions created more specialized zones and later led to significant distinction of neighborhoods as wealthier than or poorer than adjacent ones (Rees 2004, 25-30).

Specialized districts developed for such occupations as furniture makers, ship wrights, bakers, grocers, tailors, banking and warehousing, and for merchants who sold their goods at market stands along what became known as Market Street. There was even a ban placed in 1834 on steam locomotives in the district of the original city. The ban remained in effect until the 1870s, and required the cars carrying passengers and freight to switch to horses. One impact of this situation was to make the central business district around Broad and Market Streets attractive to retail stores for clothing, jewelry and other finer goods starting in the 1860s. Printing and banking also kept downtown locations (Rees 2004, 30-33).

Industry

As retailers moved from locations near the Delaware River to newer buildings closer to the city center, the vacated eastern city developed an early industrial character. The second impact of the removal of steam railroads from the original city was to shift locations of heavier manufacturing, such as lumber products and the new Baldwin locomotive works, to the less densely built areas north of Vine Street and south of South Street (Rees 2004, 33). As a result of this timing, when immigrants started arriving in greater numbers to take newly opened factory jobs, they were housed in nearby row houses in those same districts.

Philadelphia did not develop strongly demarcated and diverse industrial districts; rather, factories were dispersed around the growing city. In fact, most of them were small independent operations that only grew to larger sizes toward the end of the 1800s. This meant that, when nearby factory towns such as Germantown, Manayunk and Kensington were annexed, they did not represent a departure from the structure of the rest of city (other than downtown). Some types of factories did become clustered, especially those working with textiles. A printing and publishing district developed in the southeastern part of the original city and, later, ship-builders and refiners of oil and sugar concentrated in South Philadelphia. But Philadelphia's strength was textiles. All sorts of textiles were made and marketed in Philadelphia, from carpets to clothing and knitted socks to felt Stetson hats (Rees 2004, 31, 34).

By 1854, Philadelphia had annexed enough adjacent territory that the state declared that the entire Philadelphia County should be considered Philadelphia City. However, at the same time, it decreed that the city could not expand beyond the county. By 1880, Philadelphia was a thriving industrial city of over 840,000, including over 4000 factories (Rees 2004, 31, 34).

As recently as 1952, Murphy and Murphy could write, "Among American cities, Philadelphia makes more cigars, carpets, and rugs, refines more petroleum and sugar, makes more streamlined trains, radios, false teeth, and Bibles than any other city" (82). Since that point, the effect of suburbanization, which had its beginnings in the 1890s with the extensions of the city's trolley lines beyond the boundaries, has been draining. Additional links to Center City included subways and highways; ironically, they ended up helping many downtown workers move out to the suburbs and commute into the city (Rees 2004, 35). Unfortunately, this coincided with many of the city's factories closing down, a few at a time, starting at least with the Depression.

Neighborhoods

By the time immigration rates subsided in the middle 1900s, and the factory closings and suburban migration picked up steam, the city's population, which had reached its peak of nearly 2,000,000 in 1950, began to decline. Neighborhoods had been evolving ever since the city started expanding beyond Center City, as the highest classes built newer homes ever further from their downtown offices, and the next class in line moved into the upper crust's old neighborhood and began a chain reaction down the ranks. South Philadelphia had been immigrants' first step on that ladder for many generations. Some sections of the city functioned as small towns, forming distinctive neighborhoods (Rees 2004, 35). Figure 34.1 shows one interpretation of those neighborhoods, though both the boundaries and names are not always as well-defined as such a map might suggest.

African-Americans had been a significant minority ever since the decade before the Civil War, when Philadelphia had the largest black population north of the Mason-Dixon Line. Their numbers swelled during the Depression and afterwards. Proportionally, they grew even faster, as the majority of those leaving the city for the suburbs were whites. Today, over 40% of the city-wide population is African American, with significant numbers in many of the city's neighborhoods. Unfortunately, on average they have not kept pace financially with the rest of the city population, although there are quite a few black neighborhoods that would rank in the middle to upper-middle classes (see Figure 34.2). Hispanics are in a similar situation, though they are more recent immigrants to Philadelphia (see Figure 34.3).

In 1951, Philadelphia's Home Rule charter was approved by the state legislature. One departure from standard city government practices in Pennsylvania was a provision giving the Philadelphia City Planning Commission greater authority and responsibilities concerning land decisions. By 1960 it had issued an ambitious comprehensive plan designed to pull the city out of its growing malaise and give it a sense of cohesiveness and optimism. Unfortunately, the plan was also complex and required the cooperation, commitment and even financial resources of a diverse constituency (Macdonald 1993, 39).

Many elements of the plan ended up moving forward in rather disjointed ways. Many city slums were razed to build newer urban housing projects for the poor, but the new projects were socially dysfunctional and provided less housing than had been demolished (Rees 2004, 36). The area around Independence Hall had become a warehouse district. When the federal government acquired the land for Independence National Historical Park in 1948, it chose three city blocks connected northward from Independence Hall. However, the primary movement of traffic in that part of the city is east and west, forcing the pedestrian park visitors to cross two busy city streets in their visit (Rees 2004, 38).

Society Hill essentially renovated itself with only two sets of apartment towers funded by government money. When the new I-95 threatened its upper class residents with an on-ramp at their doorstep, they succeeded in having the ramp removed from the design. Then, to give them, and the Center City's other residents and visitors, access to coming developments

Philadelphia's Neighborhoods

#		#		#	
1	Logan Circle	31	Belmont	82	Northwood
2	Chinatown	32	Mill Creek	83	Frankford
3	Olde City	33	Haddington	84	Wissinoming
4	Society Hill	34	Overbrook	85	Mayfair
5	Washington Square	35	Carroll Park	86	Tacony
6	Rittenhouse	36	West Parkside	87	Oxford Circle
7	Schuylkill	37	Wynnefield	88	Burholme
8	SW Center City	38	Wynnfield Heights	89	Fox Chase
9	Hawthorne	39	Spring Garden	90	Rhawnhurst
10	Bella Vista	40	West Poplar	91	Lexington Park
11	Queen Village	41	Northern Liberties	92	Holmesburg
12	Pennsport	42	Francisville	93	Upper Holmesburg
13	Wharton	43	Fairmount	94	Torresdale
14	Point Breeze	44	Olde Kensington	95	Academy Gardens
15	Grays Ferry	45	Ludlow	96	Aston-Woodbridge
16	Girard Estate	46	Yorktown	97	Pennypack Woods
17	Packer Park	47	North Central	98	Winchester Park
18	Whitman	48	Sharswood	99	Pennypack
19	Eastwick	49	Brewerytown	100	Bustleton
20	Elmwood	50	Strawberry Mansion	101	West Torresdale
21	Pashcall	51	Stanton	102	Morrell Park
22	Kingsessing	52	Hartranft	103	Crestmont Farms
23	Southwest Schuylkill	53	Franklinville	104	Millbrook
24	Cobbs Creek	54	Nicetown	105	Modena
25	Walnut Hill	55	Allegheny West	106	Parkwood Manor
26	Cedar Park	56	Hunting Park	107	Mechanicsville
27	Spruce Hill	57	Fishtown	108	Byberry
28	University City	58	Kensington	109	Somerton
29	Powelton Village	59	Upper Kensington	110	Woodland Terrace
30	Mantua	60	Feltonville	111	Garden Court
		61	Juniata	112	Dunlap
		62	Richmond	113	West Powelton
		63	Bridesburg	114	East Parkside
		64	Andorra	115	Haverford North
		65	Upper Roxborough	116	SW Cedar Park
		66	Roxborough	117	Frankford Industrial Park
		67	Manayunk	118	Northeast Philadelphia Airport
		68	East Falls	119	Normandy Village
		69	Westside	120	Franklin Mills
		70	Wister	121	Port Richmond
		71	Morton	122	Southwark East
		72	East Mount Airy	123	Greenwich
		73	West Mount Airy	124	Passyunk
		74	Chestnut Hill	126	Penrose
		75	Cedarbrook	127	Bartram Village
		76	West Oak Lane	128	Angora
		77	Logan	129	Clearview
		78	East Oak Lane	130	East Poplar
		79	Olney	131	Temple University
		80	Lawndale	132	Lower Kensington
		81	Crescentville	133	West Kensington
				134	Harrowgate
				135	McGuire
				136	Rising Sun - Tioga
				137	Glenwood
				138	Fairhill
				140	Wissahickon Hills
				141	Wissahickon
				142	Germany Hill
				143	Roxborough Park
				144	Dearnley Park
				145	East Germantown
				146	SW Germantown
				147	Penn Knox
				148	West Central
				149	Melrose Park Gardens
				150	Ogontz
				151	Fern Rock
				152	Summerdale

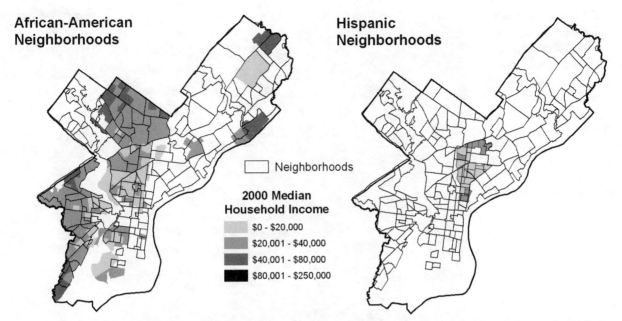

African-American Neighborhoods

Hispanic Neighborhoods

☐ Neighborhoods

2000 Median Household Income
- $0 - $20,000
- $20,001 - $40,000
- $40,001 - $80,000
- $80,001 - $250,000

Figure 34.3 **Ethnicity and Income in Philadelphia:** The areas shaded in these maps are "census tracts," which do not exactly correspond to the neighborhoods. The census tracts shaded are the ones in which at least 34% of the population delcared themselves to be African-American (on the left) or Hispanic (on the right).

along the Delaware River, the interstate was lowered below ground level and bridge-like "covers" were put over it at four key intersections (Rees 2004, 41).

Meanwhile, Philadelphia's own Chinatown was developing, too (see Figure 34.4). Chinatown can trace its roots back to the late 1800s, but its identity was solidified in the 1960s to 1970s. The Vine Street expressway was slated to demolish a major Chinese Christian church and other buildings, and potentially destroy the neighborhood. The Chinese had been hesitant to complain or get involved. A group of younger adults took the issue to City Hall, and persisted in their opposition for nearly 20 years, finally getting the expressway design modified and preserving their community (HSP 2004).

South Philadelphia similarly influenced changes to the original plan. Though a blighted area in the 1960s and early 1970s, and a mix of sometimes clashing cultures, the neighbors banded together to oppose a planned east-west expressway slated to

Figure 34.2 (Facing Page) **Philadelphia's Neighborhoods:** The boundaries shown on this map were drawn by the Philadelphia Police Department and released for public use. Many of the areas in western Philadelphia, along the Schuylkill River and in southern Philadelphia that are not identified as neighborhoods here are still residential. Other parts of those areas make up Philadelphia's largest industrial district.

replace their main commercial area. By the late 1970s, after the expressway was removed from the plan, it was beginning a transition to a lively, trendy center of *avant garde* retailers and restaurants. It is well known as the home of the Philadelphia cheese-steak sandwich as well as an active popular music and art scene.

Metropolitan Area

Philadelphia's biggest struggles have come from the loss of much of its tax base as factories closed down and many middle- and upper-class residents departed for the suburbs of Montgomery, Bucks, Chester and Delaware Counties. Drawn first to areas around the primary roadways into the city, especially the Schuylkill Expressway (Interstate 76), the Norristown and King of Prussia areas were early favorites. The US Census Bureau now considers all four suburban counties, as well as Burlington and Camden Counties, New Jersey, to be part of the Philadelphia-Camden Metropolitan Area, demonstrating that all of them have grown at Philadelphia's expense and remain strongly connected to the city.

The US Route 202 corridor, especially, is a ring road around the city which has attracted a great deal of suburban industrial development, especially since it was upgraded to limited access from West Chester to King of Prussia. So has the eastern extension of the Pennsylvania Turnpike (Interstate 276), from

Figure 34.4 **Chinatown:** Chinatown is one of Philadelphia's most visually distinctive neighborhoods. The arch, known as the Chinese Friendship Gate, spans 10th Street (four blocks east of City Hall) just north of its intersection with Arch Street (one block north of Market Street). (Photo courtesy of Joanne Mark.)

Valley Forge to its eastern end at the Delaware River. Drive along both roads, and you will almost never be out of sight of a modern suburban housing development, a shopping center, or a rather blank looking one- or two-story factory or three- to four-story office building.

Conclusions

One question presented above was whether Philadelphia is a sustainable entity. This is an issue that the city's leaders are struggling with. There are many issues competing to be the top priority, but obviously no one action, no matter how much it costs, will fix everything. The Philadelphia City Planning Commission now cooperates with the Delaware Valley Regional Planning Commission, which encompasses five Pennsylvania counties and four New Jersey counties. The challenge is to attract companies to increase the employment opportunities in the city. The unemployment rate has fluctuated between 3.6% and 6.4% between 1994 and 2004. The best scenario would be to establish the type of environment that would entice suburban middle class workers back to the city.

Such efforts cost money, however. How much should the city, the region and the state contribute towards and take responsibility for its recovery? This is a thornier issue, and one that has been particularly divisive among Pennsylvania politicians.

Bibliography

Dougherty, Percy H. 2004. "Landform Regions of Eastern Pennsylvania and their Impact on the Cultural Landscape." In Dougherty, Percy H. (ed.) Geography of the Philadelphia Region: Cradle of Democracy. pp. 1-16.

HSP 2004. Philadelphia's Chinatown: An Overview. Philadelphia, PA, Historical Society of Pennsylvania. Internet site: <http://www.hsp.org/default.aspx?id=190>, visited 12/6/04.

Macdonald, Gerald M. 1993. "Philadelphia's Penn's Landing: Changing Concepts of the Central River Front." The Pennsylvania Geographer. Vol. 31, no. 2, pp. 36-51.

Murphy, Raymond E. and Marion F. Murphy 1952. Pennsylvania Landscapes: A Geography of the Commonwealth (2nd edition). State College, PA, Penn's Valley Publishers.

Rees, Peter W. 2004. "The Spatial Evolution of Center City, Philadelphia." In Dougherty, Percy H. (ed.) Geography of the Philadelphia Region: Cradle of Democracy. pp. 25-44.

UV GSDC 1998. United States Historical Census Data Browser. Charlottesville, VA, University of Virginia, University of Virginia Geospatial and Statistical Data Center. Internet site: <http://fisher.lib.virginia.edu/census/>, visited 8/19/04.

Chapter 35

Air Pollution Issues

One consequence of the intense energy use that has characterized Pennsylvania's industrialized and urban (and suburban) past and present is air pollution. Its key difference from water pollution (Chapter 36) is air pollution's more direct tie to energy use, especially fossil fuel combustion. Some pollutants are impurities in the fuel that are released by combustion. Others are the results of chemical reactions that take place between components of the fuel and components of the atmosphere.

The US Environmental Protection Agency monitors the atmosphere for particulates (fine particles and droplets) and several gases. The particulates are very small non-combustible particles that enter the atmosphere mostly as a by-product of fuel combustion (soot, for example), and are of greatest concern if they contain elements such as mercury and lead. The gases are sulfur dioxide (SO_2), a few different nitrogen oxides (NO_x), carbon monoxide (CO), ozone (O_3) and volatile organic compounds (VOCs).

Carbon dioxide (CO_2) was also a member of this group, until the Environmental Protection Agency controversially decided that global warming, carbon dioxide's primary environmental effect, is not a national pollution problem because it is an international one. Carbon dioxide was then removed from the list of monitored and regulated pollutants (Borenstein 2003).

Sulfur dioxide is most often the result of burning coal or smelting ores. Nitrogen oxides are usually by-products of burning oil products: gasoline, diesel fuel and jet fuel, home heating oil and other fuel oils.

Carbon monoxide and carbon dioxide are released everywhere biomass and fossil fuels are burned. Ozone and the volatile organic compounds are the primary ingredients that contribute to urban smog (PA DEP 2005).

Pennsylvania has particular problems with several of these air pollutants, in part because the devices producing the pollution are older than the laws prohibiting pollution. The challenge, as we work toward the lowest acceptable pollution levels, is to reduce the pollution and improve public health without hurting the economy.

Sources and Their Locations

In Chapters 5 and 6 the general atmospheric circulation patterns that affect Pennsylvania were described. The majority of the air masses that affect us come from the west, we learned, and the specific direction of that movement influences the temperature and other characteristics of each air mass. To that set of characteristics, we can now add: pollutants.

Coal burning and the sulfur dioxide, carbon monoxide, and particulate pollution that result from it, take place mostly at factories and electric power stations. These air emissions are strongest at the largest operations, in relatively fewer locations on the state map.

The map of electric power stations in Chapter 26 (see Figure 26.2) is one representation of major sources of air pollution. In fact, Pennsylvania electric power stations have been among the top point-source polluters in the US. One environmental organization uses Environmental Protection Agency air emissions data to rank Pennsylvania as the leading source of arsenic particulates, the second leading source of dioxin (a toxin), again in particulate form, and the third leading state for lead and chromium particulates. The same source identifies Pennsylvania as the second worst state for acid gases; adds hydrochloric acid and hydrofluoric acid to the sulfuric acid that results from sulfur dioxide emissions, and identifies the Keystone coal-burning electric power station in Armstrong County as the nation's third worst single power station (Clean the Air 2005).

A map of the major cities (see Figure 1.2 in Chapter 1) is another representation of the distribution of pollution sources, especially those related to oil combustion. Each city is likely to have an industrial district, nearby electric power stations and large concentrations of buildings and automobiles. More oil is consumed in small quantities at dispersed locations by automobiles (including buses, trucks, railroad engines, airplanes, lawn mowers, tractors and construction equipment) and residential and commercial buildings heated by oil furnaces as by factories and power stations. Nitrogen oxides, carbon monoxide and carbon dioxide, and volatile organic compounds are the pollutants given off by the combustion of oil products.

Many pollutants have only local impacts because they are released lower in the atmosphere from automobiles and short smokestacks or chimneys. Pollutants that are released higher in the air travel farther with the air mass they are released into before returning to ground level. The longer distance travel also means the pollutants will stay in the atmosphere longer and have more time to react chemically with other elements there. An example of this behavior is the pollution impact known as acid precipitation, described below.

Another approach to mapping pollution levels is according to where air quality monitoring devices are located, and what pollution levels they record. The Environmental Protection Agency operates monitoring equipment at dozens of sites across Pennsylvania, and publishes the measurements recorded at each site on its Internet site. Areas which exceed the safe levels of any one of the main "criteria pollutants" are labeled as "non-attainment areas" (US EPA 2005b).

The map in Figure 35.1 shows the impacts of burning higher-sulfur coal as measured by the Environmental Protection Agency at monitoring stations near such coal burning sites as electric power stations. Fortunately, the vast majority of them have decreased the amount of pollution they emit. The set of 36 counties in which sulfur dioxide is monitored

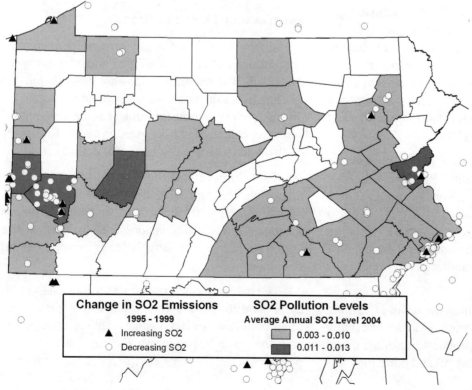

Figure 35.1 **Sulfur Dioxide Pollution around Pennsylvania:** This map shows the locations of higher emissions of Sulfur dioxide pollution as the small circles and triangles. The shades of gray represent the levels of Sulfur dioxide pollution in the state's counties where pollution monitoring takes place. The counties that are white have no data reported so their pollution levels are unknown. Sulfur dioxide is a major cause of acid precipitation and is hazardous to breathe. (Data source: US EPA 2005a)

Change in SO2 Emissions
1995 - 1999
▲ Increasing SO2
○ Decreasing SO2

SO2 Pollution Levels
Average Annual SO2 Level 2004
0.003 - 0.010
0.011 - 0.013

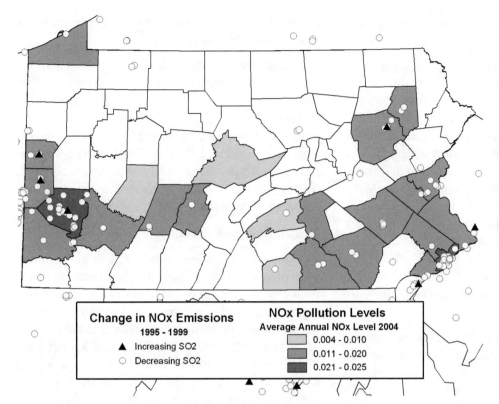

Figure 35.2 **Nitrogen Oxides Pollution around Pennsylvania:** Nitrogen oxides are again associated with acid precipitation and with the development of smog in larger cities. Again the counties that are white have no data reported so levels there are unknown. In the rest of the counties, though, pollutions levels correspond to population. With these pollutants, pollution levels are more liikely to be caused by local sources. (Data source: US EPA 2005a)

includes four in which more than one percent of the air is sulfur dioxide gas.

The distribution of elevated nitrogen oxide levels follows more closely the distribution of cities, as shown in Figure 35.2. This is expected, given that automotive sources reflect population distribution. Again, most monitors are recording smaller amounts than five years ago, so improvement is occurring. Still, Philadelphia and Pittsburgh are in the highest category and many more urban and suburban counties are just below them.

One phenomenon that can intensify pollution problems is a temperature inversion, which can affect a small area in which the *vertical* arrangement and circulation of the atmosphere is altered. In the normal vertical arrangement of air during the daytime, the air closest to the ground will be warmest and the temperature will gradually and continually decrease as you rise in altitude. During the night all of the air will cool, but the pattern will stay the same. Any warm gas released near the ground will rise through that air mass, and will cool as it rises until its temperature matches the air around it.

Temperature inversions most often occur in river valleys, such as those located in the Appalachian Plateau landscape region, and sometimes around large cities. In these settings it is possible for a new air mass featuring somewhat warmer temperatures to move in overnight, and form a lid over the lowest air with the coolest temperature. Warmer gases released in that cooler lower layer of air cannot rise through that new layer of warmer air until the entire lower layer is warmer than the air above it. If pollution is part of those gases, then the pollution can actually help to block sunlight from helping to warm the ground and the air near it. Temperature inversions have been responsible for some of the worst air pollution episodes.

Air Pollution Effects

The most dramatic consequences of our pollution are felt by people forced to breathe the pollutants. Each of the gases and the particulates are hazardous to inhale, causing toxic effects in, or more easily entering the bloodstream via, the highly sensitive tissues of the lungs. Higher concentrations of the pollutants, measured in "parts per million" or "parts per billion" as a proportion of the contents of the air, or longer exposures will result in more severe illnesses.

Sulfur dioxide, nitrogen dioxide or carbon dioxide reacts with atmospheric water vapor to create a mild sulfuric, nitric or carbonic acid, in the form of droplets. Precipitation droplets that coalesce with the acid droplets bring the acidity to the ground where it can be transferred onto trees, structures and surface waters.

Acid precipitation was not recognized as a major problem before the early 1970s, even though the potential had been noted earlier. In the 1950s and 1960s, many polluters started building taller smokestacks in order to reduce local air pollution problems. At least some of this effort resulted from a famous air pollution incident that occurred in Donora, Washington County in 1948 (explained below). Many local areas did improve. However, once scientists realized that newly degraded environments from central Pennsylvania to northeastern New York State were caused by increasing acidity, the taller smokestacks in the Ohio River valley of western Pennsylvania and other states downstream were implicated.

The acidity can have direct impacts on trees and buildings as the acid chemically reacts with the surface it reaches. Acidity entering streams and lakes is initially diluted by all the water, but over longer periods lake acidity can increase. In lakes that are located over alkaline soils and bedrock the acids are naturally "buffered," that is, neutralized. However, areas with no such limestone or related minerals are likely to become more acidic, impacting the lake ecosystem.

Fortunately, technology now exists to remove, or "scrub," the hazardous gases from the smokestack before they are released into the upper atmosphere. Unfortunately, the technology is still expensive, and although progress is being made the impacts are still being felt. The challenge is coming up with the many millions of dollars that these scrubbers cost without causing economic harm to the electricity producers and other industrial companies or to the state and federal governments.

Air pollution impacts also include the recently identified phenomenon of global warming. Though still somewhat controversial, the theory is that some pollution gases help to prevent radiant energy from leaving the Earth's atmosphere, keeping it trapped and inducing temperatures to increase. The gases that are being blamed for this climate change are generally called "greenhouse gases," and include: carbon dioxide, methane, nitrous oxide and chlorofluorocarbons. The temperature increase, as little as couple of degrees over the next fifty years, is expected to raise ocean levels and change atmospheric circulation. The impacts are not expected to be uniform over the entire Earth.

Environmental Protection Agency models of the middle 1990s projected that, by the year 2100 Pennsylvania's average annual temperature would increase 2°F to 4°F, and that precipitation would increase by ten to fifty percent, depending on the season (US EPA 1997). Now that the federal government's position on global warming is changed, we may not see the Environmental Protection Agency updating such models.

Regulation and Cleanup

The first regulations to help reduce air pollution were passed in the 1950s. These laws accomplished little. By the late 1960s there was much more public sentiment and political conviction toward environmental protection, especially with the creation of the federal Environmental Protection Agency after 1970. Early laws regulated smokestack and automobile pollution, but many pollution sources continued to pollute because of threats to the local or national economy if they were forced to make changes before they could afford it.

In the 1990 revisions to the clean air laws a new approach was adopted. Larger point-source polluters were assigned pollution "allowances," one for each ton of sulfur dioxide pollution they produced. These allowances are something like the development and mining "rights" discussed earlier, in that they separated one aspect of "business as usual" and made it a valuable commodity. While pollution limits are still in effect, polluters who make extra efforts to reduce their pollution levels can then earn money by selling the allowances they no longer need. The Environmental Protection Agency will remove allowances from that market periodically, to reflect improving air quality and to keep the pressure on the companies. The program, also known as the "cap and trade" program, produced little action in its early years, but now seems to be working.

The main benefit, from the government's perspective, is financial. The allowances market will base the value of the allowances on what the companies are willing to spend. Manufacturers of smokestack scrubbers and lower-sulfur coal will adjust their prices to what that market will bear. Total spending to achieve reductions are projected to be billions of dollars less than traditional methods of making and enforcing pollution limits (Burtraw 1996). By 2010, the Environmental Protection Agency projects, sulfur dioxide emissions should be half of what they were in 1980, the base year for the research that led to the program (US EPA 2002).

Donora, 1948

There had been many air pollution episodes before 1948, but Donora's was the first to unequivocally demonstrate the direct link between air pollution and public health. In Donora, as in many steel towns in southwestern Pennsylvania and elsewhere, people had come to accept the pollution and their breathing difficulties as better than not having a job. In Donora,

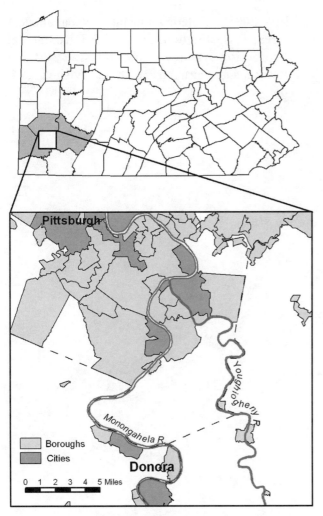

Figure 35.3 Donora, Washington County: Donora was the site of one of the nation's worst air pollution disasters in October 1948. Even though towns and cities in the immediate area experienced similar weather patterns, Donora's valley situation and larger concentration of pollution emissions made their conditions the worst.

a Monongahela River valley community about 20 miles south of Pittsburgh (see Figure 35.3), the worst combination of factors came together in 1948.

It began on a fall Friday in late October. Steel mills and coking plants up and down the Monongahela River valley were operating. In Donora several steel mills, a zinc smelter and even coal furnaces in other businesses and private homes were in full production, emitting several hazardous gases and particulates, especially sulfur dioxide. A very strong temperature inversion settled in over the valley before Friday morning. The trapped air pollutants increased their concentration all day Friday. By that evening, when the town's Halloween parade was held, visibility was reduced and people with respiratory ailments were suffering (Gammage 1998).

The weather pattern persisted throughout the day on Saturday. The zinc works continued its operations all day and into Sunday morning. At the high school's homecoming football game Saturday afternoon, the air was so thick with smoke that fans in the stands could not see the players on the field. By that evening eleven people in this town of (then) nearly 14,000 had died of respiratory illnesses (Gammage 1998).

Sunday was the worst day, even after the factory was shut down. Emergency crews carried oxygen tanks door to door in order to give residents only a few puffs. At least nine more people died and thousands, nearly half of the town's populations, were sickened. The inversion did not lift until rain started falling Sunday evening and brought in a new weather pattern on Monday (Glover 1998).

The first federal air regulations were passed in 1955, in part due to attention to Donora's tragedy. Donora's zinc works closed in 1957, and gradually the steel mills have been closing. These closings are not related to the "killer smog," but all have worked to clear the air in Donora. Unfortunately, the town's population is less than half of what it was in 1948, and unemployment is much higher.

Conclusions

Air pollution brings some visible and measurable problems, including major respiratory illnesses and deaths. Many of its impacts are more difficult to measure, though, such as minor illnesses, damage to statues and buildings, and decreasing health of vegetation. Once again, Pennsylvania has played a major role in teaching these harsh and expensive lessons. We have a long way to go before the air is clean, especially because much pollution emitted outside of the state continues to affect us, but that is the path we are on.

Bibliography

Borenstein, Seth 2003. "Bush Administration Says It Won't Regulate Carbon Dioxide." New York, NY, Natural Resources Defense Council and Knight Ridder/Tribune News Service. 8/29/03. Internet site: <http://www.nrdc.org/news/newsDetails.asp?nID=1080>, visited 3/23/05.

Burtraw, Dallas 1996. "Trading Emissions to Clean the Air: Exchanges Few but Savings Many." Resources (newsletter of Resources for the Future). Winter 1996, pp. 3-6.

Clean Air Council 2005. Beyond Mercury: Annual Power Plant Toxic Air Emissions in Pennsylvania. Philadelphia, PA, Clean Air Council. Internet site: <http://www.cleanair.org/toxicfactsheet.pdf>, visited 3/24/05.

Gammage, Jeff 1998. 20 Died. The Government Took Heed. In 1948, A Killer Fog Spurred Air Cleanup. Originally published in the Philadelphia Inquirer. Harrisburg, PA, Pennsylvania Department of Environmental Protection. Internet site: <http://www.dep.state.pa.us/dep/Rachel_Carson/dead20.htm>, visited 9/12/03.

Glover, Lynne 1998. Donora's Killer Smog Noted at 50. Originally published in the Pittsburgh Tribune-Review. Harrisburg, PA, Pennsylvania Department of Environmental Protection. Internet site: <http://www.dep.state.pa.us/dep/Rachel_Carson/killer_smog.htm>, visited 9/12/03.

PA DEP 2005. Monitoring Principal Pollutants. Harrisburg, PA, Pennsylvania Department of Environmental Protection, Bureau of Air Quality. Internet site: <http://www.dep.state.pa.us/dep/deputate/airwaste/aq/aqm/principal.htm>, visited 3/16/05.

US EPA 1997. Climate Change and Pennsylvania. Washington, DC, US Environmental Protection Agency, Office of Policy, Planning and Evaluation. September 1997. Internet site: <http://Yosemite.epa.gov/oar/globalwarming.nsf/UniqueKeyLookup/SHSU5BVMDY/$File/pa_impct.pdf>, visited 3/23/05.

US EPA 2002. Clearing the Air: The Facts About Capping and Trading Emissions. Washington, DC, US Environmental Protection Agency, Office of Air and Radiation, Clean Air Markets Division. May 2002. Internet site: <http://www.epa.gov/airmarkets/articles/clearingtheair.pdf>, visited 3/24/05.

US EPA 2005a. County Air Quality Report - Criteria Air Pollutants. 2004 Data. Washington, DC, US Environmental Protection Agency. Internet site: <http://www.epa.gov/air/data/monsum.html?st~PA~Pennsylvania>, visited 3/16/05.

US EPA 2005b. Criteria Pollutants – Nonattainment Areas. Washington, DC, US Environmental Protection Agency. Internet site: <http://www.epa.gov/airtrends/non.html>, visited 3/24/05.

Chapter 36

Water Pollution Issues

Pennsylvania is fortunate to have plentiful rainfall, most of the time, and to have plentiful water in "storage," both in surface and groundwater supplies (see Chapters 5-8). Many US states, especially in the west, are limited in one or more of these capacities.

Water quality, however, is a major concern in Pennsylvania. Once again, historical background is useful because much of our most damaging activity took place before there was very much understanding about contamination sources and processes. By the time much of that knowledge was put into the monitoring and prevention systems we now have in place (largely since the 1960s), industrial activity in the state had already begun its decline.

How Clean Is Clean?

Scientists have identified several categories of water pollutants. Sediment, which we met in Chapter 11, is particles of solids that do not readily break down. The impacts of excess sediment include water cloudiness (turbidity) and settling in calmer waters, such as bays, lakes and reservoirs. Turbidity limits sunlight penetration needed by underwater vegetation and animals. Settling of sediment gradually reduces the water capacity of the body of water. Sedimentation is most easily dealt with at its source: on the farm fields and construction sites.

Nutrients form a second category of pollutants. Nitrates and phosphates (and potassium, to a lesser extent) have their greatest impacts on lakes and bays, where they promote the growth of sun-blocking and oxygen-consuming algae. One source is lawn and farm fertilizers applied in excess of what is actually needed for the desired lawn and crop plants to grow. Another is older or poorly maintained landfills, sewage treatment plants and septic systems. Phosphates are also a cleaning agent that used to be a major ingredient in laundry detergents. Cows or other farm animals wading into streams (see Figure 36.1) also contribute to problems such as the Chesapeake Bay's ills today and Lake Erie's problems of the 1970s to 1980s (see "Lake Erie and the Chesapeake Bay" below). The Department of Environmental Protection has aggressively targeted nutrient pollution.

Organic chemicals, a third category, interact with living tissue in the animals that ingest it (including, potentially, humans). Some are pesticides, such as Chlordane, applied to kill insects and weeds. Others are industrial chemicals, such as the industrial cleaning agent polychlorinated biphenyl (PCB), or oil or grease. They may be present in the water, or they may accumulate in the tissue of fish and other water-oriented species. Reducing these chemicals requires changes in chemical formulations, in application procedures and equipment, or in the setting in which they are applied.

Inorganic chemicals such as acids, salts, chlorides and metals comprise a fourth category of pollutants. They also are toxic to sensitive animal tissues, especially in greater concentrations. Some of these compounds will react with other minerals to release additional pollutants. Sources range from acid precipitation and road/sidewalk salts to mining and industrial processes and even untouched natural sources. Prevention again requires changes in chemical formulations, in application procedures and equipment, or in the setting in which they are applied. The acids leaching out of abandoned coal mines are another major Department of Environmental Protection target (see below).

213

Figure 36.1 **Stream Pollution:** Allowing these cows such direct access to this Lancaster County stream is adding sediment and nutrients. Cows climbing the stream banks dislodge soil into the water. The lack of buffer between the pasture on the left in the picture and the stream allows any excess nutrients on that field to run off into the stream. The worst problem, though, is that the cows directly pollute the stream with their waste. Manure is very high in nutrients.

The final categories are less common, and are generally more temporary. The first is disease-causing agents, or pathogens. These can be passed from a number of different hosts into the water system. Public water supply systems can add chlorine to kill bacterial agents, though parasites, viruses and other pathogens may require different methods. The second and third temporary situations are not so much pollutants, but are physical changes to the natural water temperature or flow rate. Thermal pollution occurs when warmer water from a factory or power station mixes with surface water, changing an important biological factor in that ecosystem. When water flow is obstructed, the decreased water flow may cause shortages for residential or industrial consumers. Excessive flows following a rainstorm or a thaw can also cause temporary problems requiring the shutdown of a private or public system.

Any of these conditions that renders a water source unsuitable for drinking or a stream or river or lake unsuitable for swimming or even for fish and other wildlife water consumers, must be treated according to Environmental Protection Agency regulations. The Environmental Protection Agency also sets the levels of pollutants that can be tolerated; above those levels the state must pursue solutions.

Acid Mine Drainage

Old coal mines or piles of mining wastes can be sources of dangerous levels of acidity in streams. Any source rock that is exposed to air (via mining, usually)

and has groundwater pathways or surface channels through it is likely to be increasing the acidity of that water. Pyrite (or iron disulfide) is a common impurity in coal that gives the coal its sulfur content. Through several different chemical reactions the pyrite oxidizes in the presence of air (oxygen) and water, and creates a sulfuric acid or other form of acidity (PA DEP 2004).

The acidity in a stream near a water flow exit from source rock can be very strong. The stream is said to be suffering from Acid Mine Drainage. One indication of the presence of Acid Mine Drainage is water that has turned a yellow-brown color. The water can become so acidic as to kill much of the fish, amphibian and insect life near that point. As the polluted water flows downstream and encounters other tributaries it can become diluted enough to be no longer a problem, but that can be many miles of stream.

The solution is to create chemical reactions that will neutralize the acidity. Active mines often use tanks or channels filled with soluble limestone. This is considered an active treatment system because the limestone levels must be maintained. Experimentation has also shown that developing a wetland around the pollution point will reduce acidity if the soil in the swamp is high in certain mineral and organic matter. Even the selection of swamp vegetation can contribute to neutralizing the water (PA DEP 2004).

214

Pennsylvania has many abandoned coal mine sites that date back to before the 1970s. The responsibility for taking corrective action there falls to the state.

Lake Erie and the Chesapeake Bay

Pennsylvania has played a role in the pollution and clean-up of two very large bodies of water: Lake Erie and the Chesapeake Bay. The former was a striking success, while the latter still has a long way to go.

Lake Erie was the focus of much industrial activity from the late 1800s until at least the 1950s. The activity took place in Ohio, Michigan and even Ontario, Canada, and was accompanied by rapid urban growth in places such as Cleveland, Toledo and Detroit as well as Erie. The combination of industrial pollutants (organic and inorganic chemicals) and nutrients from agricultural fertilizers and municipal sewage created a disastrous situation. Some pollution also flowed in from the Great Lakes upstream from Lake Erie: Lakes Superior, Michigan and Huron.

Compounding the problem was that Lake Erie is much shallower than the other Great Lakes, and so cannot dilute the pollutants as effectively. By the 1960s masses of dead fish were to be found floating on the lake, and by the 1970s the newly formed Environmental Protection Agency had to ban commercial fishing on the lake. A concerted effort between the US, the bordering US states, and Canada's and Ontario's governments ensued. Sewage treatment plants were built or upgraded, phosphate-free detergents were introduced, farm runoff was restricted and factories were made to capture their pollution rather than release it.

Within ten years, the water in Lake Erie was noticeably cleaner. After about twenty years, commercial fishing was allowed again on Lake Erie. While occasional pollution episodes still occur, the lake is greatly improved.

Not so with the Chesapeake Bay. It too is shallow and surrounded by major cities (Washington, DC and Baltimore) with many more up the Susquehanna River (see Figure 36.2) and other tributaries. The Susquehanna is the source of half of the Bay's water, and a great deal of its pollution. Nutrient pollution is an especially significant problem. Nutrients cause masses of algae to cover the Chesapeake's surface and block sunlight to the underwater aquatic vegetation that many fish and shellfish depend on. The same

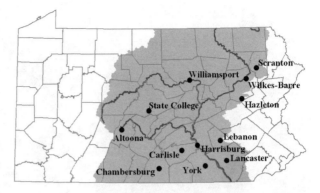

Figure 36.2 **Chesapeake Bay's Watershed in Pennsylvania:** The Susquehanna River, and a few streams that flow directly into Maryland, drain an area that represents about half of Pennsylvania and about half of the Bay's fresh water (Horton and Eichbaum 1991).

algae eventually die and, in decaying, deplete oxygen in the Bay's waters.

The consequences include reduced shellfish populations (once a mainstay of commercial Bay fishing), and contaminated, diseased and dead fish. The problems were first recognized in the 1960s, as with Lake Erie, but it has been tougher to get the states to cooperate and fund the necessary changes.

Modest progress only has been made. A recent report identified the nitrogen (nitrates) level in the Bay at 285 million pounds per year and set a target of 175 million pounds per year by 2010. 75 million of the 110 million pounds per year reduction is to come from Pennsylvania. Lancaster County in particular has been identified as a major source of the nutrients. Methods such as stream fencing to keep livestock out, planting of more vegetation along stream banks and upgrades to sewage treatment plants are to be pursued. Such measures have been voluntary or linked to loans or grants in the past, but may soon become mandatory (CBF 2003).

New Strategies

In order to follow newer Environmental Protection Agency directives, the Department of Environmental Protection has adopted a new water pollution management strategy. The goal is still to improve water quality state-wide, but the new program shifts the emphasis from individual streams (over 83,000 miles of them) to larger watersheds encompassing several connected streams. As long as the watershed is within appropriate limits, some sources of pollution can be tolerated (PA DEP 2002).

215

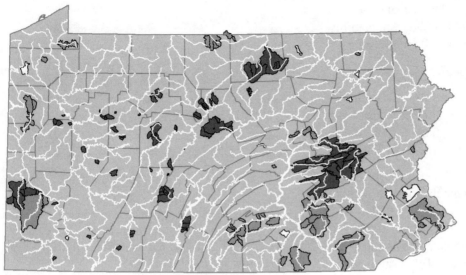

Figure 36.3 **Watersheds with Established TMDLs:** It was still early in the process when the pieces of this map were collected, but notice that several of the watersheds are very large. Acid mine drainage is a major pollutant in many Pennsylvania streams. Non-acid pollutants are more common in the southeast and other densely settled areas of the state. Solutions may require the involvement of farmers or industry, or upgrading sewage treatment facilities.

Watersheds with Specified Total Maximum Daily Loads, 2003

■ TMDL based on Acid Mine Drainage
□ TMDL for a Lake
▨ TMDL based on Non-Acid Mine Drainage Problems

The water quality measure applied to each watershed and category of pollutants is the Total Maximum Daily Load (TMDL), which should never be exceeded. An innovative method is used to calculate each watershed's permissible TMDLs. The Department of Environmental Protection gathers a group of interested citizens from each watershed, teaches them about the pollutants, and then allows them to help determine their area's TMDLs (PA DEP 2002).

The Department of Environmental Protection must identify the watersheds that violate their TMDLs, and rank the violators. The worst watersheds are dealt with first. The US EPA assists with funding and research. The research attempts to compare the costs and benefits of each strategy. Once the plan is approved by the US EPA, the citizen advisors are again called together to help decide the most workable strategy (PA DEP 2002).

The first couple of years of the strategy have succeeded in identifying initial problem watersheds (see Figure 36.3). TMDLs and improvement plans have been completed for the first 71 watersheds. The challenge is to complete the research and set strategies for hundreds of different watersheds.

Conclusions

Water pollution is a particular concern in built-up areas of Pennsylvania, but extends into many rural areas as well. The benefits from water quality improvements are many. The major reasons, of course, are to protect natural species and municipal water supplies. Benefits to the economy will also come from reduced health costs and increased recreation uses of the water.

Bibliography

CBF 2003. State Officials Commit to Reducing Nitrogen Pollution. in Bay Beginnings Pennsylvania Special Lancaster Edition (publication of the Chesapeake Bay Foundation). June 2003, page 2.

Horton, Tom and William M. Eichbaum 1991. Turning the Tide: Saving the Chesapeake Bay. Annapolis, MD, The Chesapeake Bay Foundation and Washington, DC, Island Press.

PA DEP 2002. Watershed Management and TMDLs. Harrisburg, PA, Pennsylvania Department of Environmental Protection. DEP Fact Sheet.

PA DEP 2004. The Science of Acid Mine Drainage and Passive Treatment. Harrisburg, PA, Pennsylvania Department of Environmental Protection. Internet site: <http://www.dep.state.pa.us/dep/deputate/minres/bamr/amd/science_of_amd.htm>, visited 12/8/04.

Chapter 37

Waste Management

One issue for which Pennsylvania has a particularly notorious reputation is waste management. The reputation is most strongly held among our own citizens because it involves allowing garbage from other states to be disposed of in our waste management facilities, especially landfills.

Waste management is much more than just a dumping issue, however. It is often said that Americans live in a throwaway society, meaning that creating waste has almost the status of a cultural trait. Packaging waste is certainly an integral part of our retailing-oriented economy. Wastes produced in the home are only the "tip of the iceberg;" they are also produced in commercial, institutional and industrial settings. The majority of the wastes present volume challenges, but some are concerns because they are toxic or contribute to environmental pollution. Another dimension of our waste issue is the distinction between solid and liquid wastes.

On the other hand, waste *must* be disposed of and, because of several wake-up calls since the 1960s, we are now more aware than ever of the potential impact of different types of wastes on the natural environment. The impacts of growing quantities of waste, and of improperly managed waste, are stronger than ever, for several reasons:

- We create so much more of it than ever before. Products sold at retail stores a century ago seldom came prepackaged, and required little more than a paper bag to carry home.

- Many plastic and metal products are sold as disposable, or have disposable components. Many products not marketed as disposable are intentionally made to wear out, or are cheaper to replace (buy a new one) than repair.

- Today's waste slower to break down organically in landfills. Packages made with plastics and clay-coated papers and cardboards (for brighter packaging colors) persist far longer in the underground conditions of landfills.

- Waste disposal systems common in the early 1900s (incinerators, bonfires and open dumps for solids, and stream or lake disposal for liquids, for example) are no longer permitted for environmental reasons. We are now limited to a set of restrictive and expensive disposal methods.

Solid waste disposal has only been regulated this intensely since the Solid Waste Management Act was passed by US Congress in 1968. Since then the entire system for disposing of garbage, from households, businesses and factories, has been overhauled. As a result of the intense regulation, fewer than 100 solid waste disposal facilities in Pennsylvania now handle the waste that was previously handled by over 1100 facilities (PWIA 2002a).

Types of Waste

Waste, like the air and water pollution we examined in the last two chapters, comes in a variety of types. Each type represents a different kind of environmental hazard and so forces us to take different steps to deal with it safely. Disposal in the ground, usually considered permanent, is the most common option taken. Incineration in modern systems is also making a comeback. You have probably also heard the environmentally-motivated pleas to 'reduce, reuse or recycle.' These options will be described below.

Industry produces a large amount of waste. Factories that use simpler raw materials, such as metals or wood, can usually handle the waste within the plant. Wood, paper or textile scraps, for example, can be burned for heat or to power an electric generator if they are unable to be recycled within the production process. Scrap metals are more often recycled, to such an extent that a market for scrap metals exists, and many factory products are made entirely from recycled metals.

Construction and demolition waste includes woods and metals, and also brick, concrete, vinyl siding, roofing materials and other substances. Because most projects producing such waste will create large volumes at once and most of the volume is solids that are potentially reusable in some form, it is often handled separately.

The largest amount of waste that the public encounters, termed municipal waste, is produced in households and in service and retail businesses. They are largely composed of organic materials (paper, cardboard, wood, food, lawn and garden wastes, and even plastics). This makes them easily combustible or even potentially able to be composted.

Household hazardous wastes are a different story, however. These include batteries, cleaning and polishing products, paints and thinners, pesticides and automobile fluids. Each contains organic chemicals or other toxic elements that pose great risks if they seep from a landfill into the groundwater system.

Another category is medical waste, generated at hospitals and clinics, doctors' offices, nursing homes and even in some homes. Disease agents, pharmaceutical chemicals and even materials exposed to radiation pose another set of risks.

Most radioactive waste, from hospitals as well as from laboratories and nuclear power plant maintenance work, is categorized as low-level radioactivity. These are sent to a specially designed and designated radioactive waste disposal facility in Pennsylvania. Spent fuel from the power plants is many times more dangerous. Even though the federal Department of Energy is focused on creating a permanent storage/disposal site for the nation's hundred or so nuclear power stations (and other military and civilian sources), that site has not yet been developed. All high-level radioactive waste generated at nuclear power stations is kept in long-term temporary facilities at the generation site.

All of the above wastes are considered solid waste. A separate category, with different disposal concerns and options, exists for liquid waste, generally called sewage. Sewage is handled in municipal sewage systems or by regional sewer authorities.

A third category, falling somewhere between the distinction between solid waste and sewage, is sludge. Sludge can be sewage treatment plant solid waste, septic system solid wastes, or farm manure collected and stored but not spread as fertilizer. The Department of Environmental Protection oversees a permitting system for sludge sources, waste haulers and farmers to transport sludge and apply it to non-food crops as fertilizer.

Disposal Technologies

Since the middle 1970s, federal law has required that all landfill operations be upgraded to "sanitary landfills." In a sanitary landfill the initial hole into which the solid waste will be dumped must be lined with an impermeable layer so that liquids in the waste and rainwater filtering though the site will not carry contaminants into groundwater or nearby streams. After each day's dumping is completed, the new compacted waste must be covered with several inches of soil in order to keep it from blowing or washing away and to discourage animals and birds from congregating there.

Sanitary landfills must collect the liquids that percolate through to the liner in a special drainage system, and have that treated at a sewage treatment facility. Decomposing organics in the landfill generate methane, which must be prevented from building up in explosive quantities. Many sanitary landfills have a second drainage system for drawing off the methane and use it to power electricity generators.

Any landfill has a limited lifetime capacity. At the end of its useful life it is capped, landscaped and turned over to some new land use. Some become recreational land, such as parks and ski slopes. In other cases an abandoned mine provided the original land and the filled-in area can provide a site for a factory or shopping center.

The new generation of solid waste incinerators takes advantage of the organic nature of most municipal waste, and uses the heat produced from burning it to generate electricity (see Figure 37.1). Sometimes called "trash-to-steam" facilities, the difference between these and the incinerators of old is the system that must capture all potential air pollutants. These facilities still produce waste in the form of the ashes and unburnable components of the municipal waste. This incinerator waste must be disposed of in sanitary landfills, but the advantage is that it is roughly one tenth the volume of the original waste.

Since 1988 every Pennsylvania municipality with at least 5,000 people must have a recycling program in place. The law required that each program collect at least three recyclable materials from the Department of Environmental Protection's list. The list (PA DEP 1999) includes:

Aluminum
Bi-metal/Steel cans
Corrugated paper (cardboard)
Glass containers (clear or colored)
High-grade office paper
Newsprint
Plastic bottles (any or all types)

In those communities, at least aluminum, corrugated paper and high-grade paper must be recycled. In addition, yard waste (leaves, twigs and garden trimmings) should always be recycled, as should car batteries and motor oil.

Most communities have gone much farther than the minimum because markets for recycled materials are developing. The municipality or regional solid waste authority can either earn money by selling the collected recyclables, or pay a different agent than the local solid waste authority a lower fee to take the waste.

Sewage treatment has also gone through many changes. Fifty years ago many, mostly smaller,

Figure 37.1 **Lancaster County's Trash-to-Steam Incinerator:** Lancaster County hosts one of six "resource recovery" incinerators in which municipal solid waste is burned, the heat turns water to steam, and the steam is used to generate electricity. This facility generates 36 megawatts of electricity.

communities did not treat their sewage. Many homes in Pennsylvania today still use on-lot septic tanks to hold sewage until they are periodically emptied. Sewage treatment plants are more widely required today, and part of the battle against nutrient pollution in water bodies.

Sewage treatment systems use a network of pipes to collect sewage and deliver it to the sewage treatment plant. Sewage treatment plants are required to meet much higher standards of clean water output than they once were. The sewage treatment plant uses a series of filter systems and holding tanks (in some of which biological or chemical agents are added to the sewage) to cleanse the water. The most up-to-date of these remove sediment, suspended minerals, dissolved organic matter, viruses, phosphates and nitrates, and replenished the oxygen level of the water before returning it to rivers or lakes.

Location and Transportation Issues

These wastes are ultimately the responsibility of the municipality in which they are produced, although many groups of municipalities and even entire counties have created solid waste management "authorities" as independent special purpose government units to deal with waste issues.

Most people's experience with waste disposal is the act of collecting household garbage and putting out for the local trash hauler. What follows is a transportation system that may have several components, depending on how large the municipality or authority and how many components the integrated solution has.

Pennsylvania has 63 landfills, of which six are designated for construction/demolition waste only and six are resource recovery facilities (see Figure 37.2). The landfills can be part of a larger solid waste authority's operations, or they can be privately owned businesses. All landfills must be licensed by the state Department of Environmental Protection and inspected periodically.

The locations of these landfills show how competing factors have to be weighed in location decisions. Obviously, the larger an area's population, the more waste it will generate. There are some interesting regional differences in the rate at which solid waste is generated (see Figure 37.3). Factors involved probably include the some measure of how often people shop and what they buy, the relative affluence of the population, the level of recycling in the community, and the local political and cultural attitudes toward environmental concerns.

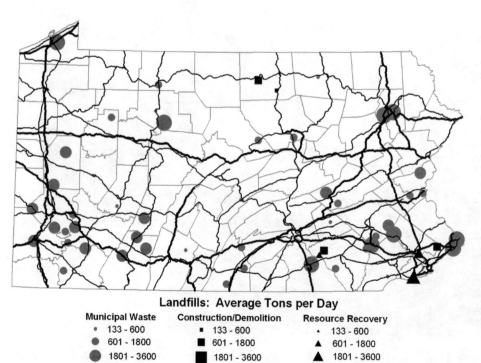

Figure 37.2 **Landfills Across Pennsylvania and Their Capacities:** 51 municipal waste landfills, 6 construction/demolition landfills, and 6 resource recovery (incinerator-based) facilities serve Pennsylvania. Some are too small for this map. The largest are the two units in the southeastern corner of the state northeast of Philadelphia. Located near the FairlessSteel works in Bucks County, The GROWS landfill takes 10,000 tons of trash per day, and the Tullytown landfill takes another 8,333 tons per day. (Data source: PA DEP 2004)

Landfills: Average Tons per Day

Municipal Waste	Construction/Demolition	Resource Recovery
133 - 600	133 - 600	133 - 600
601 - 1800	601 - 1800	601 - 1800
1801 - 3600	1801 - 3600	1801 - 3600
3601 - 10000	3601 - 10000	3601 - 10000

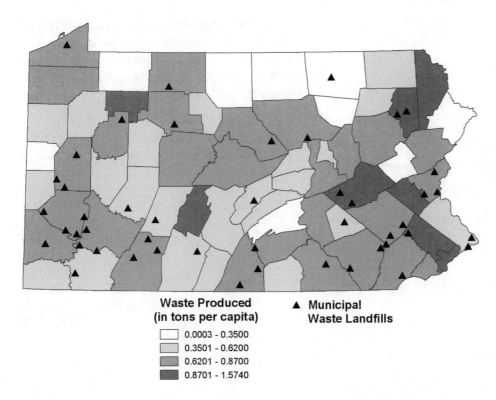

Figure 37.3 **Municipal Waste Generation Across Pennsylvania:** While the map in Figure 37.2 shows where the trash ends up, this map shows where it comes from. Our lowest level of trash production comes from our rural counties, but the highest levels come from urban and rural counties alike (Data source: PA DEP 2002).

Waste Produced (in tons per capita)

▲ **Municipal Waste Landfills**

- 0.0003 - 0.3500
- 0.3501 - 0.6200
- 0.6201 - 0.8700
- 0.8701 - 1.5740

Trash trucks, with either compaction systems or open tops, are a very visible part of the municipal solid waste disposal system. Since 1999 Pennsylvania and several surrounding states have been holding simultaneous "Trashnets." Trash trucks are pulled over on the road, inspected by the Pennsylvania Department of Transportation, the State Police and the Department of Environmental Protection, and cited for any vehicle safety violations or load containment violations. Typically thousands of trucks are inspected, and hundreds to thousands of trucks are issued at least one citation (PA DEP 2001).

An even more controversial issue is the practice of Pennsylvania landfills accepting out-of-state solid waste, a practice in which Pennsylvania ranks number one by a large margin in the US. The list of states from which the waste is brought extends from Florida to Maine and as far west as Kansas and Louisiana (PEN 2001). Approximately half of the solid waste that ends up in Pennsylvania landfills comes from out-of-state (PA DEP 2004). The sizes of the different landfills and their locations relative to the interstate highways (shown in Figure 37.2) may give some clues as to which landfills are accepting out-of-state solid waste.

The quantity of out-of-state solid waste is seen as a problem by Pennsylvania's Department of

Environmental Protection, but as an opportunity by the state's landfill owners. In fact, the issue has already been decided by the US Supreme Court. Their decision was that for any state to restrict out-of-state trucks delivering waste to Pennsylvania landfills would violate interstate commerce laws. Ironically, it was a lawsuit brought by the City of Philadelphia against New Jersey landfills in the late 1970s that prompted the Supreme Court to make this decision. Philadelphia was running out of nearby landfills at the time and the New Jersey sites were its only option (PWIA 2002b.

Several consequences result from the importation of waste into Pennsylvania. We potentially lose capacity for our own waste, our state government incurs costs to inspect both the landfills and the contents of the out-of-state trucks to make sure that nothing illegal is being dumped, and these heavier-than-average vehicles cause wear and tear on our major highways. Municipal wastes represent potential air and water pollution. On the other hand, the landowners are participating in a viable business venture which makes use of land that may otherwise be idle, former strip mines or quarries, for example. The state earns tax income from these businesses as well as from the out-of-state trucks that buy gasoline and other services while in Pennsylvania. The potential exists for imposing extra fees or taxes on out-of-state haulers.

The question is whether the benefits outweigh the costs, both short-term and long-term.

Conclusion

The waste disposal issue is connected to nearly all of the other topics included in this book. It is an environmental, economic and cultural concern. It has impacts on the visible landscape as well as on air quality and water quality. In addition to falling into the category of service businesses, it is a need for every other business in the state. Transportation, energy and urban-vs.-rural problems and benefits can result from it. Above all, as a geographical issue, questions of how much waste gets generated where and why can stimulate interesting cultural and economic speculations and studies. That is what geographers thrive on.

Bibliography

PA DEP 1999. What Can Be Recycled? Harrisburg, PA, Department of Environmental Protection, Waste Management Bureau. Internet site: <http://www.dep.state.pa.us/dep/deputate/airwaste/wm/RECYCLE/Recywrks/recywrks2.htm>, visited 3/29/05.

PA DEP 2001. Ridge Administration Record on Waste Issues. Harrisburg, PA, Pennsylvania Department of Environmental Protection Air/Waste Deputate. Internet site: <http://www.dep.state.pa.us/dep/deputate/airwaste/wm/TargetingTrash/2001_trash_record.htm>, visited 12/9/04.

PA DEP 2002. Municipal Waste Report: 2002. Harrisburg, PA, Department of Environmental Protection, Bureau of Waste Management, Division of Reporting and Fee Collection. Internet site: <http://www.dep.state.pa.us/dep/deputate/airwaste/wm/drfc/reports/repinfo.htm>, visited 12/9/04.

PA DEP 2004. Pennsylvania Municipal Waste Disposal Facilities. Harrisburg, PA, Department of Environmental Protection, Bureau of Waste Management. Internet site: <http://www.dep.state.pa.us/dep/deputate/airwaste/wm/MRW/Docs/Landfill_list.htm>, visited 12/9/04.

PEN 2001. Welcome to Pennsylvania, America Dumps Here. Pennsylvania Environmental Network. Internet site: <http://www.penweb.org/waste/importation/>, visited 5/1/02.

PWIA 2002a. The Evolution of the Solid Waste Management Industry in Pennsylvania. The Pennsylvania Waste Industries Association. Internet site: <http://www.pawasteindustries.org/indhistory.htm>, visited 12/9/04.

PWIA 2002b. Out-Of-State Waste. The Pennsylvania Waste Industries Association. Internet site: <http://www.pawasteindustries.org/oos.htm>, visited 3/29/05.

Bibliography

Abler, Ronald F. 1995. "Services." Chapter 17 in Miller, E. Willard (ed.) A Geography of Pennsylvania. pp. 296-311.

Ackroyd-Kelly, Ian 1987. "The Near Country: The Historical Geography of the Pocono Resorts." The Pennsylvania Geographer. Vol. 25, nos. 1 & 2 (Spring/Summer), pp. 18-23.

Agnes in Northeastern Pennsylvania 2004. Flood Facts. Internet site: <http://www.agnesinnepa.org/modules.php?op=modload&name=News&file=article&sid=3>, visited 2/4/04.

Anonymous 2005. Pittsburghese. Pittsburgh, PA. Internet site: <http://www.pittsburghese.com/>, visited 1/19/05.

Bailey, Robert G. 1995. Descriptions of the Ecoregions of the United States, second edition. Washington, DC, US Department of Agriculture, US Forest Service. Miscellaneous Publication number 1391.

Barnes, John H. and William D. Sevon 2002. The Geological Story of Pennsylvania (3rd edition). 4th series, Educational Series 4. Harrisburg, Pennsylvania Geological Survey.

Barnes, John H. and Robert C. Smith, II 2001. The Nonfuel Mineral Resources of Pennsylvania. Educational Series publication 12. Harrisburg, Department of Conservation and Natural Resources, Bureau of Topographic and Geologic Survey.

Bauman, John F. 2002. "Philadelphia (city, Pennsylvania)." in Microsoft Encarta Encyclopedia 2002. Microsoft Corporation.

Becher, Albert E. 1999. "Groundwater." Chapter 44 in Shultz, Charles H. (ed.) The Geology of Pennsylvania. pp. 666-677.

Beck, Bill 1995. PP&L 75 Years of Powering the Future: An Illustrated History of Pennsylvania Power and Light Co. Allentown, PA, PP&L.

Boas, Charles W. 1987. "The Railroads of York County, PA: Rail Center to Recreation." The Pennsylvania Geographer. Vol. 25, nos. 1&2 (Spring/Summer), pp. 12-17.

Borenstein, Seth 2003. "Bush Administration Says It Won't Regulate Carbon Dioxide." New York, NY, Natural Resources Defense Council and Knight Ridder/Tribune News Service. 8/29/03. Internet site: <http://www.nrdc.org/news/newsDetails.asp?nID=1080>, visited 3/23/05.

Briggs, Reginald P. 1999. "Appalachian Plateaus Province and the Eastern Lake Section of the Central Lowland Province." Chapter 30 in Shultz, Charles H. (ed.) The Geology of Pennsylvania. pp. 362-377.

Brotherswar.com 2004. The Battle of Gettysburg and the American Civil War. Internet site: <http://www.brotherswar.com>, visited 9/15/04.

Brown, Andrew 1962. Geology and the Gettysburg Campaign. Educational Series 5. Harrisburg, Pennsylvania Geological Survey.

Burtraw, Dallas 1996. "Trading Emissions to Clean the Air: Exchanges Few but Savings Many." Resources (news-letter of Resources for the Future). Winter 1996, pp. 3-6.

CBF 2003. State Officials Commit to Reducing Nitrogen Pollution. in Bay Beginnings Pennsylvania Special Lancaster Edition (publication of the Chesapeake Bay Foundation). June 2003, p. 2.

Chamberlin, Clint 2004. Canals in Pennsylvania. Northeast Rails. Internet site: <http://www.northeast.railfan.net/canal.html>, visited 11/7/04.

City-data.com 2004. Gettysburg, Pennsylvania. Internet site: <http://www.city-data.com/city/Gettysburg-Pennsylvania.html>, visited 9/14/04.

City-data.com 2004. Warren, Pennsylvania. Internet site: <http://www.city-data.com/city/Warren-Pennsylvania.html>, visited 9/14/04.

City of Warren 2004. City of Warren, Pennsylvania. Warren, PA, City of Warren. Internet site <http://www.cityofwarrenpa.org/>, visited 9/12/04.

Clean Air Council 2005. Beyond Mercury: Annual Power Plant Toxic Air Emissions in Pennsylvania. Philadelphia, PA, Clean Air Council. Internet site: <http://www.cleanair.org/toxicfactsheet.pdf>, visited 3/24/05.

Cox, Harold E. 2004. Pennsylvania Presidential Election Returns. Wilkes-Barre, PA, Wilkes University: The Wilkes University Election Statistics Project: Dr. Harold E. Cox, Director. Internet site: <http://wilkes-fs1.wilkes.edu/~hcox>, visited 10/25/04.

Dougherty, Percy H. (ed.) 2004. Geography of the Philadelphia Region: Cradle of Democracy. Prepared for the 100th Annual Meeting of the Association of American Geographers. Washington, DC, Association of American Geographers.

Dougherty, Percy H. 2004. "Landform Regions of Eastern Pennsylvania and their Impact on the Cultural Landscape." in Dougherty, Percy H. (ed.) Geography of the Philadelphia Region: Cradle of Democracy. pp. 1-16.

Dubin, Murray 1996. South Philadelphia: Mummers, Memories, and the Melrose Diner. Philadelphia, Temple University Press.

Dudden, Arthur P. 1982. "The City Embraces 'Normalcy' 1919-1929." in Weigly, Russell F. (ed.), Philadelphia: A 300-Year History. pp. 566-600.

Dunne, Thomas and Luna B. Leopold 1978. Water in Environmental Planning. New York, W.H. Freeman and Co.

Dykstra, Anne Marie 1989. "Pennsylvania's Past: Exploration and Settlement." in Cuff et al.: The Atlas of Pennsylvania. pp. 80-81.

Edmunds, William E. 1999. "Bituminous Coal." Chapter 37 in Shultz, Charles H. (ed.) The Geology of Pennsylvania. pp. 470-481.

Eggleston, Jane R., Thomas M. Kehn, and Gordon H. Wood, Jr. 1999. "Anthracite." Chapter 36 in Shultz, Charles H. (ed.) The Geology of Pennsylvania. pp. 458-469.

Erickson, Rodney A. 1995. "The Internal Spatial Structure of Pennsylvania's Metropolitan Areas." Chapter 19 in Miller, E. Willard (ed.) A Geography of Pennsylvania. pp. 336-355.

Erickson, Rodney A. 1995. "The Location and Growth of Pennsylvania's Metropolitan Areas." Chapter 18 in Miller, E. Willard (ed.) A Geography of Pennsylvania. pp. 315-335.

Fergus, Chuck 2003. Elk. Harrisburg, PA, Pennsylvania Game Commission. Internet site: <http://sites.state.pa.us/PA_Exec/PGC/x_notes/elk.htm>, visited 5/28/04.

Fischer, John 2002. 2002 Mummers Parade Part 1: Quaker City, Downtowners, Golden Sunrise and Murray Win… But at What Price? Philadelphia, About.com. Internet site: <http://philadelphia.about.com/library/gallery/aa010302a.htm>, visited 1/20/05.

Fleeger, Gary M. 1999. The Geology of Pennsylvania's Groundwater. 4th series, Educational Series 3. Harrisburg, Pennsylvania Geological Survey.

Frederick, Paul 2004. Take a Look at Cook Forest State Park. Allegheny-online. Internet site: <http://www.allegheny-online.com/cookforest.html>, visited 9/24/04.

Gammage, Jeff 1998. 20 Died. The Government Took Heed. In 1948, A Killer Fog Spurred Air Cleanup. Originally published in the Philadelphia Inquirer. Harrisburg, PA, Pennsylvania Department of Environmental Protection.

Internet site: <http://www.dep.state.pa.us/dep/Rachel_Carson/dead20.htm>, visited 9/12/03.

Geiger, Charles 2004. "Historical Electricity Production in Pennsylvania." The Pennsylvania Geographer, Vol. 42, no. 2 (Fall/Winter), forthcoming.

Geiger, Charles and Kent Barnes 1994. "Indoor Radon Hazard: A Geographical Assessment and Case Study." Applied Geography, 14(4): 350-371.

Geisinger Health System 2004. About Geisinger. Danville, PA, Geisinger Health System. Internet site: <http://www.geisinger.edu/about/history.shtml>, visited 12/13/04.

Getis, Arthur, Judith Getis and I.E. Quastler 2001. The United States and Canada: The Land and the People. Boston, McGraw-Hill.

GE Transportation 2005. Facts and History. Erie, PA, General Electric Transportation Rail. Internet site: <http://www.getransportation.com/general/locomotives/fast_facts.asp>, visited 3/15/05.

Gibson, Campbell J. and Emily Lennon 1999. "Table 13: Nativity of the Population, for Regions, Divisions, and States: 1850 to 1990." from: Historical Census Statistics on the Foreign-born Population of the United States: 1850-1990. Washington, DC, U.S. Department of Commerce, Census Bureau. Internet site: <http://www.census.gov/population/www/documentation/twps0029/tab13.html>, visited 7/14/04.

Glass, Joseph W. 1979. "Be Ye Separate, Saith the Lord: Old Order Amish in Lancaster County." In Cybriwsky, Roman A. (ed.): The Philadelphia Region: Selected Essays and Field Trip Itineraries. Washington, DC, Association of American Geographers.

Glass, Joseph W. 1986. The Pennsylvania Culture Region: A View from the Barn. Ann Arbor, MI, UMI Research Press.

Glover, Lynne 1998. Donora's Killer Smog Noted at 50. Originally published in the Pittsburgh Tribune-Review. Harrisburg, PA, Pennsylvania Department of Environmental Protection. Internet site: <http://www.dep.state.pa.us/dep/Rachel_Carson/killer_smog.htm>, visited 9/12/03.

GPU Nuclear 1991. The TMI-2 Story. Middletown, PA, GPU (originally General Public Utilities) Nuclear, Public Affairs Department.

Gray, Carlyle 1999. "Cornwall-Type Iron Deposits." Chapter 40B in Shultz, Charles H. (ed.) The Geology of Pennsylvania. pp. 566-573.

Harper, John A., Derek B. Tatlock and Robert T. Wolfe, Jr. 1999. "Petroleum-Shallow Oil and Natural Gas."

Chapter 38A in Shultz, Charles H. (ed.) The Geology of Pennsylvania. pp. 484-505.

Hartmann, Edward George 1978. Americans from Wales. New York, NY, Octagon Books: Farrar, Strauss and Giroux.

Helms, Douglas 1998. Natural Resources Conservation Service Brief History. Washington, DC, US Department of Agriculture, Natural Resources Conservation Service. Internet site: <http://www.nrcs.usda.gov/about/history/articles/briefhistory.html>, visited 10/4/04.

Holechek, Jerry L., Richard A. Cole, James T. Fisher, and Raul Valdez 2003. Natural Resources: Ecology, Economics, and Policy. Upper Saddle River, NJ, Prentice-Hall.

Horton, Tom and William M. Eichbaum 1991. Turning the Tide: Saving the Chesapeake Bay. Annapolis, MD, The Chesapeake Bay Foundation and Washington, DC, Island Press.

HSP 2004. Philadelphia's Chinatown: An Overview. Philadelphia, PA, Historical Society of Pennsylvania. Internet site: <http://www.hsp.org/default.aspx?id=190>, visited 12/6/04.

Ingham, John N. 1991. Making Iron and Steel: Independent Mills in Pittsburgh, 1820-1920. Columbus, Ohio State University Press.

Inners, Jon D. 1999. "Sedimentary and Metasedimentary Iron Deposits." Chapter 40A in Shultz, Charles H. (ed.) The Geology of Pennsylvania. pp. 556-565.

IPAA 2004. Oil and Gas in Your State: Pennsylvania. Independent Petroleum Association of America, Washington, DC. Internet site: <http://www.ipaa.org/info/In Your State/default.asp?State=Pennsylvania>, visited 8/10/04.

Jacobson, Mike and Cathy Seyler 2004. Economic Contribution of Forestry to Pennsylvania. University Park, PA, The Pennsylvania State University, College of Agricultural Sciences, School of Forest Resources. Internet site: <http://rnrext.cas.psu.edu/counties/extmap.htm>, visited: 7/29/04.

Johnstone, Barbara 2004. Pittsburgh Speech and Society. Pittsburgh, PA, Carnegie-Mellon University. Internet site: <http://English.cmu.edu/pittsburghspeech>, visited 1/19/05.

Klein, Frederic Shriver 1964. Old Lancaster: Historic Pennsylvania Community from its Beginnings to 1865. Lancaster, PA, Early America Series, Inc.

Kline, Hibberd V.B. 1981. "Welcome to Pittsburgh." The Pennsylvania Geographer. Vol. 19, no. 3 (October), pp. 1-6.

Kochanov, William E. 2002. Sinkholes in Pennsylvania. 4th series, Educational Series 11. Harrisburg, Pennsylvania Geological Survey.

Kosack, Joe 2001. History of the Pennsylvania Elk. Harrisburg, PA, Pennsylvania Game Commission. Internet site: <http://sites.state.pa.us/PA_Exec/PGC/elk/history.htm>, visited 9/27/02.

Lancaster Farmland Trust 2004. What's New. Lancaster, PA, Lancaster Farmland Trust. Internet site: <http://www.savelancasterfarms.org/savelanc/cwp/view.asp?a=3&q=570958>, visited 10/28/04.

Lapsansky 2002. "Building Democratic Communities: 1800-1850." Chapter 4 in Miller and Pencak (eds.) Pennsyl-vania: A History of the Commonwealth. pp. 151-202.

Lavine, Mary P. 1990. "The Legacy of the Johnstown Flood." The Pennsylvania Geographer. Vol. 28, no. 2, 68-80.

LC APB 2004. Agricultural Preserve Board. Lancaster, PA, Lancaster County Agricultural Preserve Board. Internet site: <http://www.co.lancaster.pa.us/lanco/cwp/view.asp? a=371&Q=384772&tx=1>, visited 10/28/04.

LC APB 2005. Personal communication. Lancaster, PA, Lancaster County Agricultural Preserve Board. 1/27/05.

Lemon, James T. 1972. The Best Poor Man's Country: A Geographical Study of Early Southeastern Pennsylvania. Baltimore, Johns Hopkins University Press.

Levittown 2002. The History of Levittown. Levittown, PA. Internet site: <http://www.levittownpa.org/Levittown.html>, visited 10/19/04.

Lewis, Peirce 1995. "American Roots in Pennsylvania Soil." Chapter 1 in Miller, E. Willard (ed.) A Geography of Pennsylvania. pp. 1-13.

Lewis, Tom 1997. Divided Highways: Building the Interstate Highways, Transforming American Life. New York, NY, Penguin Putnam.

Licht, Walter 2002. "Civil Wars: 1850-1900." Chapter 5 in Miller and Pencak (eds.) Pennsylvania: A History of the Commonwealth. pp. 203-256.

Liebhold, Sandy 2003. Gypsy Moth in North America. Morgantown, WV, US Department of Agriculture: US Forest Service: Forest Service Northeastern Research Station. Internet site: <http://www.fs.fed.us/ne/morgantown/4557/gmoth/>, visited 9/29/04.

Little League 2002. Little League Baseball Historical Timeline. Williamsport, PA, Little League Baseball,

Inc. Internet site: <http://www.littleleague.org/history/index.htm>, visited 1/25/05.

Lukacs, John 1981. Philadelphia: Patricians and Philistines 1900-1950. New York, Farrar, Straus, Giroux.

Lycoming County 2005. Internet site <http://www.williamsport.org/visitors/mill_row.htm>, visited 1/28/05.

Magda, Matthew S. 1986. Welsh in Pennsylvania. Harrisburg, PA, Pennsylvania Historical and Museum Commission. Internet site: <http://www.phmc.state.pa.us/ppet/welsh/page1.asp?secid=31>, visited 7/14/04.

MARFC 2004. The Life of Hurricane Agnes. State College, PA, Middle Atlantic River Forecast Center. Internet site: <http://www.erh.noaa.gov/marfc/Flood/agnes.html>, visited 2/12/03.

Marsh, Ben and Peirce Lewis 1995. "Landforms and Human Habitat." Chapter 2 in Miller, E. Willard (ed.) A Geography of Pennsylvania. pp. 17-43.

Marvel 2005. Radio Roadtrip.com – Baseball. Marvel Company, LLC. Internet site: <http://radioroadtrip.com/baseball.htm>, visited 3/5/05.

Macdonald, Gerald M. 1993. "Philadelphia's Penn's Landing: Changing Concepts of the Central River Front." The Pennsylvania Geographer. Vol. 31, no. 2, pp. 36-51.

McNab, W. Henry and Peter E. Avers 1994. Ecological Subregions of the United States. Washington, DC, US Department of Agriculture, US Forest Service. Internet site: <http://www.fs.fed.us/land/pubs/ecoregions/index.html>, visited 8/10/04.

MC RPC 2004. Mercer County's Comprehensive Plan – Trends and Data. Hermitage, PA, Mercer County Regional Planning Commission. Internet site: <http://www.mcrpc.com/trends.htm>, visited 11/30/04.

Meyers, Thomas J. 1990. Amish. Canada, Mennonite Historical Society of Canada. Internet site, created 1998, copyrighted 2004: <http://www.mhsc.ca/index.asp ?content=http://www.mhsc.ca/encyclopedia/contents/A4574ME.html>, visited 7/25/04.

Miller, E. Willard 1981. "Pittsburgh: Patterns of Evolution." The Pennsylvania Geographer. Vol. 19, no. 3 (October), pp. 6-20.

Miller, E. Willard (ed.) 1995. A Geography of Pennsylvania. University Park, PA, The Pennsylvania State University Press.

Miller, E. Willard 1995. "Pittsburgh: An Urban Region in Transition." Chapter 21 in Miller, E. Willard (ed.) A Geography of Pennsylvania. pp. 374-395.

Miller, E. Willard 1995. "Soil Resources." Chapter 5 in Miller, E. Willard (ed.) A Geography of Pennsylvania. pp. 67-73.

Miller, E. Willard 1995. "Transportation." Chapter 14 in Miller, E. Willard (ed.) A Geography of Pennsylvania. pp. 234-251.

Miller, Randall M. and William Pencak (eds.) 2002. Pennsylvania: A History of the Commonwealth. University Park and Harrisburg, The Pennsylvania State University Press and the Pennsylvania Historical and Museum Commission.

Mulhollem, Jeff 2003. Pennsylvania Forests Changing from Red Oak to Red Maple Dominated. University Park, PA, The Pennsylvania State University. News Release. Internet site: <http://aginfo.psu.edu/News.march03/forest.html>, site visited 3/26/04.

Muller, Edward K. 2000. "Downtown Pittsburgh Renaissance and Renewal." in Patrick and Scarpaci (eds.) A Geographic Perspective of Pittsburgh and the Alleghenies: From Precambrian to Post-Industrial, pp. 17-27.

Murphy, Raymond E. and Marion F. Murphy 1952. Pennsylvania Landscapes: A Geography of the Commonwealth (2nd edition). State College, PA, Penn's Valley Publishers.

NOAA 1992. Tornadoes: Nature's Most Violent /Storms. Washington, DC, National Oceanic and Atmospheric Administration. Internet site: <http://www.nws.noaa.gov/om/brochures/tornado.htm>, visited 9/14/04.

NOAA 2002. Monthly Station Normals of Temperature, Precipitation, and Heating and Cooling Degree Days 1971-2000: Pennsylvania. Climatography of the United States series, Publication No. 81. Asheville, NC, National Oceanic and Atmospheric Administration: National Climatic Data Center.

NOAA 2004. SNOW1413.shp (digital map file). Internet site: <http://www5.ncdc.noaa.gov/cgi-bin/climaps/climaps.pl>, visited 7/13/04.

NPWA 2003. Water Currents. Lansdale, PA, North Penn Water Authority. January 2003 newsletter. Internet site: <http://www.northpennwater.org/pdf/newsletter0103.pdf>, visited 12/30/04.

NukeWorker.com 2005. Saxton Nuclear Experimental Facility. Powell, TN, NukeWorker.com. Internet site: <http://nukeworker.com/nuke_facilities/North_America/usa/NRC_Facilities/Region_1/saxton/index.shtml>, visited 1/13/05.

Oil City Area Chamber of Commerce 2004. City of Oil City: Historical Overview. Oil City, PA. Internet site:

<http://www.oilcitychamber.org/history.htm>, visited 9/1/04.

Oser, David R. 2004. Bryn Mawr Historical Information. Bryn Mawr, PA, Main Line Real Estate/REMAX Executive Realty. Internet site: <http://www.mainlinerealestate.com/bryn_mawr_history.htm>, visited 7/14/04.

PA APSS 2004. Hazleton: Pennsylvania State Soil. Harrisburg, PA, Pennsylvania Association of Professional Soil Scientists. Internet site: <http://www.papss.org/hazleton.htm>, visited 7/19/04.

PA AOPC 2002. "The Structure of Pennsylvania's Unified Judicial System." in Report of the Administrative Office of Pennsylvania Courts 2001 [annual report]. Harrisburg, PA, Administrative Office of Pennsylvania's Courts. Internet site: <http://www.courts.state.pa.us/Index/Aopc/AnnualReport/annual01/07struct.pdf>, visited 10/21/04.

PA AOPC 2005. Pennsylvania's Unified Judicial System. Harrisburg, PA, Administrative Office of Pennsylvania's Courts. Internet site: <http://www.courts.state.pa.us/index.asp>, visited 1/21/05.

PA BCEL 2003. Official Voter Registration Data, November 2003. Harrisburg, PA, Pennsylvania Department of State, Bureau of Commissions, Elections and Legislation. Internet site: <http://www.dos.state.pa.us/bcel/cwp/view.asp?a=1099&Q=441857&PM=1>, visited 8/5/04.

PA Bureau of Forestry 1975. A Chronology of Events in Pennsylvania Forestry Showing Things As They Happened to Penn's Woods. Harrisburg, PA, PA Department of Environmental Resources (now the Department of Environmental Protection): Bureau of Forestry. Internet site: <http://www.dep. state.pa.us/dep/pa_env-her/historycalforestry.htm>, visited 9/29/04.

PA Canal Society 2004. Pennsylvania's Canal Era 1792-1931. Easton, PA, Pennsylvania Canal Society. Internet site: <http://www.pa-canal-society.org/>, visited 7/25/04.

PA DA 2004. PA's Forest Product Industry. Harrisburg, PA, Pennsylvania Department of Agriculture: Hardwoods Development Council. Internet site: <http://www.agriculture.state.pa.us/agriculture/cwp/view.asp?a=3&q=128894>, visited 7/19/04.

PA DCED 2003. Home Rule in Pennsylvania (7th ed.). Harrisburg, PA, Pennsylvania Department of Community and Economic Development, Governor's Center for Local Government Services. Internet site: <http://www.inventpa.com/docs/Document/application/pdf/a882c48f-b26e-4cc0-9b3d-0681c5662df4/home-rule.pdf>, visited 10/18/04.

PA DCNR 2003. State Forest Resource Management Plan: Executive Summary. Harrisburg, PA, Pennsylvania Department of Conservation and Natural Resources, Bureau of Forestry. Internet site: <http://www.dcnr.state.pa.us/forestry/sfrmp/execsummary.htm>, visited 1/27/05.

PA DCNR 2004. "The Grand Canyon of Pennsylvania." Pennsylvania Department of Conservation and Natural Resources State Parks, Harrisburg, PA. Internet site: <http:www.dcnr.state.pa.us/stateparks/prks/leonardharrison.aspx>, visited 8/29/04.

PA DEP 1999. What Can Be Recycled? Harrisburg, PA, Department of Environmental Protection, Waste Management Bureau. Internet site: <http://www.dep.state.pa.us/dep/deputate/airwaste/wm/RECYCLE/Recywrks/recywrks2.htm>, visited 3/29/05.

PA DEP 2001. 2000 Annual Report on Mining Activities in the Commonwealth of Pennsylvania. Harrisburg, PA, Pennsylvania Department of Environmental Protection.

PA DEP 2001. Ridge Administration Record on Waste Issues. Harrisburg, PA, Pennsylvania Department of Envi-ronmental Protection Air/Waste Deputate. Internet site: <http://www.dep.state.pa.us/dep/deputate/airwaste/wm/TargetingTrash/2001_trash_record.htm>, visited 12/9/04.

PA DEP 2002. Comprehensive Stormwater Management Policy. Harrisburg, PA, PA Department of Environmental Protection. Document ID 392-0300-002.

PA DEP 2002. Municipal Waste Report: 2002. Harrisburg, PA, Department of Environmental Protection, Bureau of Waste Management, Division of Reporting and Fee Collection. Internet site: <http://www.dep.state.pa.us/dep/deputate/airwaste/wm/drfc/reports/repinfo.htm>, visited 12/9/04.

PA DEP 2002. Watershed Management and TMDLs. Harrisburg, PA, Pennsylvania Department of Environmental Protection. DEP Fact Sheet.

PA DEP 2004. DSAW Dam Graphs. Harrisburg, PA, Pennsylvania Department of Environmental Protection: Bureau of Waterways Engineering. Internet site: <http://www.dep.state.pa.us/deputate/watermgt/we/damprogram/ndsad/main/graphs.htm>, visited 6/28/04.

PA DEP 2004. Pennsylvania Municipal Waste Disposal Facilities. Harrisburg, PA, Department of Environmental Protection, Bureau of Waste Management. Internet site: <http://

www.dep.state.pa.us/dep/deputate/airwaste/wm/MRW/ Docs/Landfill_list.htm>, visited 12/9/04.

PA DEP 2004. The Science of Acid Mine Drainage and Passive Treatment. Harrisburg, PA, Pennsylvania Department of Environmental Protection. Internet site: <http://www.dep.state.pa.us/dep/deputate/minres/bamr/ amd/science_of_amd.htm>, visited 12/8/04.

PA DEP 2005. Monitoring Principal Pollutants. Harrisburg, PA, Pennsylvania Department of Environmental Protection, Bureau of Air Quality. Internet site: <http://www.dep.state.pa.us/dep/deputate/airwaste/aq/aqm/ principal.htm>, visited 3/16/05.

PA DGS 2003. The Pennsylvania Manual. Harrisburg, PA, Department of General Services. Internet site: <http://www.dgs.state.pa.us/pamanual/site/ default.asp>, visited 1/21/05.

PA Game Commission 2004. About the Pennsylvania Game Commission. Harrisburg, PA, Pennsylvania Game Commission. Internet site: <http://www.pgc. state.pa.us/pgc/cwp/view.asp?a=481 &q=151287&pgcNav=|>, visited 10/4/04.

PA Highways 2004. Breezewood Services. Pennsylvania Highways. Internet site: <http://www.pahighways.com/ interstates/bwoodservices_files/bwoodservices_ vml_1.htm>, visited 11/9/04.

PA HMC 2004. Pennsylvania State History: Maturity 1945-1995. Harrisburg, PA, Pennsylvania Historical and Museum Commission. Internet site: <http:// www.phmc.state.pa.us/bah/pahist/mature.asp ?secid=31>, visited 11/9/04.

PA State Climatologist 2004. [No Title: a climatological description of Pennsylvania]. State College, PA, Office of the Pennsylvania State Climatologist. Internet site: <http://pasc.met.psu.edu/PA_Climatologist/state/ index.html>, visited 7/6/04.

PA State Climatologist 2004. Number of Tornadoes: 1881-Present. State College, PA, Office of the Pennsylvania State Climatologist. Internet site: <http:// pasc.met.psu.edu/PA_Climatologist/state/misc/ tornado00.jpg>, visited 7/6/04.

PA State Museum 2003. Levittown, PA: Building the Suburban Dream. Harrisburg, PA, The State Museum of Pennsylvania. Internet site: <http:// server1.fandm.edu/levittown/>, visited 10/19/04.

PA Turnpike Commission 2004. Mon/Fayette Expressway and Southern Beltway. Harrisburg, PA, Pennsylvania Department of Transportation, Pennsylvania Turnpike Commission. Internet site: <http:// www.paturnpike.com/monfaysb/>, visited 12/5/04.

Patrick, Kevin J. and Joseph L. Scarpaci, Jr. (eds.) 2000. A Geographic Perspective of Pittsburgh and the Alleghenies: From Precambrian to Post-Industrial. Washington, DC, Association of American Geographers.

Pearson, Brooks C. 1997. "18th and 19th Century Welsh Migration to the United States." Pennsylvania Geographer. Vol. 35, no. 1, pp. 38-54.

PEL-East 2004. Pennsylvania's Geo-Politics. Philadelphia, PA, Pennsylvania Economy League, Inc., Eastern Division. Internet site: <http://www.peleast.org/ geo.htm>, visited 7/21/04.

PEN 2001. Welcome to Pennsylvania, America Dumps Here. Pennsylvania Environmental Network. Internet site: <http://www.penweb.org/waste/importation/>, visited 5/1/02.

Philadelphia 2005. Mummers Parade History. Philadelphia, PA, Philadelphia Recreation Department. Internet site: <http://www.phila.gov/recreation/ mummers/mummers_history.html>, visited 1/20/05.

Pillsbury, Richard R. 1970. "The Urban Street Pattern as a Cultural Indicator: Pennsylvania, 1682-1815." Annals of the Association of American Geographers. Vol. 60 (September), pp. 428-446.

Pomerantz, Joanne T. and Joan M. Welch 1996. "Utilization of Woody Browse by White-Tailed Deer (Odocoileus virginianus) in Valley Forge National Historical Park." Pennsylvania Geographer. Vol. 34, no. 2, pp. 87-97.

Potter, Noel, Jr. 1999. "Southeast of Blue Mountain." Chapter 28 in Shultz, Charles H. (ed.) The Geology of Pennsylvania. pp. 344-351.

Public Broadcast System 2004. They Made America: John Wanamaker. Internet site: <http://www.pbs.org/wgbh/ theymadeamerica/whomade/wanamaker_ hi.html>, visited 11/19/04.

PWIA 2002. The Evolution of the Solid Waste Management Industry In Pennsylvania. The Pennsylvania Waste Industries Association. Internet site: <http:// www.pawasteindustries.org/indhistory. htm>, visited 12/9/04.

PWIA 2002. Out-Of-State Waste. The Pennsylvania Waste Industries Association. Internet site: <http:// www.pawasteindustries.org/oos.htm>, visited 3/29/05.

Quaker State 2004. Quaker State: A History of Industry Leadership. Shell Oil Products US, Houston TX. Internet site: <http://www.quakerstate.com/pages/ about/history.asp>, visited 9/2/04.

Rees, Peter W. 2004. "The Spatial Evolution of Center City, Philadelphia." In Dougherty, Percy H. (ed.)

Geography of the Philadelphia Region: Cradle of Democracy. pp. 25-44.

Reisinger, Mark E. 2003. "The Origins and Development of the York Industrial District in South Central Pennsylvania: 1700-1920." The Pennsylvania Geographer. Vol. 41, no. 2 (Fall/Winter), pp. 67-94.

RHC 2005. About the Railroaders Memorial Museum. Altoona, PA, Railroaders Heritage Corporation. Internet site: <http://www.railroadcity.com/museum.htm>, visited 2/8/05.

Richter, Daniel K. 2002. "The First Pennsylvanians." Chapter 1 in Miller, Randall M. and William Pencak (eds.) Pennsylvania: A History of the Commonwealth. pp. 3-46.

Rodgers, Allan L. 1995. "The Rise and Decline of Pennsylvania's Steel Industry." Chapter 16 in Miller, E. Willard (ed.) A Geography of Pennsylvania. pp. 285-295.

Rose, Arthur W. 1999. "Radon." Chapter 55B in Shultz, Charles H. (ed.) The Geology of Pennsylvania. pp. 786-793.

Rossi, Theresa 1999. "Climate." Chapter 43 in Shultz, Charles H. (ed.) The Geology of Pennsylvania. pp. 658-665.

RRM PA 2005. About Us. Strasburg, PA, Railroad Museum of Pennsylvania. Internet site: <http://www.rrmuseumpa.org/about/welcome/aboutus.htm>, visited 2/8/05.

Rusk 2003. "'Little Boxes' – Limited Horizons: A Study of Fragmented Local Governance in Penn-sylvania: Its Scope, Consequences, and Reforms." Background paper for the project: Back to Prosperity: A Competitive Agenda for Renewing Pennsylvania. Washington, DC, Brookings Institution, Center on Urban and Metropolitan Policy. Internet version: <http://www.brookings. edu/pennsylvania>, visited 10/18/04.

SAGE 2003. River Discharge Database. Madison, WI, University of Wisconsin-Madison: Gaylord Nelson Institute for Environmental Studies: Center for Sustainability and the Global Environment. Internet site: <http://www.sage.wisc.edu/riverdata/scripts/keysearch.php?numfiles=50&startnum=2000>, visited 7/8/04.

Scharnberger, Charles K. 2003. Earthquake Hazard in Pennsylvania. 4th series, Educational Series 10. Harrisburg, Pennsylvania Geological Survey.

Schein, Richard D. and E. Willard Miller 1995. "Forest Resources." Chapter 6 in Miller, E. Willard (ed.) A Geography of Pennsylvania. pp. 74-83.

Schuyler, David 2002. A City Transformed: Redevelopment, Race, and Suburbanization in Lancaster, Pennsylvania 1940-1980. University Park, PA, The Pennsylvania State University Press.

Sevon, William D. and Gary M. Fleeger 1999. Pennsylvania and the Ice Age (2nd edition). 4th series, Educational Series 6. Harrisburg, Pennsylvania Geological Survey.

Shultz, Charles H. (ed.) 1999. The Geology of Pennsylvania. Harrisburg and Pittsburgh, The Pennsylvania Geological Survey and the Pittsburgh Geological Society.

Shultz, Charles H. 1999. "Overview." Chapter 2 in Shultz, Charles H. (ed.) The Geology of Pennsylvania. pp. 12-21.

Smith, Joe Bart 1987. "The Growth and Development of York City from 1740 to 1890." The Pennsylvania Geographer. Vol. 25, nos. 1&2 (Spring/Summer), pp. 8-11.

Stanitski-Martin, Diane and Kay R.S. Williams 1999. "A Climatological Summary of Pennsylvania's State System of Higher Education Universities." The Pennsylvania Geographer. Vol. 37, no. 2 (Fall/Winter), pp. 80-101.

Stranahan, Susan Q. 1993. Susquehanna, River of Dreams. Baltimore, MD, The Johns Hopkins University Press.

Strausbaugh, Judy A. 2004. "Using Horse Sense in Buggy Safety." Sunday News, August 29, 2004. Lancaster, PA, pp. A1, A4.

Superbrands 2005. Woolworth's. London, UK, Superbrands, Ltd. Internet site: <http://www.superbrands.org/21843>, visited 3/2/05.

Thompson, Glenn H. and J. Peter Wilshusen 1999. "Geological Influences on Pennsylvania's History and Scenery." Chapter 57 in Shultz, Charles H. (ed.) The Geology of Pennsylvania. pp. 810-819.

Toker, Franklin 1986. Pittsburgh: An Urban Portrait. University Park, PA, Pennsylvania State University Press.

Trainweb 2004. Horseshoe Curve: Background. Trainweb. Internet site: <http://www.trainweb.org/horseshoecurve-nrhs/Guide.htm> visited 8/26/04.

Trifonoff, Karen 2000. "The Mine Fire in Centralia, Pennsylvania." The Pennsylvania Geographer. Vol. 32, no. 2, pp. 3-24.

US ACE 1999. National Inventory of Dams (1999 update). Alexandria, VA, US Army Corps of Engineers: Army Topographic Engineering Center. Internet site: <http://crunch.tec.army.mil/nid/webpages/nid.cfm>, visited 9/17/04.

US Airways 2004. Economic Impact Study Finds Pittsburgh Air Hub Generates $3.1 Billion for Region, Supports 33,300 Jobs. Arlington, VA, US Airways, Inc. Internet site: <www.usairways.com/about/press_2003/nw_03_0917.htm>, visited 2/11/05.

US Airways 2004. US Airways History. Arlington, VA, US Airways, Inc. Internet site: <www.usairways.com/about/corporate/profile/history/company_history.htm>, visited 2/11/05.

US CDC c. 1992. Public Health Assessment: North Penn-Area 1, Souderton, Montgomery County, Pennsylvania. Atlanta, GA, US Department of Health and Human Services, Centers for Disease Control, Agency for Toxic Substances and Disease Registry. Internet site: <http://www.atsdr.cdc.gov/HAC/PHA/penn/npa_p1.html>, visited 12/31/04.

US Census Bureau 1999. 1997 Economic Census: Retailing. Washington, DC, US Department of Commerce, Bureau of the Census. Internet site: <http://www.census.gov/epcd/www/97EC_US.HTM>, visited multiple times.

US Census Bureau 2000. 1997 Economic Census: Wholesaling. Washington, DC, US Department of Commerce, Bureau of the Census. Internet site: <http://www.census.gov/epcd/www/97EC_US.HTM>, visited multiple times.

US Census Bureau 2001. 1997 Economic Census: Manufacturing. Washington, DC, US Department of Commerce, Bureau of the Census. Internet site: <http://www.census.gov/epcd/www/97EC_US.HTM>, visited multiple times.

US Census Bureau 2002. Census 2000 Summary File 1 100-Percent Data. Washington, DC, US Department of Commerce, Bureau of the Census. Internet site: <http://www.census.gov>, visited multiple times.

US Census Bureau 2002. Census 2000 Summary File 3 (SF 3) 100-Percent Data. Washington, DC, US Department of Commerce, Census Bureau. Internet site: <http://www.census.gov>, multiple visits.

US Census Bureau 2002. Measuring America: Long Form Questionnaire. This longer publication has been separated into several documents for posting on the Census Bureau Web site. Washington, DC, US Department of Commerce, Census Bureau. Internet site: <www.census.gov/prod/2002pubs/pol02marv-pt4.pdf>, visited 8/14/04.

US Census Bureau 2004. Table 1: Annual Estimates of the Population for the United States and States, and for Puerto Rico: April 1, 2000 to July 1, 2004 (NST-EST2004-01). Washington, DC, US Department of Commerce, Census Bureau, Population Division. Internet site: <http://www.census.gov/popest/states/NST-ann-est.html>, visited 1/6/05.

US Census Bureau 2004. Table 5: Annual Estimates of the Components of Population Change for the United States and States: July 1, 2003 to July 1, 2004 (NST-EST2004-05). Washington, DC, US Department of Commerce, US Census Bureau, Population Division. Internet site: <http://www.census.gov/popest/states/NST-comp-chg.html>, visited 1/6/05.

US Census Bureau 2004. North American Industry Classification System (NAICS). Washington, DC, US Department of Commerce, Bureau of the Census. Internet site: <http://www.census.gov/epcd/www/naics.html>, visited multiple times.

USDA 1975. Soil Survey of Franklin County, Pennsylvania. US Department of Agriculture: Soil Conservation Service in cooperation with Pennsylvania State University: College of Agriculture and with Pennsylvania Department of Environmental Resources: State Conservation Commission.

USDA 2002. Food Consumption (per capita) Data System. Washington, DC, US Department of Agriculture, Economic Research Service. Internet site: <http://www.ers.usda.gov/Data/foodcon-sumption/>, visited 6/24/04.

US DOE 1991. Inventory of Electric Power Plants in the United States 1990. Washington, DC, US Department of Energy, Energy Information Administration.

US DOT 2003. Pennsylvania Transportation Profile. Washington, DC, US Department of Transportation, Bureau of Transportation Statistics. Internet site: <http://www.bts.gov/publications/state_transportation_profiles/pennsylvania/index.html>, visited 8/23/04.

US EPA 1997. Climate Change and Pennsylvania. Washington, DC, US Environmental Protection Agency, Office of Policy, Planning and Evaluation. September 1997. Internet site: <http://Yosemite.epa.gov/oar/globalwarming.nsf/UniqueKeyLookup/SHSU5BVMDY/$File/pa_impct.pdf>, visited 3/23/05.

US EPA 2002. Clearing the Air: The Facts About Capping and Trading Emissions. Washington, DC, US Environmental Protection Agency, Office of Air and Radiation, Clean Air Markets Division. May 2002. Internet site: <http://www.epa.gov/airmarkets/articles/clearingtheair.pdf>, visited 3/24/05.

US EPA 2005a. County Air Quality Report - Criteria Air Pollutants. 2004 Data. Washington, DC, US Environmental Protection Agency. Internet site: <http://www.epa.gov/air/data/monsum.html?st~PA~Pennsylvania>, visited 3/16/05.

US EPA 2005b. Criteria Pollutants – Nonattainment Areas. Washington, DC, US Environmental Protection Agency. Internet site: <http://www.epa.gov/airtrends/non.html>, visited 3/24/05.

US EPA Region III 2000. Slag from Sharon Steel to be Reused Under Innovative Agreement. Environmental News, September 20, 2000. Philadelphia, PA, US Environmental Protection Agency Region III, Mid-Atlantic Hazardous Site Cleanup. Internet site: <http://www.epa.gov/reg3hwmd/super/sites/PAD001933175/pr/2000-09-20.htm>, visited 12/1/04.

US Forest Service 2004. Allegheny National Forest: Forest Facts. Milwaukee, WI, US Department of Agriculture: US Forest Service: Eastern Region – R9. Internet site: <http://www.fs.fed.us/r9/forests/allegheny/about/forest_facts/>, visited 9/29/04.

US Forest Service 2004. Allegheny National Forest: History of the Allegheny National Forest. Milwaukee, WI, US Department of Agriculture: US Forest Service: Eastern Region – R9. Internet site: <http://www.fs.fed.us/r9/forests/allegheny/about/history/>, visited 9/29/04.

USGS 1999. Digital Ortho Quarter Quad (aerial photograph): [various locations]. Downloaded in TIFF format from http://www.pasda.psu.edu/access/doq99list.cgi, multiple dates. US Geological Survey, Washington, DC.

USGS 2004. Geographic Names Information System. Washington, DC, US Department of the Interior, Geological Survey. Internet site: <http://www.geonames.usgs.gov>, visited 10/8/04.

US NARA 2004. Historical Election Results: Electoral Votes by State. Washington, DC, US National Archives and Records Administration, Office of the Federal Register. Internet site: <http://www.archives.gov/federal_register/electoral_college/votes/votes_by_state.html>, visited 10/25/04.

US NASS 2004. 2002 Census of Agriculture – County Data: Pennsylvania. Washington, DC, US Department of Agriculture: National Agricultural Statistics Service. Internet site: <http://www.nass.usda.gov/census/census02/volume1/pa/index2.htm>, visited 7/22/04.

US NASS 2004. 2002 Census of Agriculture – State Data: Pennsylvania. Washington, DC, US Department of Agriculture: National Agricultural Statistics Service. Internet site: <http://www.nass.usda.gov/census/census02/volume1/pa/index1.htm>, visited 10/28/04.

US NPS 1998. A Roar Like Thunder. US National Park Service. Internet site: <http://www.cr.nps.gov/nr/twhp/wwwlps/lessons/5johnstown/5facts1.htm>, visited 10/8/02.

US NPS 2003. Steamtown/Chamber of Commerce Release Annual Report: Economic Impact Favorable Upon Region. Washington, DC, US Department of Interior, National Park Service, Steamtown National Historic Site. Internet site: <http://www.nps.gov/stea/economicimpact.htm>, visited 2/8/05.

US NRCS 2000. "Table 10 – Estimated Average Annual Sheet and Rill Erosion on Nonfederal Land, By State and Year," from Summary Report, 1997 National Resources Inventory (revised 2000). Washington, DC, US Department of Agriculture: Natural Resources Conservation Service. Internet site: <http://www.nrcs.usda.gov/technical/NRI/1997/summary_report/table10.html>, visited 10/1/04.

US NRCS 2004. Highly Erodible Land Conservation Compliance Provisions. Washington, DC, US Department of Agriculture: Natural Resources Conservation Service. Internet site: <http://www.nrcs.usda.gov/programs/helc/>, visited 10/1/04.

Van Trump, James 2004. The Controversial Spelling of "Pittsburgh," or Why The "H?" Pittsburgh, PA, Pittsburgh History and Landmarks Foundation. Internet site: <http://www.phlf.org/phlfnews/essays/pittsburg.html>, visited 12/3/04.

Wallingford, Bret D. 2002. "Deer Management and the Concept of Change." Pennsylvania Game News, 73(7). Internet site: <http://sites.state.pa.us/PA_Exec/PGC/deer/GN0207.htm>, visited 9/27/02.

Warren Chamber 2004. Largest Employers. Warren, PA, Warren County Chamber of Business and Industry. Internet site <http://www.wcda.com/>, visited 9/12/04.

Warren, Kenneth 2001. Big Steel: The First Century of the United States Steel Corporation 1902-2001. Pittsburgh, University of Pittsburgh Press.

Welch's 2005. Welch's Company Information. Concord, MA, Welch Foods, Inc. Internet site: <http://www.welchs.com/company/general_co_info.html>, visited 1/23/05.

Wildlife Information Center 1996. White-tailed Deer in The Kittatinny Raptor Corridor. Wildlife Bulletin No. 17. Slatington, PA, Wildlife Information Center, Inc. Internet site: <http://www.wildlifeinfo.org/Bulletins/wildlifebulletin17.htm>, visited 10/4/04.

Williams, Anthony V. 1995. "Political Geography." Chapter 10 in Miller, E. Willard (ed.) A Geography of Pennsylvania. pp. 154-164.

Woolrich 2004. About Woolrich. Woolrich, PA, Woolrich, Inc. Internet site: <http://www.woolrich.com>, visited 8/23/04.

Yarnal, Brent 1995. "Climate." Chapter 3 in Miller, E. Willard (ed.) A Geography of Pennsylvania. pp. 44-55.

Yergin, Daniel 1991. The Prize: The Epic Quest for Oil, Money and Power. New York, Simon and Schuster.

Zelinsky, Wilbur 1977. "The Pennsylvania Town: An Overdue Geographical Account." Geographical Review. Vol. 67, no. 2 (April), pp. 127-147.

Zelinsky, Wilbur 1995. "Cultural Geography." Chapter 9 in Miller, E. Willard (ed.) A Geography of Pennsylvania. pp. 132-153.

Zelinsky, Wilbur 1995. "Ethnic Geography." Chapter 8 in Miller, E. Willard (ed.) A Geography of Pennsylvania. pp. 113-131.

Index to Places in Pennsylvania

A

Adams County 42, 126
Allegheny County 86, 171, 175,
    176, 180
Allegheny Front 8, 13, 14, 35, 39,
    40, 63, 144
Allegheny National Forest 38, 48,
    64, 134, 183
Allegheny Portage Railroad National
    Historic Site 148, 183
Allegheny Reservoir 38, 47, 64
Allegheny River 14, 29, 37, 38, 46,
    48, 64, 195, 196, 197
Allentown 13, 18, 186
Altoona 14, 15, 146, 147, 148
Appalachian Mountains 13
Appalachian Plateau 13, 15, 24, 26,
    34, 35, 36, 37, 40, 45, 47, 52,
    55, 58, 59, 61, 63, 104, 126,
    134, 138, 149, 156, 186, 194,
    197, 201
Appalachian Trail 183
Armstrong County 208
Atlantic Coastal Plain 8, 14, 19, 37,
    43, 52, 61, 63, 96, 151, 201

B

Beaver County 158, 159
Beaver Valley 159
Bedford County 158
Berks County 10, 21, 24, 28, 88,
    137, 164
Berwick 159
Bethlehem 13, 146, 168, 170
Blair County 73
Bloomsburg 111
Blue Mountain 13
Boyertown 28
Bradford County 103
Breezewood 152
Broad Top 13
Bryn Mawr 90
Bucks County 10, 24, 54, 88, 95,
    172, 205

C

Caernarvon Township 88
Cambria County 89
Cameron County 72, 73

Camp Hill 118
Carbon County 73
Central Lowland 13, 14, 58, 126
Centralia 141
Centre County 28, 73, 80, 180
Chalfont 54
Chambersburg 13
Chesapeake Bay 45, 70, 108, 131,
    213, 215
Chester 191
Chester County 88, 123, 130, 131,
    205
Chinatown 104, 205
Clairton 172
Clarion River 64
Clearfield County 73, 133
Clinton County 73, 164
Codorus Creek 165
Colton Point State Park 21
Columbia 134, 145, 166
Columbia County 141
Conestoga River 144
Connellsville 168
Conowingo Dam 46
Cook Forest State Park 64, 65
Cornwall 24
Crawford County 35
Cumberland County 24, 176
Cumberland Valley 13, 24

D

Danville 180
Dauphin County 159
Delaware Bay 8, 70
Delaware County 34, 135, 205
Delaware National Wild and Scenic
    River 183
Delaware River 8, 10, 29, 34, 40,
    45, 46, 48, 53, 63, 138, 145,
    182, 196, 201, 202, 205, 206
Delaware Water Gap 186
Delaware Water Gap National
    Recreation Area 183
Donora 210, 211

E

East Conemaugh 49
Easton 13
Edinboro 35
Eisenhower National Historic Site

    183
Elk County 72, 73
Endless Mountains 14, 194
Enterprise 30
Erie 14, 18, 79, 117, 151, 191, 192,
    215
Erie County 35, 126, 135, 163, 191

F

Farrell 193
Forest County 73
Fort Necessity National Battlefield
    183
Franklin 49
Franklin County 24
Friendship Hill National Historic Site
    183

G

Genesee River 45
Germantown 202
Gettysburg 10, 42, 43, 126
Gettysburg National Military Park
    183
Golden Triangle 197
Grand Canyon of Pennsylvania 20
Great Valley 12, 13, 24, 43
Grove City 193
Gwynedd Township 88

H

Hanover 166
Harrisburg 13, 19, 89, 104, 117,
    118, 150, 151, 166, 169, 176,
    193
Hazleton 79, 157
Hermitage 193
Hershey 164
Homestead 171, 172
Hopewell Furnace National Historic
    Site 183
Horseshoe Curve 14, 146, 147, 149
Horseshoe Curve National Historic
    Landmark 148

I

Independence National Historical
    Park 183, 203
Irwin 150, 172

J

Jefferson County 135
Jersey Shore 164
Johnstown 14, 48, 49, 145, 168, 169, 196
Johnstown Flood National Memorial 183

K

Kensington 100, 202
King of Prussia 175, 205
Kinzua Dam 38, 47, 48, 64

L

Lackawanna County 180
Lake Conemaugh 49
Lake Erie 13, 14, 34, 35, 37, 45, 63, 126, 151, 191, 192, 213, 215
Lancaster 117, 166, 174, 192, 193
Lancaster County 10, 21, 24, 88, 122, 123, 124, 126, 130, 131, 144, 145, 166, 215
Lansdale 53
Lebanon County 10, 24, 137, 164, 168
Lehigh County 10, 21, 24, 29, 180
Lehigh River 145, 168
Lehigh Valley 12, 24, 29, 168
Leonard Harrison State Park 21
Levittown 95, 96, 172
Limerick 159
Little Conemaugh River 49
Luzerne County 159, 180
Lycoming County 130, 135, 164

M

Main Line 14, 87, 89, 90, 98
Manayunk 202
Marietta 134
McKean County 34, 35, 135
Meadowcroft Rock Shelter 75
Mercer 193
Mercer County 193
Middlesex 150
Middletown 159
Mifflin County 122
Mineral Point 49
Monongahela River 171, 195, 196, 197, 211
Monroe County 73, 181
Montgomery County 53, 88, 159, 175, 205
Montour County 180

Montrose 191, 194
Mount Carmel 142, 157
Mount Davis 7, 13, 35

N

New England 10, 12, 24, 28, 58, 63, 126
New Stanton 152
Nittany Valley 13
Norristown 150, 205
North Branch (of the Susquehanna River) 13
North East 163
North Wales 88
Northampton County 10, 29
Northern Tier 103, 104

O

Oakland 199
Ohio River 45, 46, 158, 168, 193, 195, 196, 210
Oil City 29, 30, 140
Oil Creek 29, 30

P

Peach Bottom 159
Petroleum Center 30
Philadelphia 9, 14, 18, 19, 35, 40, 43, 45, 53, 76, 79, 81, 85, 86, 87, 88, 89, 91, 93, 96, 99, 101, 102, 103, 104, 105, 107, 109, 111, 112, 117, 118, 119, 127, 137, 138, 144, 145, 146, 150, 151, 152, 164, 165, 168, 173, 174, 175, 176, 177, 179, 180, 184, 186, 187, 189, 191, 192, 193, 194, 195, 201, 202, 203, 205, 206, 209, 221
Piedmont 8, 9, 10, 11, 19, 21, 24, 28, 35, 37, 42, 43, 52, 58, 61, 63, 126, 164, 201
Pike County 79, 181, 182
Pine Creek 21
Pine Creek Gorge 20
Pine Grove Furnace 24
Pithole 30, 140
Pittsburgh 14, 18, 24, 30, 38, 48, 76, 85, 86, 89, 95, 104, 109, 111, 115, 117, 118, 145, 146, 150, 151, 152, 156, 158, 164, 168, 169, 170, 171, 172, 175, 176, 189, 191, 192, 193, 195, 196, 197, 198, 199, 200, 209
Pocono Mountains 14, 186, 187

Point State Park 197
Portland 29
Potomac Heritage Trail 183
Potter County 45, 73, 74
Pottstown 159
Pottsville 168
Presque Isle 14, 191

R

Reading 117, 164, 165
Ridge and Valley 10, 11, 12, 13, 19, 24, 25, 26, 28, 34, 35, 39, 41, 43, 46, 52, 58, 59, 63, 126, 138, 141, 149, 201
Rivers of Steel National Heritage Area 172

S

Saxton 158
Schuylkill County 138
Schuylkill River 8, 19, 46, 138, 144, 145, 168, 201
Scranton 13, 35, 76, 79, 111, 147, 148, 150, 180, 186, 194
Shamokin 41, 157
Sharon 193
Shenango River 193
Shippensburg 13
Shippingport 158
Society Hill 203
Somerset County 7, 35, 122, 123, 130, 160
Souderton 53, 54
South Fork 49
South Fork (town) 49
South Mountain 12, 43
South Philadelphia 95, 99, 100, 203, 205
State College 13, 180
Steamtown National Historic Site 148, 183
Steelton 169
Strasburg 15, 98, 147, 148
Sullivan County 103
Sunbury 157
Susquehanna (nuclear power station) 159
Susquehanna County 103, 191, 194
Susquehanna River 13, 19, 21, 45, 46, 47, 48, 118, 133, 134, 145, 165, 166, 176, 196, 201, 215

T

Thaddeus Kosciuszko National

Memorial 183
The Golden Triangle 197
The Hill 199
The Point 197, 199
The Strip 199
Three Mile Island 159
Tioga County 103
Tioga State Forest 21
Titusville 30, 140, 156

U

Upper Delaware National Wild and
    Scenic River 183

V

Valley Forge 206
Valley Forge National Historical
    Park 72, 183
Venango County 29, 140, 156

W

Warren 37, 38, 47
Warren County 35, 37, 47
Washington County 210
Wayne County 34
Welsh Mountains 88
West Branch (of the Susquehanna
    River) 133, 135
West Chester 205
Wilkes-Barre 13, 35, 48, 79, 180,
    186
Williamsport 134, 135, 136, 157
Woolrich 164
Wrightsville 166
Wyoming County 103
Wyoming Valley 13, 156

Y

York 117, 165, 166, 192
York County 24, 47, 159, 165, 166
York Haven Dam 47
Youghiogheny River 14, 168

Map Credits

Each map element in the list below is a computer file used in ESRI, Inc.'s ArcGIS software program to add that element to computer-drawn maps. Each element is followed by its source in parentheses. After the colon are listed the figures (with page numbers) in which it appears. The source ESRI, Inc. is used to indicate files that came with the ArcGIS software. Sources followed by [PASDA] were downloaded from the Pennsylvania Spatial Data Access Internet site: <http://www.pasda.psu.edu>.

Air quality monitoring sites (US Environmental Protection Agency, 2005 <http://www.epa.gov/air/data>: 35.1 (208), 35.2 (209).

Airports (Pennsylvania Department of Health, 2002 [PASDA]): 25.2 (152).

Bedrock geology (Pennsylvania Department of Conservation and Natural Resources Bureau of Topographic and Geologic Survey, 2001 [PASDA]): 4.7 (31).

Canals (created by Geiger, 2004): 24.3 (145).

Carbonate bedrock (Pennsylvania Department of Transportation, 2001 [PASDA]): 4.5 (28).

Census tracts (City of Philadelphia, 2001 [PASDA]): 34.3 (205).

Cities: as point locations, see "Places" (below); as defined areas, see "Municipality boundaries" (below).

Cities, international (ESRI, Inc.): 14.4 (89).

Coal deposits (Pennsylvania Department of Transportation, 2001 [PASDA]): 4.1 (25), 23.1 (138).

Coal mined areas (Pennsylvania Department of Conservation and Natural Resources, 2001 [PASDA]): 23.1 (138).

Cooling degree day isolines (created by Geiger, 2004): 6.3 (41).

Counties (Pennsylvania Department of Transportation, 2004 [PASDA]): 1.1 (2), 1.2 (3), 1.3 (4), 2.1 (8), 3.2 (21), 4.1 (25), 4.3 (26), 4.4 (27), 4.5 (28), 4.6 (30), 5.1 (34), 5.3 (36), 5.4 (36), 5.5 (37), 6.2 (40), 6.3 (41), 6.4 (42), 6.5 (42), 7.2 (47), 7.4 (49), 8.3 (53), 8.4 (54), 9.2 (57), 9.3 (58), 9.4 (59), 10.1 (62), 10.2 (65), 10.3 (66), 11.1 (68), 12.1 (73), 13.2 (80), 13.3 (80), 13.4 (81), 13.5 (81), 14.1 (86), 14.2 (87), 14.3 (88), 15.3 (94), 16.1 (99), 16.2 (100), 17.1 (102), 18.1 (108), 18.2 (109),

18.3 (109), 18.4 (110), 19.1 (115), 19.2 (116), 19.3 (116), 19.4 (117), 21.2 (127), 21.4 (128), 21.5 (128), 21.6 (129), 21.8 (130), 22.1 (134), 22.2 (136), 23.1 (138), 23.4 (142), 24.4 (146), 24.6 (147), 25.1 (150), 25.2 (152), 25.3 (153), 26.2 (157), 27.1 (162), 27.2 (163), 27.3 (164), 27.4 (165), 28.1 (169), 28.2 (170), 28.3 (171), 29.2 (175), 29.3 (176), 30.1 (179), 30.2 (180), 30.3 (181), 31.1 (184), 31.2 (185), 31.3 (185), 31.4 (186), 32.1 (192), 32.2 (192), 32.3 (193), 32.4 (194), 35.1 (208), 35.2 (209), 35.3 (211), 36.2 (215), 36.3 (216), 37.2 (220), 37.3 (221).

Countries/regions (ESRI, Inc.): 14.4 (89).

Dams (US Army Corps of Engineers, 2004: "National Inventory of Dams <http://crunch.tec.army.mil/nid/webpages/nid.cfm>): 7.2 (47).

Electric power stations (created by Geiger, 2002): 26.2 (157).

Forest/ecoregions US Environmental Protection Agency, 1994 [PASDA]): 10.1 (62).

Glaciers (Pennsylvania Department of Conservation and Natural Resources Bureau of Topographic and Geologic Survey, 1995 [PASDA]): 3.1 (20).

Heating degree day isolines (created by Geiger, 2004): 5.4 (36).

Highways and major roads (Pennsylvania Department of Transportation, 2004 [PASDA]): 1.3 (4), 3.2 (21), 8.4 (54), 12.1 (73), 14.3 (88), 14.5 (89), 15.2 (93), 19.4 (117), 21.8 (130), 23.5 (142), 25.1 (150), 25.2 (152), 25.3 (153), 27.4 (165), 28.3 (171), 31.4 (186), 32.1 (192), 32.2 (192), 32.3 (193), 32.4 (194), 37.2 (220).

Lancaster County agricultural preservation designated areas (Lancaster County GIS Office and Lancaster County Agricultural Preserve Board, 2005): 21.8 (130).

Landfills (created by Geiger, 2004, from zip codes in Landfills database published by Pennsylvania Department of Environmental Protection, <http://www.dep.state.pa.us/dep/deputate/airwaste/wm/MRW/Docs/Landfill_list.htm>): 37.2 (220), 37.3 (221).

Landform regions (Pennsylvania Department of Conservation and Natural Resources Bureau of Topographic and Geologic Survey, 1999 [PASDA]): 2.1 (8), 2.3 (9), 2.5 (11), 2.6 (12), 4.1 (25), 6.5

(42), 9.2 (57), 9.4 (59), 11.1 (68), 31.4 (186), 16.2 (100), 19.4 (117), 21.8 (130), 24.3 (145), 27.4 (165), 28.1 (169), 28.3 (171), 31.4 (186), 32.1 (192), 32.2 (192), 32.3 (193), 32.4 (194), 33.1 (196), 34.2 (204), 34.3 (205), 35.3 (211), 36.2 (215).

Metropolitan Statistical Areas (created by Geiger, 2004): 15.3 (94).

Municipality boundaries, including cities, boroughs and/ or townships (Pennsylvania Department of Transportation, 2004 [PASDA]): 2.8 (15), 3.2 (21), 4.6 (30), 5.5 (37), 6.5 (42), 7.4 (49), 8.4 (54), 9.4 (59), 12.1 (73), 14.3 (88), 15.2 (93), 15.5 (95), 15.1 (92), 15.2 (93), 18.4 (110), 21.8 (130), 23.5 (142), 25.2 (152), 27.4 (165), 28.3 (171), 31.4 (186), 32.1 (192), 32.2 (192), 32.3 (193), 32.4 (194), 34.2 (204), 34.3 (205), 35.3 (211).

National Fish Hatcheries (Pennsylvania Department of Environmental Protection, 2001 [PASDA]): 31.1 (184).

National Forest (Pennsylvania Department of Environmental Protection, 2001 [PASDA]): 5.5 (37), 10.2 (65), 31.1 (184).

National Landmarks (Pennsylvania Department of Environmental Protection, 2001 [PASDA]): 31.1 (184).

National Parks (Pennsylvania Department of Environmental Protection, 2001 [PASDA]): 6.5 (42), 31.1 (184).

National Wilderness Preserves (Pennsylvania Department of Environmental Protection, 2001 [PASDA]): 3.2 (21), 31.1 (184).

Oil regions (Pennsylvania Department of Environmental Protection, 2001 [PASDA]): 4.3 (26), 23.4 (140).

Oil wells (Pennsylvania Department of Environmental Protection, 2001 [PASDA]): 23.4 (140).

Pennsylvania House of Representatives districts (Pennsylvania Department of Transportation, 2004 [PASDA]): 18.2 (109).

Pennsylvania Senate districts (Pennsylvania Department of Transportation, 2004 [PASDA]): 18.3 (109).

Philadelphia neighborhoods (City of Philadelphia, 1998 [PASDA]): 34.2 (204), 34.3 (205).

Pittsburgh neighborhoods (created by Jonathan Egger, Millersville University Geo-Graphics Lab, 2004): 33.3 (198).

Places, including cities and towns as point locations (ESRI, Inc.): 7.4 (49), 8.4 (54), 15.2 (93), 16.1 (99), 17.1 (102), 24.3 (145), 31.3 (185), 33.1 (196), 36.2 (215).

Precipitation isolines (created by Geiger): 6.2 (40).

Railroads, active and inactive (Pennsylvania Department of Environmental Protection, 2001 [PASDA]): 2.8 (15), 7.4 (49), 14.3 (88), 14.5 (89), 24.4 (146), 24.6 (147), 28.3 (171).

Rivers and lakes (Pennsylvania Department of Environmental Protection, 2001 [PASDA]): 1.2 (3), 3.1 (20), 3.2 (21), 4.6 (30), 5.5 (37), 7.1 (46), 7.2 (47), 7.4 (49), 8.4 (54), 10.2 (65), 12.1 (73), 15.1 (92), 16.2 (100), 19.4 (117), 21.8 (130), 24.3 (145), 27.4 (165), 28.1 (169), 28.3 (171), 31.4 (186), 32.1 (192), 32.2 (192), 32.3 (193), 32.4 (194), 33.1 (196), 34.2 (204), 34.3 (205), 35.3 (211), 36.2 (215).

Roads, local (Pennsylvania Department of Transportation, 2001 [PASDA]): 15.1 (92), 15.4 (94), 16.2 (100), 19.4 (117), 32.1 (192), 32.2 (192), 32.3 (193), 32.4 (194).

Snowfall (US National Oceanic and Atmospheric Administration, National Climatic Data Center, 2001 <http://ww1.ncdc.noaa.gov/pub/data/climaps/shapefiles/>): 5.3 (36).

Soils (US Natural Resources Conservation Service STATSGO Program, 1995): 9.2 (57), 9.3 (58), 9.4 (54), 11.1 (68).

State Forests (Pennsylvania Department of Environmental Protection, 2001 [PASDA]): 3.2 (21), 10.2 (65), 12.1 (73), 31.2 (185).

State Parks (Pennsylvania Department of Environmental Protection, 2001 [PASDA]): 3.2 (21), 10.2 (65), 12.1 (73), 31.2 (185).

State Gamelands (Pennsylvania Department of Environmental Protection, 2001 [PASDA]): 10.2 (65), 12.1 (73), 31.2 (185).

States (ESRI, Inc.): 5.2 (35), 6.1 (39), 17.1 (102), 31.3 (185), 33.1 (196), 35.1 (208) 35.2 (209).

Steel mills (created by Geiger, 2004): 28.1 (169).

Temperature isolines (created by Geiger, 2004): 5.1 (34).

US House of Representatives districts (Pennsylvania Department of Transportation, 2004 [PASDA]): 18.1 (108).

Watersheds (Pennsylvania Department of Environmental Protection, 2001 [PASDA]): 7.1 (46), 36.2 (215), 36.3 (216).

Weather Stations (created by Geiger, 2004, from US National Oceanic and Atmospheric Administration

Climatology of the US Publication 81 "Monthly Station Normals 1971-2000): 5.1 (34), 5.4 (36), 6.2 (40), 6.3 (41).

Welsh Tract boundary (created by Geiger, 2004): 14.3 (88).